GEOTECHNICAL ENGINEERING

A Practical Problem Solving Approach

N. Sivakugan | Braja M. Das

J.ROSS

PUBLISHING

Copyright © 2010 by J. Ross Publishing, Inc.

ISBN-13: 978-1-60427-016-7

Printed and bound in the U.S.A. Printed on acid-free paper

Library of Congress Cataloging-in-Publication Data

Sivakugan, Nagaratnam, 1956-
Geotechnical engineering : a practical problem solving approach / by
Nagaratnam Sivakugan and Braja M. Das.
    p. cm.
  Includes bibliographical references and index.
  ISBN 978-1-60427-016-7 (pbk. : alk. paper)
  1. Soil mechanics. 2. Foundations. 3. Earthwork. I. Das, Braja M.,

TA710.S536 2009
624.1′5136—dc22

2009032547

Direct all inquiries to J. Ross Publishing, Inc., 5765 N. Andrews Way, Fort Lauderdale, FL 33309.

Phone: (954) 727-9333

Fax: (561) 892-0700

Web: www.jrosspub.com

To our parents, teachers, and wives

Contents

Preface

We both have been quite successful as geotechnical engineering teachers. In *Geotechnical Engineering: A Practical Problem Solving Approach*, we have tried to cover every major geotechnical topic in the simplest way possible. We have adopted a hands-on approach with a strong, practical bias. You will learn the material through several worked examples that take geotechnical engineering principles and apply them to realistic problems that you are likely to encounter in real-life field situations. This is our attempt to write a straightforward, no-nonsense, geotechnical engineering textbook that will appeal to a new generation of students. This is said with no disrespect to the variety of geotechnical engineering textbooks already available—each serves a purpose.

We have used a few symbols to facilitate quick referencing and to call your attention to key concepts. This symbol appears at the end of a chapter wherever it is necessary to emphasize a particular point and your need to understand it.

There are a few thoughtfully selected review exercises at the end of each chapter, and answers are given whenever possible. Remember, when you practice as a professional engineer you will not get to see the solutions! You will simply design with confidence and have it checked by a colleague. The degree of difficulty increases with each review exercise. The symbol shown here appears beside the most challenging problems.

We also try to nurture the habit of self-learning through exercises that relate to topics not covered in this book. Here, you are expected to surf the Web; or even better, refer to library books. The knowledge obtained from both the research activity and the material itself will complement the material from this book and is an integral part of learning. Such research-type questions are identified by the symbol shown here. Today, the *www* is at your fingertips, so this should not be a problem. There are many dedicated Web sites for geotechnical resources and reference materials (e.g., Center for Integrating Information on Geoengineering at http://www.geoengineer.org). Give proper references for research-type questions in your short essays. Sites like Wikipedia (http://en.wikipedia.org) and YouTube (http://www.youtube.com) can provide useful information including images and video clips. To obtain the best references, you must go to the library and conduct a proper literature search using appropriate key words.

We have included eight quizzes to test your comprehension. These are closed-book quizzes that should be completed within the specified times. They are designed to make you think and show you what you have missed.

The site investigation chapter has a slightly different layout. The nature of this topic is quite descriptive and less reliant on problem solving. It is good to have a clear idea of what the different in situ testing devices look like. For this reason, we have included several quality photographs. The purpose of the site investigation exercise is to derive the soil parameters from the in situ test data. A wide range of empirical correlations that are used in practice are summarized in this chapter. Tests are included that are rarely covered in traditional textbooks—such as the borehole shear test and the K_0 stepped blade test—and are followed by review questions that encourage the reader to review other sources of literature and hence nurture the habit of research.

Foundation Engineering is one of the main areas of geotechnical engineering; therefore, considerable effort was directed toward Chapters 12 and 13, which cover the topics of bearing capacity and settlements of shallow and deep foundations.

This is not a place for us to document everything we know in geotechnical engineering. We realize that this is your first geotechnical engineering book and have endeavored to give sufficient breadth and depth covering all major topics in soil mechanics and foundation engineering.

A free DVD containing the *Student Edition* of *GeoStudio* is included with this book. It is a powerful software suite that can be used for solving a wide range of geotechnical problems and is a useful complement to traditional learning. We are grateful to Mr. Paul Bryden and the GeoStudio team for their advice and support.

We are grateful to the following people who have contributed either by reviewing chapters from the book and providing suggestions for improvement: Dr. Jay Ameratunga, Coffey Geotechnics; Ms. Julie Lovisa, James Cook University; Kirralee Rankine, Golder Associates; and Shailesh Singh, Coffey Geotechnics; or by providing photographs or data: Dr. Jay Ameratunga, Coffey Geotechnics; Mr. Mark Arnold, Douglas Partners; Mr. Martyn Ellis, PMC, UK; Professor Robin Fell, University of New South Wales; Dr. Chris Haberfield, Golder Associates; Professor Silvano Marchetti, University of L'Aquila, Italy; Dr. Kandiah Pirapakaran, Coffey Geotechnics; Dr. Kirralee Rankine, Golder Associates; Dr. Kelda Rankine, Golder Associates; Dr. Ajanta Sachan, IIT Kanpur, India; Mr. Leonard Sands, Venezuela; Dr. Shailesh Singh, Coffey Geotechnics; Mr. Bruce Stewart, Douglas Partners; Professor David White, Iowa State University.

We wish to thank Mrs. Janice Das and Mrs. Rohini Sivakugan, who provided manuscript preparation and proofreading assistance. Finally, we wish to thank Mr. Tim Pletscher of J. Ross Publishing for his prompt response to all our questions and for his valuable contributions at various stages.

N. Sivakugan and B. M. Das

About the Authors

Dr. Nagaratnam Sivakugan is an associate professor and head of Civil and Environmental Engineering at the School of Engineering and Physical Sciences, James Cook University-Australia. He graduated from the University of Peradeniya–Sri Lanka, with First Class Honours and received his MSCE and PhD from Purdue University. As a chartered professional engineer and registered professional engineer of Queensland, he does substantial consulting work for geotechnical and mining companies throughout Australia and the world. He is a Fellow of Engineers, Australia. Dr. Sivakugan has supervised eight PhD candidates to completion and has published more than 50 scientific and technical papers in refereed international journals, and 50 more in refereed international conference proceedings. He serves on the editorial board of the *International Journal of Geotechnical Engineering* (IJGE) and is an active reviewer for more than 10 international journals. In 2000, he developed a suite of fully animated Geotechnical PowerPoint slideshows that are now used worldwide as an effective teaching and learning tool. An updated version is available for free downloads at http://www.jrosspub.com.

Dr. Braja M. Das, Professor and Dean Emeritus, California State University–Sacramento, is presently a geotechnical consulting engineer in the state of Nevada. He earned his MS in civil engineering from the University of Iowa and his PhD in geotechnical engineering from the University of Wisconsin–Madison. He is a Fellow of the American Society of Civil Engineers and is a registered professional engineer. He is the author of several geotechnical engineering texts and reference books including *Principles of Geotechnical Engineering, Principles of Foundation Engineering, Fundamentals of Geotechnical Engineering, Introduction to Geotechnical Engineering, Principles of Soil Dynamics, Shallow Foundations: Bearing Capacity and Settlement, Advanced Soil Mechanics, Earth Anchors,* and *Theoretical Foundation Engineering*. Dr. Das has served on the editorial boards of several international journals and is currently the editor in chief of the *International Journal of Geotechnical Engineering*. He has authored more than 250 technical papers in the area of geotechnical engineering.

Web
Added
Value™

This book has free material available for download from the
Web Added Value™ resource center at *www.jrosspub.com*

At J. Ross Publishing we are committed to providing today's professional with practical, hands-on tools that enhance the learning experience and give readers an opportunity to apply what they have learned. That is why we offer free ancillary materials available for download on this book and all participating Web Added Value™ publications. These online resources may include interactive versions of material that appears in the book or supplemental templates, worksheets, models, plans, case studies, proposals, spreadsheets, and assessment tools, among other things. Whenever you see the WAV™ symbol in any of our publications, it means bonus materials accompany the book and are available from the Web Added Value™ Download Resource Center at www.jrosspub.com.

Downloads for *Geotechnical Engineering: A Practical Problem Solving Approach* include PowerPoint slides to assist in classroom instruction and learning.

Introduction

1

1.1 GENERAL

What is *Geotechnical Engineering*? The term *geo* means earth or soil. There are many words that begin with *geo*—*geology*, *geodesy*, *geography*, and *geomorphology* to name a few. They all have something to do with the earth. Geotechnical engineering deals with the engineering aspects of *soils* and *rocks*, sometimes known as *geomaterials*. It is a relatively young discipline that would not have been part of the curriculum in the earlier part of the last century. The designs of every building, service, and infrastructure facility built on the ground must give due consideration to the engineering behavior of the underlying soil and rock to ensure that it performs satisfactorily during its design life. A good understanding of *engineering geology* will strengthen your skills as a geotechnical engineer.

Mechanics is the physical science that deals with forces and equilibrium, and is covered in subjects like Engineering Mechanics, Strength of Materials, or Mechanics of Materials. In *Soil Mechanics* and *Rock Mechanics*, we apply these principles to soils and rocks respectively. Pioneering work in geotechnical engineering was carried out by Karl Terzaghi (1882–1963), acknowledged as the father of soil mechanics and author of *Erdbaumechanik auf bodenphysikalischer grundlage (1925),* the first textbook on the subject.

Foundation Engineering is the application of the soil mechanics principles to design earth and earth-supported structures such as foundations, retaining structures, dams, etc. Traditional geotechnical engineering, which is also called *geomechanics* or *geoengineering*, includes soil mechanics and foundation engineering. The escalation of human interference with the environment and the subsequent need to address new problems has created a need for a new branch of engineering that will deal with hazardous waste disposal, landfills, ground water contamination, potential acid sulphate soils, etc. This branch is called *environmental geomechanics* or *geoenvironmental engineering*.

1.2 SOILS

Soils are formed over thousands of years through the weathering of parent rocks, which can be *igneous*, *sedimentary*, or *metamorphic* rocks. Igneous rocks (e.g., granite) are formed by the cooling of magma (underground) or lava (above the ground). Sedimentary rocks (e.g., limestone,

1

shale) are formed by gradual deposition of fine soil grains over a long period. Metamorphic rocks (e.g., marble) are formed by altering igneous or sedimentary rocks by pressure or temperature, or both.

Soils are primarily of two types: residual or transported. *Residual soils* remain at the location of their geologic origin when they are formed by weathering of the parent rock. When the weathered soils are transported by glacier, wind, water, or gravity and are deposited away from their geologic origin, they are called *transported soils*. Depending on the geologic agent involved in the transportation process, the soil derives its special name: glacier—*glacial*; wind—*aeolian*; sea—*marine*; lake—*lacustrine*; river—*alluvial*; gravity—*colluvial*. Human beings also can act as the transporting agents in the soil formation process, and the soil thus formed is called a *fill*.

Soils are quite different from other engineering materials, which makes them interesting and at the same time challenging. Presence of water within the voids further complicates the picture. Table 1.1 compares soils with other engineering materials such as steel.

We often simplify the problem so that it can be solved using soil mechanics principles. Sometimes soil is assumed to be a homogeneous isotropic elastic continuum, which is far from reality. Nevertheless, such approximations enable us to develop simple theories and arrive at some solutions that may be approximate. Depending on the quality of the data and the degree of simplification, appropriate *safety factors* are used.

Geotechnical engineering is a science, but its practice is an art. There is a lot of judgment involved in the profession. The same data can be interpreted in different ways. When there are limited data available, it becomes necessary to make assumptions. Considering the simplifications in the geotechnical engineering fundamentals, uncertainty, and scatter in the data, it may not always make sense to calculate everything to two decimal places. All these make the field of geotechnical engineering quite different from other engineering disciplines.

Table 1.1 Soils vs. other engineering materials

Soils	Others (e.g., steel)
1. *Particulate medium*—consists of grains	*Continuous medium*—a continuum
2. Three phases—solid grains, water, and air	Single phase
3. Heterogeneous—high degree of variability	Homogeneous
4. High degree of anisotropy	Mostly isotropic*
5. No tensile strength	Significant tensile strength
6. Fails mainly in shear	Fails in compression, tension, or shear

*Isotropic = same property in all directions

1.3 APPLICATIONS

Geotechnical engineering applications include foundations, retaining walls, dams, sheet piles, braced excavations, reinforced earth, slope stability, and ground improvement. Foundations such as *footings* or *piles* are used to support buildings and transfer the loads from the super-structure to the underlying soils. *Retaining walls* are used to provide lateral support and maintain stability between two different ground levels. *Sheet piles* are continuous impervious walls that are made by driving interlocking sections into the ground. They are useful in dewatering work. *Braced excavation* involves bracing and supporting the walls of a narrow trench, which may be required for burying a pipeline. Lately, *geosynthetics* are becoming increasingly popular for reinforcing soils in an attempt to improve the stability of footings, retaining walls, etc. When working with natural or man-made slopes, it is necessary to ensure their stability. The geotechnical characteristics of weak ground are often improved by *ground improvement* techniques such as compaction, etc.

Figure 1.1a shows a *soil nailing* operation where a reinforcement bar is placed in a drill hole and surrounded with concrete to provide stability to the neighboring soil. Figure 1.1b shows the Itaipu Dam in Brazil, the largest hydroelectric facility in the world. Figure 1.1c shows treated timber piles. Figure 1.1d shows steel sheet piles being driven into the ground. Figure 1.1e shows a gabion wall that consists of wire mesh cages filled with stones. Figure 1.1f shows a containment wall built in the sea for dumping dredged spoils in Brisbane, Australia.

1.4 SOIL TESTING

Prior to any design or construction, it is necessary to understand the soil conditions at the site. Figure 1.2a shows a *trial pit* that has been made using a backhoe. It gives a clear idea of what is lying beneath the ground, but only to a depth of 5 m or less. The first 2 m of the pit shown in the figure are *clays* that are followed by *sands* at the bottom. Samples can be taken from these trial pits for further study in the laboratory. Figure 1.2b shows the drill rig set up on a barge for some offshore site investigation. To access soils at larger depths, boreholes are made using drill rigs (Figure 1.2c) from which samples can be collected. The boreholes are typically 75 mm in diameter and can extend to depths exceeding 50 m. In addition to taking samples from boreholes and trial pits, it is quite common to carry out some *in situ* or *field tests* within or outside the boreholes. The most common in situ test is a *penetration test* (e.g., standard penetration test, cone penetration test) where a probe is pushed into the ground, and the resistance to penetration is measured. The penetration resistance can be used to identify the soil type and estimate the soil strength and stiffness.

Figure 1.1 Geotechnical applications: (a) soil nailing (b) Itaipu Dam (c) timber piles (d) sheet piles (e) gabion wall (Courtesy of Dr. Kirralee Rankine, Golder Associates) (f) sea wall to contain dredged spoils

1.5 GEOTECHNICAL LITERATURE

Some of the early geotechnical engineering textbooks were written by Terzaghi (1943), Terzaghi and Peck (1948, 1967), Taylor (1948), Peck et al. (1974), and Lambe and Whitman (1979). They are classics and will always have their place. While the content and layout may not appeal to

Figure 1.2 Soil testing: (a) a trial pit (Courtesy of Dr. Shailesh Singh) (b) drill rig mounted on a barge (Courtesy of Dr. Kelda Rankine, Golder Associates) (c) a drill rig (Courtesy of Mr. Bruce Stewart, Douglas Partners)

the present generation, they serve as useful references. *Geotechnical journals* provide reports on recent developments and any innovative, global research that is being carried out on geotechnical topics. *Proceedings of conferences* can also be a good reference source. Through universities and research organizations, some of the literature can be accessed online or ordered through an interlibrary loan. There are still those who do not place all their work on the Web, so you may not find everything you need simply by surfing. Nevertheless, there are a few dedicated geotechnical Web sites that have good literature, images, and videos.

When writing an essay or report, it is a good practice to credit the source when referring to someone else's work, including the data. A common practice is to include in parentheses both the name of the author or authors and the year of the publication. At the end of the report, include a complete list of references in alphabetical order. Each item listed should include the names of the authors with their initials, the year of the publication, the title of the publication, the publishing company, the location of the publisher, and the page numbers. The style of referencing and listing differs between publications. In this book (See References), we have followed the style adapted by the American Society of Civil Engineers (ASCE).

Professional engineers often have a modest collection of *handbooks* and design aids in their libraries. These include the *Canadian Foundation Engineering Manual (2006),* the *Naval Facility Design Manual* (U.S. Navy 1971), and the design manuals published by U.S. Army Corps of Engineers. These handbooks are written mainly for practicing engineers and will have limited coverage of the theoretical developments and fundamentals.

1.6 NUMERICAL MODELING

Numerical modeling involves *finite element* or *finite difference* techniques that are implemented on micro or mainframe computers. Here, the soil is often represented as a continuum with an appropriate *constitutive model* (e.g., linear elastic material obeying Hooke's law) and *boundary conditions.* The constitutive model specifies how the material deforms when subjected to specific loading. The boundary conditions define the loading and displacements at the boundaries. A problem without boundary conditions cannot be solved; the boundary conditions make the solution unique.

Figure 1.3 shows a coarse mesh for an embankment underlain by two different soil layers. Due to symmetry, only the right half of the problem is analyzed, thus saving computational time. Making the mesh finer will result in a better solution, but will increase computational time. The bottom and right boundaries are selected after some trials to ensure that the displacements are negligible and that the stresses remain unaffected by the embankment loading.

The model geometry is discretized into hundreds or thousands of elements, each element having three or four nodes. Equations relating loads and displacements are written for every node, and the resulting simultaneous equations are solved to determine the unknowns. *ABAQUS, PLAXIS, FLAC,* and *GeoStudio 2007* are some of the popular software packages that are being used in geotechnical modeling worldwide.

To give you a taste of numerical modeling, we have included a free DVD containing the *Student Edition* of *GeoStudio 2007,* a software suite developed by *GEO-SLOPE International* (http://www.geo-slope.com) to perform numerical modeling of geotechnical and geoenvironmental problems. It is quite popular worldwide and is being used in more than 100 countries; not only in universities, but also in professional practices by consulting engineers. It includes eight stand-alone software modules: *SLOPE/W* (slope stability), *SEEP/W* (seepage), *SIGMA/W* (stresses and deformations), *QUAKE/W* (dynamic loadings), *TEMP/W* (geothermal), *CTRAN/W* (contaminant transport), *AIR/W* (airflow), and *VADOSE/W* (vadose zone and soil cover), which are integrated to work with each other. For example, the output from one program can be imported into another as input. There are tutorial movies that are downloadable from the Web site. Press *F1* for help. You can subscribe to their free monthly electronic newsletter, *Direct Contact,* which has some useful tips that will come in handy when using these programs.

The *GeoStudio 2007 Student Edition* DVD included with this book contains all eight programs with limited features (e.g., 3 materials, 10 regions, and 500 elements, when used with

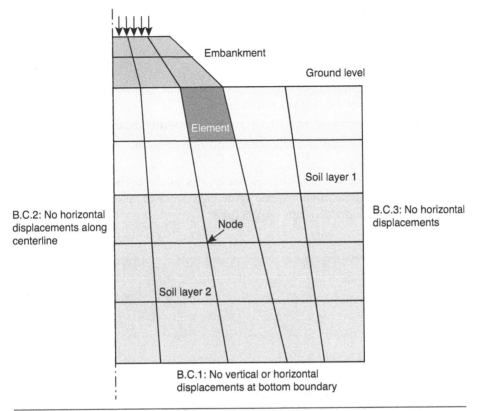

Figure 1.3 A simple mesh for an embankment underlain by two different soil layers

finite element analyses). It also contains a comprehensive engineering manual (e.g., *Stability Modeling with SLOPE/W 2007*) for each of the programs. *SLOPE/W* works on the basis of limit equilibrium theory using the method of slices. The other programs within the suite use finite element analysis. *SEEP/W*, *SIGMA/W*, and *SLOPE/W* have been used extensively in Chapters 6, 7, and 15 for solving problems. Once you become proficient with the *Student Edition*, you will require very little start-up time with the professional versions in the workplace.

It is uncommon to teach numerical modeling of geotechnical engineering during the first degree of a civil engineering program; it is more commonly viewed as a postgraduate subject with firm grounding in finite element and finite difference methods, constitutive models, etc. Nevertheless, in the professional engineering practice, fresh and recent graduates get to do some simple numerical modeling work. Numerical modeling is a very powerful tool when used correctly. No matter how sophisticated the model is, the output can only be as good as the input. Therefore, realistic results can be obtained only by using the right soil parameters.

- ❖ Geotechnical engineering, geomechanics, geoengineering, and soil mechanics are more or less the same.
- ❖ Soils are quite different from other engineering materials.
- ❖ Soils are tested to derive the engineering properties that can be used in designs.
- ❖ Try all sources of references: books, journals, conference proceedings, and the mighty World Wide Web. You will be surprised to see some good video clips on YouTube.

REVIEW EXERCISES

1. List five geotechnical Web Sites.

2. List 10 geotechnical applications and write two or three sentences about each.

3. List 10 geotechnical textbooks.

4. List five geotechnical journals.

5. List five names of those who made significant contributions to the early developments in geotechnical engineering.

6. List five different rock types.

Quiz 1. Introduction

Duration: 20 minutes

You have not started learning geotechnical engineering. Nevertheless, you will be able to answer most of the questions. Each question is worth one point.

1. What would be the mass of a 1 m by 1 m by 1 m rock?

2. What is *permeability*?

3. What is the difference between *gravel* and *clay*? Which is more permeable?

4. What is *water content* of a soil?

5. What is *porosity* of a soil?

6. What is *factor of safety*?

7. Why do we compact the soil in earthwork?

8. What is the difference between *mass* and *weight*?

9. What is the difference between *density* and unit *weight*?

10. What is the difference between *strength* and *stiffness*?

Phase Relations

2

2.1 INTRODUCTION

Soils generally contain *air*, *water*, and *solid grains*, known as the three phases. The relative proportions of these three phases play an important role in the engineering behavior of soils. The two extreme cases here are *dry* soils and *saturated* soils, both having only two phases. Dry soils have no water, and the voids are filled with only air. Saturated soils have no air, and the voids are filled with only water. Soils beneath the water table are often assumed to be saturated. Very often in geotechnical problems (e.g., earthworks) and in laboratory tests on soils, it is required to compute masses (or weights) and volumes of the different phases present within the soil.

In this chapter, you will learn how to compute masses and volumes of the different phases in a soil. We will define some simple terms and develop expressions that relate them, which will help in the computations that appear in most chapters. The definitions are quite logical, and although it is important that you *understand* them, it is not necessary that you memorize them.

2.2 DEFINITIONS

Let's consider the soil mass shown in Figure 2.1a, where all three phases are present. For simplicity, let's separate the three phases and stack them as shown in Figure 2.1b, which is known as a *phase diagram*. Here, the volumes are shown on the left and the masses on the right. M and V denote mass (or weight) and volume respectively. The subscripts are: a = air, w = water, s = soil grains (solids), v = voids, and t = total quantity of the soil under consideration. Since the mass of air M_a is negligible, $M_t = M_s + M_w$. Also, $V_v = V_w + V_a$, and $V_t = V_s + V_w + V_a$.

Water content w is a mass ratio that is used to quantify the amount of water present within the soil and is defined as:

$$w = \frac{M_w}{M_s} \times 100\%$$ (2.1)

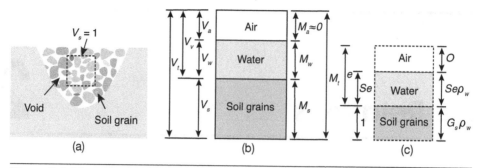

Figure 2.1 (a) a soil mass (b) phase diagram (c) phase diagram with $V_s = 1$

This is generally expressed as a percentage. Drying the soil in the oven at 105°C for 24 hours is the standard method for determining water content. The natural water content of most soils would be well below 100%, but organic soils and some marine clays can be at water contents greater than 100%.

Example 2.1: A soil sample of 26.2 g was placed in a 105°C oven for 24 hours. The dry mass of the sample turned out to be 19.5 g. What is the water content?

Solution:

$$M_t = 26.2 \text{ g}, M_s = 19.5 \text{ g}$$
$$\therefore M_w = 26.2 - 19.5 = 6.7 \text{ g}$$
$$\therefore w = (6.7/19.5) \times 100\% = 34.4\%$$

Void ratio e and porosity n are two volumetric ratios used to quantify the voids that are present within the soil. Generally, void ratio is expressed as a decimal number (e.g., 0.82) and porosity is expressed as a percentage (e.g., 45.1%) ranging from 0% to 100%. They are defined as:

$$e = \frac{V_v}{V_s} \tag{2.2}$$

$$n = \frac{V_v}{V_t} \times 100\% \tag{2.3}$$

Void ratios typically lie between 0.4 and 1 for sands, and 0.3 to 1.5 for clays. For organic soils and soft clays, the void ratio can be even more.

The degree of saturation S is a measure of the void volume that is occupied by water, expressed as a percentage ranging from 0% to 100%. It is defined as:

$$S = \frac{V_w}{V_v} \times 100\% \tag{2.4}$$

For dry soils $S = 0$ and for saturated soils (e.g., below the water table) $S = 100\%$.

Density ρ of the soil is simply the mass per unit volume. However, because of the different phases present within the soil, there are several forms of densities used in geotechnical engineering. The most common one is the bulk density ρ_m, also known as total, moist, or wet density. It is the total mass divided by total volume ($\rho_m = M_t/V_t$). Dry density ρ_d is the density of the soil at the same volume, assuming there is no water (i.e., $\rho_d = M_s/V_t$). Saturated density ρ_{sat} is the bulk density when the voids are filled with water (i.e., $\rho_{sat} = M_t/V_t$ when $S = 100\%$). Submerged density ρ' is the effective density of the soil when submerged (considering buoyancy effects) and is defined as:

$$\mathbf{r}' = \mathbf{r}_{sat} - \mathbf{r}_w \tag{2.5}$$

When weight (e.g., kN) is used instead of mass (e.g., g, kg, t), density becomes unit weight γ. You may remember that $\gamma = \rho\,g$. Never mix densities and unit weights. The definitions of bulk unit weight γ_m, dry unit weight γ_d, saturated unit weight γ_{sat}, and submerged unit weight γ' are similar to those of corresponding densities. Density of water ρ_w is 1.0 g/cm³, 1.0 t/m³, or 1000 kg/m³, and its unit weight γ_w is 9.81 kN/m³.

Specific gravity of a soil grain G_s is the ratio of the density of the soil grain to the density of the water. We know that specific gravity of mercury = 13.6, steel = 7.5, and water = 1.0. For most soils, specific gravity varies little—ranging from 2.6 to 2.8. If G_s is not known, it is reasonable to assume a value in this range. There are exceptions, where mine tailings rich in minerals have G_s values as high as 4.5. For organic soils or fly ash, it can even be lower than 2 (See Worked Example 11). The specific gravity of soil grains is generally measured using *pycnometers* (density bottles of fixed volume).

Example 2.2: A 90 g sample of dry sands was placed in a pycnometer (a density bottle used for determining the specific gravity of soil grains), and the pycnometer was filled with water; its mass is 719.3 g. A clean pycnometer was filled with water and has a mass of 663.2 g. Find the specific gravity of the sand grains.

Solution: $M_s = 90$ g. Let's find the mass of the water displaced by the sand (i.e., same volume) using Archimedes' principle. It is given by (think!!) $90 + 663.2 - 719.3 = 33.9$ g.

$$\therefore G_s = 90/33.9 = 2.65$$

2.3 PHASE RELATIONS

All the terms introduced above (e.g., w, e, S, γ_d) are ratios and therefore do not depend on the quantity of soil under consideration. In a homogeneous soil mass, they should be the same anywhere. Let's consider a portion of the soil where the volume of the soil grains is unity (i.e., $V_s = 1$) and develop the phase diagram as shown in Figure 2.1c. Here, we have simply used the given definitions and the fact that $V_s = 1$ to compute the other masses and volumes. The weights (shown on the right) are obtained simply by multiplying the volumes (shown on the left) by the corresponding densities. Now let's develop some simple and useful expressions for water content, porosity, and the different densities and unit weights. Here, we express water content (w) and degree of saturation (S) as decimal numbers instead of percentages:

$$w = \frac{M_w}{M_s} = \frac{Se}{G_s} \tag{2.6}$$

$$n = \frac{V_v}{V_t} = \frac{e}{1+e} \tag{2.7}$$

$$\rho_m = \frac{M_t}{V_t} = \left(\frac{G_s + Se}{1+e}\right)\rho_w \tag{2.8}$$

The expressions for ρ_d and ρ_{sat} can be deduced from Equation 2.8 by substituting $S = 0$ and 1 respectively. They are:

$$\rho_d = \frac{M_s}{V_t} = \left(\frac{G_s}{1+e}\right)\rho_w \tag{2.9}$$

$$\rho_{sat} = \frac{M_t}{V_t} = \left(\frac{G_s + e}{1+e}\right)\rho_w \tag{2.10}$$

From Equations 2.5 and 2.10:

$$\rho' = \left(\frac{G_s - 1}{1+e}\right)\rho_w \tag{2.11}$$

Similar equations hold for unit weights too, where ρ is replaced by γ.

Example 2.3: A saturated soil sample has water content of 24.2% and the specific gravity of the soil grains is 2.73. What are the dry and saturated unit weights?

Solution: $S = 1$, $w = 0.242$, $G_s = 2.73$

$\therefore$ From Equation 2.6 $\rightarrow e = (0.242)(2.73) = 0.661$

$$\gamma_d = \left(\frac{G_s \gamma_w}{1+e}\right) = \left(\frac{2.73 \times 9.81}{1+0.661}\right) = 16.12 \text{ kN/m}^3$$

$$\gamma_{sat} = \left(\frac{G_s + e}{1+e}\right)\gamma_w = \left(\frac{2.73 + 0.661}{1+0.661}\right) \times 9.81 = 20.03 \text{ kN/m}^3$$

It is not necessary to memorize the different equations relating the phases. From the definitions and the phase diagram for $V_s = 1$ (Figure 2.1c), one can derive them quickly. It is a good practice to go from the fundamentals.

The densities (or unit weights), water content, and specific gravity are the ones that are measured in the laboratory. Void ratio, porosity, and degree of saturation are generally not measured, but are calculated from the phase relations.

Example 2.4: The unit weight of a dry sandy soil is 15.5 kN/m³. The specific gravity of the soil grains is 2.64. If the soil becomes saturated, at the same void ratio, what would be the water content and unit weight?

Solution:

$$\gamma_d = \frac{G_s \gamma_w}{1+e} \rightarrow 15.5 = \frac{2.64 \times 9.81}{1+e} \rightarrow e = 0.6781$$

If the soil gets saturated, $S = 1 \rightarrow w = \dfrac{Se}{G_s} = \dfrac{1 \times 0.671}{2.64} = 0.254$ or 25.4%

$$\gamma_{sat} = \frac{G_s + e}{1+e} \gamma_w = \frac{2.64 + 0.671}{1 + 0.671} \times 9.81 = 19.3 \text{ kN/m}^3$$

- ❖ Do not try to memorize the equations. *Understand* the definitions and develop the phase relations from the phase diagram with $V_s = 1$. If you are determined to memorize some of the equations, you would benefit most from Equations 2.6 and 2.8.
- ❖ You can work with weights (and unit weights) or masses (and densities), but you should never mix them.
- ❖ Assume G_s (2.6 to 2.8) when required.
- ❖ Soil grains are incompressible. Their mass M_s and volume V_s remain the same at any void ratio.
- ❖ γ (N/m³) = ρ (kg/m³) g (m/s²).
- ❖ γ_w = 9.81 kN/m³; ρ_w = 1.0 g/cm³ = 1.0 t/m³ = 1000 kg/m³.

Reminder

WORKED EXAMPLES

1. Show that bulk density, dry density, and water content are related by $\rho_m = \rho_d(1 + w)$.

   Solution:

   $$\rho_m = \left(\frac{G_s + Se}{1+e}\right)\rho_w = \left(\frac{G_s + wG_s}{1+e}\right)\rho_w = \left(\frac{G_s(1+w)}{1+e}\right)\rho_w = \rho_d(1+w)$$

2. 5 kg of soil is at natural water content of 3%. How much water would you add to the above soil to bring the water content to 12%?

   Solution: Let's find the dry mass M_s (kg) of soil grains first.

   $$w = 0.03 = \frac{5 - M_s}{M_s} \rightarrow M_s = 4.854 \text{ kg and } M_w = 0.146 \text{ kg}$$

   $$\text{At } w = 12\%, M_w = 0.12 \times 4.854 = 0.583 \text{ kg}$$

   $$\therefore \text{Quantity of water to add} = 0.583 - 0.146 = 0.437 \text{ kg or 437 ml}$$

3. A 38 mm diameter and 76 mm long cylindrical clay sample has a mass of 174.2 g. After drying in the oven at 105°C for 24 hours, the mass is reduced to 148.4 g. Find the dry density, bulk density, and water content of the clay.

   Assuming the specific gravity of the soil grains as 2.71, find the degree of saturation of the clay.

   Solution: Sample volume $= \pi(1.9)^2(7.6) = 86.2 \text{ cm}^3$; $M_t = 174.2$ g;
   $M_s = 148.4$ g.

   $$\therefore \rho_d = 148.4/86.2 = 1.722 \text{ g/cm}^3$$
   $$\rho_m = 174.2/86.2 = 2.201 \text{ g/cm}^3$$
   $$w = M_w/M_s = (174.2 - 148.4)/148.4 = 0.174 \text{ or } 17.4\%$$

   Substituting in Equation 2.9:

   $$1.722 = \frac{(2.71)(1.0)}{1+e} \rightarrow e = 0.574$$

   Substituting in Equation 2.6:

   $$0.174 = \frac{S(0.574)}{2.71} \rightarrow S = 0.822 \text{ or } 82.2\%$$

4. Soil excavated from a borrow area is being used to construct an embankment. The void ratio of the in situ soil at the borrow area is 1.14, and it is required that the soil in the embankment be compacted to a void ratio of 0.70. With 200,000 m³ of soil removed from the borrow area, how many cubic meters of embankment can be made?

Solution: The volume of the soil grains V_s remain the same in the borrow area and in the embankment.

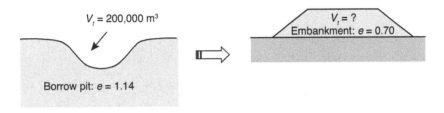

At the borrow area:

$$e = 1.14 = \frac{200,000 - V_s}{V_s} \rightarrow V_s = 93,457.9 \text{ m}^3$$

At the embankment:

$$e = 0.70 = \frac{V_t - 93,457.9}{93,457.9} \rightarrow V_t = 158,879 \text{ m}^3$$

5. A saturated, undisturbed clay sample collected below the water table has a wet mass of 651 g. The volume of the sample was determined to be 390 cm³. When dried in the oven for 24 hours, the sample has a mass of 416 g. What is the specific gravity of the soil grains?

Solution:

$$M_t = 651 \text{ g}; M_s = 416 \text{ g}; \text{ and } V_t = 390 \text{ cm}^3$$
$$\therefore w = 56.5\%, \rho_d = 1.067 \text{ g/cm}^3, \rho_{sat} = 1.669 \text{ g/cm}^3$$
$$S = 100\% \text{ (Given)}$$

Substituting in Equation 2.6:

$$0.565 = \frac{(1.0)(e)}{G_s} \rightarrow e = 0.565 \, G_s$$

$$\rho_d = \frac{G_s \rho_w}{1+e} \rightarrow 1.067 = \frac{(G_s)(1)}{1+0.565 G_s} \rightarrow G_s = 2.69$$

6. A 200 m long section of a 15 m wide canal is being deepened 1.5 m by means of a dredge. The effluent from the dredge has a unit weight of 12.4 kN/m³. The soil at the bottom of the canal has an in situ unit weight of 18.7 kN/m³. The specific gravity of the soil grains is 2.72. If the effluent is being pumped at a rate of 400 L per minute, how many operational hours will be required to complete the dredge work?

Solution: Let's find the volume of solid grains ($V_s = x$) to be removed by dredging.

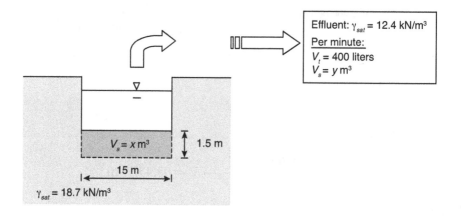

Volume of soil to be removed $= (1.5)(15)(200) = 4500 \ \text{m}^3$.

In situ unit weight (saturated) $= 18.7 \ \text{kN/m}^3$.

$$\gamma_{sat} = \left(\frac{G_s + e}{1 + e}\right)\gamma_w \rightarrow 18.7 = \left(\frac{2.72 + e}{1 + e}\right)9.81 \rightarrow e_{\text{in situ}} = 0.898$$

$$e = 0.898 = \frac{4500 - x}{x} \rightarrow x = 2370.9 \ \text{m}^3 \text{ of soil grains to be dredged}$$

Now, let's see how much soil grains ($V_s = y$) are being pumped out every minute, where $V_t = 400 \ \text{L} = 0.400 \ \text{m}^3$

$$\gamma_{sat} \text{ (effluent)} = 12.4 \ \text{kN/m}^3 \rightarrow e_{\text{effluent}} = 5.515$$

$$e = 5.515 = \frac{0.400 - y}{y} \rightarrow y = 0.0614 \ \text{m}^3 \text{ of soil grains per minute}$$

$$\therefore \text{Operational hours required} = \frac{2370.9}{0.0614 \times 60} = 644 \ \text{hours}$$

7. A 1 m-thick fill is compacted by a roller, and the thickness reduced by 90 mm. If the initial void ratio of the fill was 0.94, what is the new void ratio after the compaction?

Solution: Let's consider a 1 m² area in plan. Find the volume of soil grains V_s.

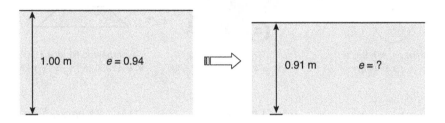

$$\therefore V_t = 1.0 \text{ m}^3 \rightarrow e = \frac{1-V_s}{V_s} \rightarrow 0.94 = \frac{1-V_s}{V_s} \rightarrow V_s = 0.516 \text{ m}^3$$

The new volume after the compaction = 0.91 m × 1.0 m² = 0.91 m³, where $V_s = 0.516$ m³ and $V_v = 0.394$ m³:

$$\therefore e = \frac{0.394}{0.516} = 0.764$$

8. The undisturbed soil at a borrow pit has a bulk unit weight of 19.1 kN/m³ and water content of 9.5%. The soil from this borrow will be used to construct a compacted fill with a finished volume of 42,000 m³. The soil is excavated by machinery and placed in trucks, each with a capacity of 4.50 m³. When loaded to the full capacity, each load of soil weighs 67.5 kN.

In the construction process, the trucks dump the soil at the site, then the soil is spread and broken up. Water is then sprinkled to bring the water content to 15%. Finally, the soil is compacted to a dry unit weight of 17.1 kN/m³.

  a. Assuming each load is to the full capacity, how many truckloads are required to construct the fill?
  b. What would be the volume of the pit in the borrow area?
  c. How many liters of water should be added to a truckload?

Solution: The water content of the borrow pit and the truck must be the same. In addition, the mass of the soil grains at the fill and the borrow pit is the same.

a.

At the borrow pit	In the truck	At the fill

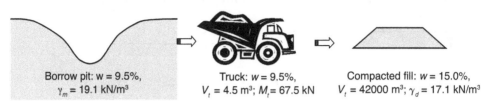

Borrow pit: $w = 9.5\%$, $\gamma_m = 19.1$ kN/m³   Truck: $w = 9.5\%$, $V_t = 4.5$ m³; $M_t = 67.5$ kN   Compacted fill: $w = 15.0\%$, $V_t = 42000$ m³; $\gamma_d = 17.1$ kN/m³

$w = 9.5\%$

$\gamma_m = 19.1$ kN/m³

$V_t = 4.50$ m³

$M_t = 67.5$ kN

$w = 9.5\%$ (same as in borrow)

$V_t = 42{,}000$ m³

$w = 15\%$

$\gamma_d = 17.1$ kN/m³

$$0.095 = \frac{67.5 - M_s}{M_s}$$

$$\therefore M_s = 61.64 \text{ kN}$$

$$M_s = (17.1)(42{,}000)$$

$$= 718{,}200 \text{ kN}$$

$\therefore$ Number of truck loads required $= 718{,}200/61.64 = 11{,}652$

b. At the borrow area, $w = 9.5\%$ and $M_s = 718{,}200$ kN (same as at the fill).

$\therefore M_w = 0.095 \times 718{,}200 = 68{,}229.0$ kN $\rightarrow M_t = 786{,}429.0$ kN
$\therefore V_t = 786{,}429.0/19.1 = 39{,}920.3$ m³

c. M_s per truckload is 61.64 kN, and the water content is increased from 9.5% to 15%. Therefore, the quantity of water that has to be added per truckload $= 61.64 \times 0.055 = 3.39$ kN or 345.6 L.

9. An irregularly shaped, undisturbed soil lump has a mass of 4074 g. To measure the volume, it was required to *thinly* coat the sample with wax (the mass and volume of which can be neglected) and weigh it submerged in water when suspended by a string. The submerged mass of the sample is 1991 g. Later, the water content of the sample and the specific gravity of the soil grains were determined to be 12.4% and 2.75 respectively. Determine the void ratio and the degree of saturation of the sample.

Solution: Mass of the water displaced $=$ upthrust $= 4074 - 1991$ g $= 2083$ g

$\therefore$ Volume of the soil specimen $= 2083$ cm³

$$w = \frac{Se}{G_s} \rightarrow Se = (0.124)(2.75) = 0.341$$

$$\rho_m = \left(\frac{G_s + Se}{1+e}\right)\rho_w \rightarrow \frac{4074}{2083} = \left(\frac{2.75 + 0.341}{1+e}\right) \rightarrow e = 0.580$$

$$w = \frac{Se}{G_s} \rightarrow S = (0.124)(2.75)/(0.580) = 0.588 \text{ or } 58.8\%$$

10. A sample of an irregular lump of *saturated* clay with a mass of 605.2 g was coated with wax. The total mass of the coated lump was 614.2 g. The volume of the coated lump was determined to be 311 cm³ by the water displacement method as used in Worked Example 9. After carefully removing the wax, the lump of clay was oven dried to a dry mass of 479.2 g. The specific gravity of the wax is 0.90. Determine the water content, dry unit weight, and the specific gravity of the soil grains.

Solution:

$$M_t = 605.2 \text{ g}, M_s = 479.2 \text{ g} \rightarrow M_w = 126.0 \text{ g}, V_w = 126 \text{ cm}^3 \text{ and } w = 26.3\%$$
$$M_{wax} = 614.2 - 605.2 = 9.0 \text{ g} \rightarrow V_{wax} = 9.0/0.9 = 10 \text{ cm}^3$$
$$V_{soil \ grains} = 311 - 126 - 10 = 175 \text{ cm}^3 \rightarrow G_s = 479.2/175 = 2.74$$
$$\rho_d = 479.2/(175 + 126) = 1.592 \text{ g/cm}^3 \rightarrow \gamma_d = 1.592 \times 9.81 = 15.62 \text{ kN/m}^3$$

11. A series of experiments are being conducted in a laboratory where fly ash ($G_s = 2.07$) is being mixed with sand ($G_s = 2.65$) at various proportions by weight. If the suggested mixes are 100/0, 90/10, 80/20...10/90, and 0/100, compute the average values of the specific gravities for all the mixes. Show the results graphically and in tabular form.

Solution: Let's show here a specimen calculation for a 70/30 mix, which contains 70% fly ash and 30% sand by weight. Let's consider 700 g of fly ash and 300 g of sand.

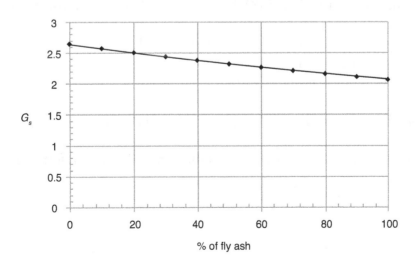

Volume of fly ash = $700/2.07 = 338.2 \text{ cm}^3$

Volume of sand = $300/2.65 = 113.2 \text{ cm}^3$

Total mass = 1000 g

Total volume = $338.2 + 113.2 = 451.4 \text{ cm}^3$

∴ Density = $1000/451.4 = 2.22 \text{ g/cm}^3 \rightarrow G_s = 2.22$

Mix	Fly ash (g)	Sand (g)	Fly ash (cm³)	Sand (cm³)	G_s
100/0	1000	0	483.09	0.00	2.07
90/10	900	100	434.78	37.74	2.12
80/20	800	200	386.47	75.47	2.16
70/30	700	300	338.16	113.21	2.22
60/40	600	400	289.86	150.94	2.27
50/50	500	500	241.55	188.68	2.32
40/60	400	600	193.24	226.42	2.38
30/70	300	700	144.93	264.15	2.44
20/80	200	800	96.62	301.89	2.51
10/90	100	900	48.31	339.62	2.58
0/100	0	1000	0.00	377.36	2.65

REVIEW EXERCISES

1. State whether the following are true or false.
   a. A porosity of 40% implies that 40% of the total volume consists of voids
   b. A degree of saturation of 40% implies that 40% of the total volume consists of water
   c. Larger void ratios correspond to larger dry densities
   d. Water content cannot exceed 100%
   e. The void ratio cannot exceed 1

2. From the expressions for ρ_m, ρ_{sat}, ρ_d, and ρ', deduce that $\rho' < \rho_d \le \rho_m \le \rho_{sat}$.

3. Tabulate the specific gravity values of different soil and rock forming minerals (e.g., quartz).

4. A thin-walled sampling tube of a 75 mm internal diameter is pushed into the wall of an excavation, and a 200 mm long undisturbed sample with a mass of 1740.6 g was obtained. When dried in the oven, the mass was 1421.2 g. Assuming that the specific gravity of the soil grains is 2.70, find the void ratio, water content, degree of saturation, bulk density, and dry density.

Answer: 0.679, 22.5%, 89.5%, 1.97 t/m³, 1.61 t/m³

5. A large piece of rock with a volume of 0.65 m³ has 4% porosity. The specific gravity of the rock mineral is 2.75. What is the weight of this rock? Assume the rock is dry.

Answer: 16.83 kN

6. A soil-water suspension is made by adding water to 50 g of dry soil, making 1000 ml of suspension. The specific gravity of the soil grains is 2.73. What is the total mass of the suspension?

Answer: 1031.7 g

7. A soil is mixed at a water content of 16% and compacted in a 1000 ml cylindrical mold. The sample extruded from the mold has a mass of 1620 g, and the specific gravity of the soil grains is 2.69. Find the void ratio, degree of saturation, and dry unit weight of the compacted sample. If the sample is soaked in water at the same void ratio, what would be the new water content?

Answer: 0.926, 46.5%, 1.397 t/m³, 34.4%

8. A sample of soil is compacted into a cylindrical compaction mold with a volume of 944 cm³. The mass of the compacted soil specimen is 1910 g and its water content is calculated at 14.5%. Specific gravity of the soil grains is 2.66. Compute the degree of saturation, density, and unit weight of the compacted soil.

Answer: 76.4%, 2.023 g/cm³, 19.85 kN/m³

9. The soil used in constructing an embankment is obtained from a borrow area where the in situ void ratio is 1.02. The soil at the embankment is required to be compacted to a void ratio of 0.72. If the finished volume of the embankment is 90,000 m³, what would be the volume of the soil excavated at the borrow area?

Answer: 105,698 m³

10. A subbase for an airport runway 100 m wide, 2000 m long, and 500 mm thick is to be constructed out of a clayey sand excavated from a nearby borrow where the in situ water content is 6%. This soil is being transported into trucks having a capacity of 8 m³, where

each load weighs 13.2 metric tons (1 metric ton = 1000 kg). In the subbase course, the soil will be placed at a water content of 14.2% to a dry density of 1.89 t/m³.

a. How many truckloads will be required to complete the job?

b. How many liters of water should be added to each truckload?

c. If the subbase becomes saturated, what would be the new water content?

Answer: 15,177, 1021 L, 15.9%

11. The bulk unit weight and water content of a soil at a borrow pit are 17.2 kN/m³ and 8.2% respectively. A highway fill is being constructed using the soil from this borrow at a dry unit weight of 18.05 kN/m³. Find the volume of the borrow pit that would make one cubic meter of the finished highway fill.

Answer: 1.136 m³

12. A soil to be used in the construction of an embankment is obtained by hydraulic dredging of a nearby canal. The embankment is to be placed at a dry density of 1.72 t/m³ and will have a finished volume of 20,000 m³. The in situ saturated density of the soil at the bottom of the canal is 1.64 t/m³. The effluent from the dredging operation, having a density of 1.43 t/m³, is pumped to the embankment site at the rate of 600 L per minute. The specific gravity of the soil grains is 2.70.

a. How many operational hours would be required to dredge sufficient soil for the embankment?

b. What would be the volume of the excavation at the bottom of the canal?

Answer: 1396 hours, 33,841 m³

13. A contractor needs 300 m³ of aggregate base for a highway construction project. It will be compacted to a dry unit weight of 19.8 kN/m³. This material is available in a stockpile at a local material supply yard at a water content of 7%, but is sold by the metric ton and not by cubic meters.

a. How many tons of aggregate should the contractor purchase?

b. A few weeks later, an intense rainstorm increased the water content of the stockpile to 15%. If the contractor orders the same quantity for an identical section of the highway, how many cubic meters of compacted aggregate base will he produce?

Answer: 648 t, 279.2 m³

14. A sandy soil consists of perfectly spherical grains of the same diameter. At the loosest possible packing, the particles are stacked directly above each other. Show that the void ratio is 0.910.

There are few possible arrangements for a denser packing. You can (with some difficulty) show that the corresponding void ratios are 0.654, 0.433, and 0.350 (densest). Use the diagram shown below to visualize this. See how the void ratio decreases with the increasing number of contact points.

Note: This is not for the fainthearted!

Loosest   Dense

Soil Classification

3

3.1 INTRODUCTION

Soils can behave quite differently depending on their geotechnical characteristics. In *coarse-grained soils* where the grains are larger than 0.075 mm (75 μm), the engineering behavior is influenced mainly by the relative proportions of the different grain sizes present within the soil, the density of their packing, and the shapes of the grains. In *fine-grained soils* where the grains are smaller than 0.075 mm, the mineralogy of the soil grains and the water content have greater influence on the engineering behavior than do the grain sizes. The borderline between coarse- and fine-grained soils is 0.075 mm, which is the smallest grain size one can distinguish with the naked eye. Based on the grain sizes, soils can be grouped as *clays, silts, sands, gravels, cobbles,* and *boulders* as shown in Figure 3.1. This figure shows the borderline values as per the *Unified Soil Classification System (USCS),* the *British Standards (BS),* and the *Australian Standards (AS).* Within these major groups, soils can still behave differently, and we will look at some systematic methods of classifying them into distinct subgroups.

3.2 COARSE-GRAINED SOILS

The major factors that influence the engineering behavior of a coarse-grained soil are: (a) relative proportions of the different grain sizes, (b) packing density, (c) grain shape. Let's discuss these three separately.

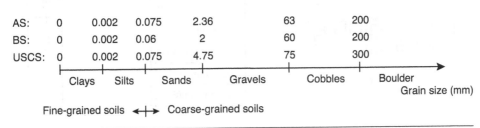

Figure 3.1 Major soil groups

3.2.1 Grain Size Distribution

The relative proportions of the different grain sizes in a soil are quantified in the form of *grain size distribution*. They are determined through *sieve analysis* (ASTM D6913; AS 1289.3.6.1) in coarse-grained soils and through *hydrometer analysis* (ASTM D422; AS 1289.3.6.3) in fine-grained soils.

In sieve analysis, a coarse-grained soil is passed through a set of sieves stacked with opening sizes increasing upward. Figure 3.2a shows a sieve with 0.425 mm diameter openings. When 1.2 kg soil was placed on this sieve and shaken well (using a *sieve shaker*), 0.3 kg passes through the openings and 0.9 kg is retained on it. Therefore, 25% of the grains are finer than 0.425 mm and 75% are coarser. The same exercise is now carried out on another soil with a stack of sieves (Figure 3.2b) where 900 g soil was sent through the sieves, and the masses retained are shown in the figure. The percentage of soil finer than 0.425 mm is given by $[(240 + 140 + 60)/900] \times 100\% = 48.9\%$. Sometimes in North America, sieves are specified by a sieve number instead of by the size of the openings. A 0.075 mm sieve is also known as No. 200 sieve, implying that there are 200 openings per inch. Similarly, No. 4 sieve = 4.75 mm and No. 40 sieve = 0.425 mm.

In the case of fine-grained soils, a *hydrometer* is used to determine the grain size. A hydrometer is a floating device used for measuring the density of a liquid. It is placed in a soil-water suspension where about 50 g of fine-grained soil is mixed with water to make 1000 ml of suspension (Figure 3.2c). The hydrometer is used to measure the density of the suspension at different times for a period of one day or longer. As the grains settle, the density of the suspension decreases. The time-density record is translated into grain size percentage passing data using Stokes' law. The hydrometer data can be merged with those from sieve analysis for the complete grain size distribution. *Laser sizing*, a relatively new technique, is becoming more popular for determining the grain size distributions of the fine-grained soils. Here, the soil grains are sent through a laser beam where the rays are scattered at different angles depending on the grain sizes.

The grain size distribution data is generally presented in the form of a *grain size distribution curve* shown in the figure in Example 3.1, where percentage passing is plotted against the corresponding grain size. Since the grain sizes vary in a wide range, they are usually shown on a logarithmic scale.

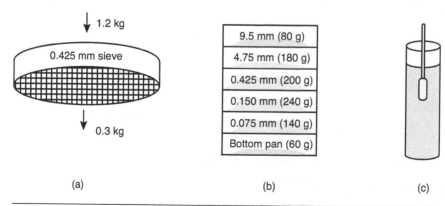

Figure 3.2 Grain size analysis: (a) a sieve (b) stack of sieves (c) hydrometer test

Example 3.1: Using the data from sieve analysis shown in Figure 3.2b, plot the grain size distribution data with grain size on the x-axis using a logarithmic scale and percentage passing on the y-axis.

Solution: Let's compute the *cumulative* percent passing each sieve size and present as:

Size (mm)	9.5	4.75	0.425	0.150	0.075
% passing	91.1	71.1	48.9	22.2	6.7

The grain size distribution curve is shown:

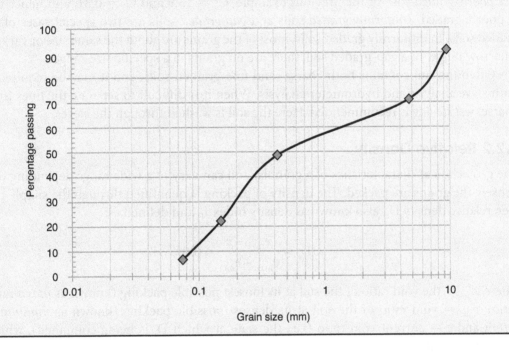

The grain size distribution gives a complete and quantitative picture of the relative proportions of the different grain sizes within the soil mass. At this stage, let's define some important grain sizes such as D_{10}, D_{30}, and D_{60}, which are used to define the shape of the grain size distribution curve. D_{10} is the grain size corresponding to 10% passing; i.e., 10% of the grains are smaller than this size. Similar definitions hold for D_{30}, D_{60}, etc. In Example 3.1, $D_{10} = 0.088$ mm, $D_{30} = 0.195$ mm, and $D_{60} = 1.4$ mm. The shape of the grain size distribution curve is described through two simple parameters: the coefficient of uniformity (C_u) and the coefficient of curvature (C_c). They are defined as:

$$C_u = \frac{D_{60}}{D_{10}}$$ (3.1)

and

$$C_c = \frac{D_{30}^2}{D_{10}D_{60}} \qquad (3.2)$$

A coarse-grained soil is said to be *well-graded* if it consists of soil grains representing a wide range of sizes where the smaller grains fill the voids created by the larger grains, thus producing a dense packing. A sand is described as well-graded if $C_u > 6$ and $C_c = 1\text{-}3$. A gravel is well-graded if $C_u > 4$ and $C_c = 1\text{-}3$. A coarse-grained soil that cannot be described as well-graded is a *poorly graded* soil. In the previous example, $C_u = 15.9$ and $C_c = 0.31$, and hence the soil is poorly graded. *Uniformly graded* soils and *gap-graded* soils are two special cases of poorly graded soils. In uniformly graded soils, most of the grains are about the same size or vary within a narrow range. In a gap-graded soil, there are no grains in a specific size range.

Often the soil contains both coarse- and fine-grained soils, and it may be required to do both sieve analysis and hydrometer analysis. When it is difficult to separate the fines from the coarse, *wet sieving* is recommended. Here the soil is washed through the sieves.

3.2.2 Relative Density

The geotechnical characteristics of a granular soil can vary in a wide range depending on how densely the grains are packed. The density of packing is quantified through the simple parameter, relative density D_r, also known as density index I_D and defined as:

$$D_r = \frac{e_{max} - e}{e_{max} - e_{min}} \times 100\% \qquad (3.3)$$

where e_{max} = the void ratio of the soil at its loosest possible packing (known as *maximum void ratio*); e_{min} = void ratio of the soil at its densest possible packing (known as *minimum void ratio*); and e = current void ratio (i.e., the state at which D_r is being computed), which lies between e_{max} and e_{min}.

The loosest state is achieved by raining the soil from a small height (ASTM D4254; AS1289.5.5.1). The densest state is obtained by compacting a moist soil sample, vibrating a moist soil sample, or both (ASTM D4253; AS 1289.5.5.1) in a rigid cylindrical mold. Relative density varies between 0% and 100%; 0% for the loosest state and 100% for the densest state. Terms such as *loose* and *dense* are often used when referring to the density of packing of granular soils. Figure 3.3 shows the commonly used terms and the suggested ranges of relative densities.

In terms of unit weights, relative density can be expressed as:

$$D_r = \frac{(\gamma_d - \gamma_{d,min})}{(\gamma_{d,max} - \gamma_{d,min})} \frac{\gamma_{d,max}}{\gamma_d} \qquad (3.4)$$

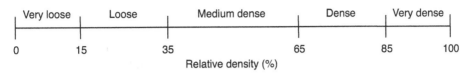

Figure 3.3 Granular soil designations based on relative densities

3.2.3 Grain Shape

Shapes of the grains can be *angular, subangular, subrounded,* or *rounded* (Figure 3.4a–d). When the grains are angular there is more interlocking among the grains, and therefore the strength and stiffness of the soils would be greater. For example, in roadwork, angular aggregates would provide better interlocking and resistance against dislodgement.

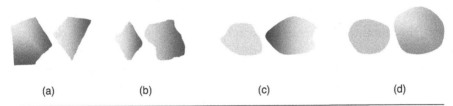

Figure 3.4 Grain shapes: (a) angular (b) subangular (c) subrounded (d) rounded

---

Example 3.2: Maximum and minimum dry density tests were carried out on sand (G_s = 2.67), using a one liter compaction mold. In the loosest state, 1376 g of dry sand filled the mold. At 8% water content with vibratory compaction, 1774 g of wet sand was placed in the mold at its densest state. If the void ratio of this sand at the site is 0.72, what is the relative density?

Solution:
$$\rho_d = \frac{G_S \rho_w}{1+e} \rightarrow e_{max} = \frac{2.67 \times 1}{1.376} - 1 = 0.940$$

At densest state:
$$\rho_{d,max} = \frac{\rho_{m,max}}{1+w} = \frac{1.774}{1+0.08} = 1.643 \text{ g/cm}^3 \rightarrow e_{min} = 0.625$$

$$D_r = \frac{e_{max} - e}{e_{max} - e_{min}} \times 100\% = \frac{0.940 - 0.720}{0.940 - 0.625} \times 100\% = 69.8\% \ldots \text{ dense sand}$$

3.3 FINE-GRAINED SOILS

While gravel, sand, and silt grains are equidimensional (i.e., same order of dimensions in the three orthogonal directions), clay *particles* (or grains) are generally two-dimensional or sometimes one-dimensional. They look like flakes or needles. Their surfaces are electrically charged due to a charge imbalance between the cations and anions in their atomic structures. Since the particles are flakey and finer than 2 μm, they have larger *specific surfaces* (surface area per unit mass) than do silts, sands, or gravels. Large specific surfaces and the electric charges make the clays sticky when wet, and make them *cohesive,* which makes them behave differently than *noncohesive soils* do, such as sands and gravels. Clays are also known as *cohesive soils.* To understand the behavior of clays, it is necessary to have some knowledge about clay mineralogy.

3.3.1 Clay Mineralogy

Earth is about 12,500 km in diameter, and most geotechnical engineering work is confined to the top few hundred meters of the *crust,* which is comprised essentially of oxygen (49.2%), silicon (25.7%), and aluminum (7.5%) present in the form of oxides, with some Fe^{3+}, Ca^{2+}, Na^+, K^+, Mg^{2+}, etc. The atomic structure of a clay mineral is made of one of the two structural units: *tetrahedrons* containing a silicon atom at the center surrounded by four oxygen atoms at the corners, and *octahedrons* containing aluminum or magnesium ions at the center surrounded by six hydroxyl or oxygen ions at the corners, as shown in Figures 3.5a and 3.5c. When several of these units are joined together along a common base, they make tetrahedral and octahedral sheets, which are represented schematically with the symbols shown in Figures 3.5b and 3.5d. An octahedral sheet containing aluminum cations is called *gibbsite,* and when it contains magnesium cations, it is called *brucite.*

Different clay minerals are produced by stacking tetrahedral and octahedral sheets in different ways. Three of the most common minerals, *kaolinite, illite,* and *montmorillonite,* are shown schematically in Figure 3.6a–c. Kaolinite is formed by stacking several layers of alternating tet-

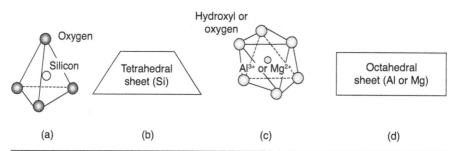

Figure 3.5 Atomic structural units of clay minerals: (a) tetrahedron (b) tetradedral sheet (c) octahedron (d) octahedral sheet

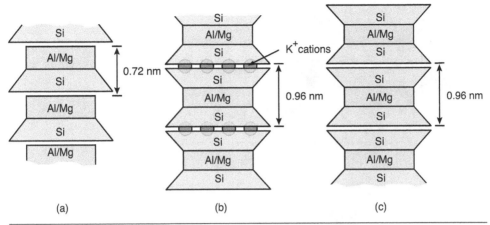

Figure 3.6 Three major clay minerals: (a) kaolinite (b) illite (c) montmorillonite

rahedral and octahedral sheets, each 0.72 nm in thickness, stacked on top of each other (Figure 3.6a). They are held together by strong hydrogen bonds that prevent them from separating. Kaolinite is used in ceramics, paper, paint, and medicine. Illite is formed by stacking several layers 0.96 nm thick that consist of an octahedral sheet sandwiched between two tetrahedral sheets (one inverted) as shown in Figure 3.6b. They are held together by potassium ions, where the bonds are not as strong as in kaolinite. Montmorillonites (Figure 3.6c), also known as *smectites*, have the same atomic structure as illite, but the layers are held together by weak van der Waals forces. When water gets between the layers, they are easily separated and there will be a substantial increase in volume, known as *swelling*. Montmorillonitic clays are called *expansive* or *reactive clays*. They expand in the presence of water and shrink when dried. This shrink-swell behavior causes billions of dollars worth of damage to buildings and roads across the globe. Other clay minerals that are of some interest in geotechnical engineering are *chlorite*, *halloysite*, *vermiculite*, and *attapulgite*.

The specific surfaces of these three major clay minerals are kaolinite = 15 m^2/g, illite = 80 m^2/g, and montmorillonite = 800 m^2/g. There is always a charge imbalance within a clay particle due to substitution of cations within the *pore water*, and the net effect is to make the clay particle negatively charged. The charge deficiency (i.e., the negative charge) is significantly larger for montmorillonites than for kaolinites. Depending on the mineralogy of the clay particles and chemistry of the *pore water*, the clay particles can form different *fabrics*. Two of the extreme situations are *dispersed* (also known as *oriented*) and *flocculated* fabrics. In a dispersed fabric, most of the clay particles are oriented in the same direction. In a flocculated fabric, they are randomly oriented. Clay microfabric can be examined using a scanning electron microscope (SEM) or atomic force microscope (AFM). The scanning electron micrograph of a dispersed kaolinite clay fabric is shown in Figure 3.7.

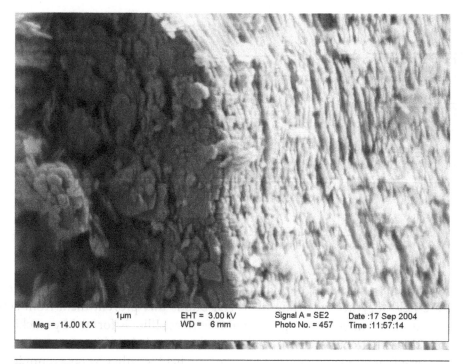

Figure 3.7 Scanning electron micrograph of a dispersed clay fabric (Courtesy of Dr. Ajanta Sachan, IIT Kanpur, India)

3.3.2 Atterberg Limits

Atterberg limits were developed by A. Atterberg, a Swedish scientist, in 1911 for pottery and were later modified to suit geotechnical engineering needs by Arthur Casagrande in 1932. When a dry fine-grained soil is mixed with water in small increments, the soil will pass through distinct states known as *brittle solid*, *semi-solid*, *plastic solid* and *liquid*, as shown in Figure 3.8. Atterberg limits are simply borderline water contents that separate the different consistencies the fine-grained soils can have. These borderline water contents are *shrinkage limit*, *plastic limit* and *liquid limit*.

Shrinkage limit (SL or w_s) is the highest water content below which there will be no reduction in volume when the soil is dried. Plastic limit (PL or w_p) is the lowest water content at which the soil shows plastic behavior. Above the liquid limit (LL or w_L), the soil flows like a liquid. When the water content is between PL and LL, the soil remains plastic and the difference between LL and PL is known as the *plasticity index* (PI or I_p). Silts have little or no plasticity, and their PI $\approx$ 0. These original definitions of the Atterberg limits are rather vague and are not reproducible, especially by inexperienced operators. Casagrande (1932) standardized the test procedures which are discussed below.

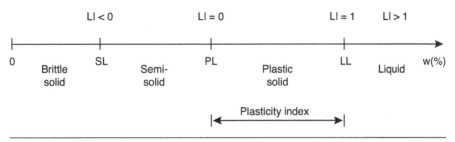

Figure 3.8 Atterberg limits

Soil fraction smaller than 0.425 mm is used in the laboratory tests for LL and PL. Liquid limit is determined by two different methods: *Casagrande's percussion cup method* (ASTM D4318; AS 1289.3.1.1; Figure 3.9a) and *Swedish fall cone method* (ASTM D4318; AS 1289.3.9.1; Figure 3.9b). In Casagrande's percussion cup method, the moist soil pat is placed in the cup and a standard groove is cut using a grooving tool (Figure 3.9a). The cup is raised and dropped over a height of 10 mm, hitting a hard rubber or micarta plastic base, and the number of blows required to make the groove close over 12.5 mm ($\frac{1}{2}$ inch) is recorded at different water contents. Liquid limit is defined as the water content at which such closure occurs at 25 blows. In a Swedish fall cone test, a stainless steel cone, having a mass of 80 g and angle of 30°, is initially positioned to touch the moist soil sample in a standard cup (Figure 3.9b). It is released to fall freely and penetrate the moist soil for 5 seconds, and the penetration is recorded at different water contents. The water content at which the penetration is 20 mm is the liquid limit. Plastic limit is defined as the lowest water content at which the soil can be rolled into a 3 mm ($\frac{1}{8}$ in.) thread (ASTM D4318; AS 1289.3.2.1). In geotechnical engineering, LL and PL are used more than SL.

Liquidity index (LI or I_L) is a measure of how close the natural water content (w_n) is to the liquid limit, and is defined as:

$$LI = \frac{w_n - PL}{LL - PL} \qquad (3.5)$$

It takes the value of 1.0 at LL and 0 at PL. At water content greater than LL, LI is greater than 1.

Linear shrinkage (LS) is a simple test to measure the potential of the clay to shrink, which is also an indirect measure of the plasticity. Here, a soil pat mixed at water content near the liquid limit is placed in a standard mold (Figure 3.9c) and in an oven for 24 hours (AS 1289.3.4.1). The percentage reduction in the length of the soil is known as linear shrinkage, which is approximately equal to 40–50% of PI.

Let's consider two different fine-grained soils X (20% clay and 80% silts) and Y (80% clay and 20% silts), having the same plasticity index of 40. In X, the 20% clay contributes to all the plasticity, whereas in Y, there is a significantly larger quantity of clay contributing to the same

(a)

(b)

(c)

Figure 3.9 Liquid limit and linear shrinkage test devices: (a) Casagrande's percussion cup with grooving tool (b) fall cone device (c) shrinkage limit mold

degree of plasticity. Understandably, the clay component in X is more plastic than the one in Y. This is quantified by the term *activity* (A), defined as:

$$A = \frac{\text{PI}}{\text{\% clay fraction}} \qquad (3.6)$$

Thus, the activities of clays X and Y are 2 and 0.5 respectively. Larger activity values (e.g., > 1.5) generally suggest potential swell-shrink problems.

3.4 SOIL CLASSIFICATION

The person at the site classifying the samples is different from the one who will do the designs in the office. Therefore, it is necessary to communicate the soil description as precisely as possible, from the site to the design office. A *soil classification system* does just that. It is a systematic method that groups soils of similar behavior, describes them, and classifies them. The strict guidelines and the standard terms proposed eliminate any ambiguity and make it a universal language among geotechnical engineers. There are several soil classification systems currently in use. The *Unified Soil Classification System* (ASTM D2487) is the most popular one that is used in geotechnical engineering worldwide. The American Association of State Highway Transportation Officials (AASHTO) classification system is quite popular for roadwork where soils are grouped according to their suitability as subgrade or embankment materials. There are also country-specific standards such as Australian Standards (AS), British Standards (BS), Indian Standards (IS), etc.

3.4.1 Unified Soil Classification System (USCS)

The Unified Soil Classification System (USCS) was originally developed by Casagrande (1948), and revised in 1952 by Casagrande, the U.S. Bureau of Reclamation (USBR), and the U.S. Army Corps of Engineers to make it suitable for wider geotechnical applications. The coarse-grained soils are classified based on their grain size distribution and the fine-grained soils based on Atterberg limits. The four major soil groups in the USCS, defined on the basis of the grain size, are gravel (G), sand (S), silt (M), and clay (C). Two other special groups are *organic soils* (O) and *peat* (Pt). Organic soils are mostly clays containing organic material that may have come from decomposed living organisms, plants, and animals. When the liquid limit reduces by more than 25% upon over-drying, the soil can be classified as an organic soil.

When the coarse fraction within a soil is greater than 50%, it is classified as a coarse-grained soil. Within a coarse-grained soil, if the gravel fraction is more than the sand fraction, then it is classified as gravel and vice versa. When the fine fraction is greater than 50%, it is classified as a fine-grained soil.

The USCS recommends a symbol in the form of *XY* for a soil where the prefix X is the *major soil group* and suffix Y is the *descriptor*. Coarse-grained soils (G or S) are described as well-graded W, poorly graded P, silty M, or clayey C. Fine-grained soils are described on the basis of plasticity as low L or high H. These are summarized in Table 3.1.

Table 3.1 Major soil groups, descriptors, and symbols

Major soil group (X)	Descriptor (Y)	Possible symbols (XY)
Gravel (G) Sand (S)	Well graded (W) Poorly graded (P) Silty (M) Clayey (C)	GW, GP, GM, GC SW, SP, SM, SC
Silt (M) Clay (C) Organic (O)	Low plasticity (L) High plasticity (H)	ML, MH CL, CH OL, OH

Fine-grained soils are classified based on Atterberg limits, irrespective of the relative proportions of clays and silts, which are of little value in classification. Casagrande (1948) proposed the PI-LL chart shown in Figure 3.10 where the A-line separates the clays from silts. Most fine-grained soils plot near the A-line. The U-line is the upper limit for any fine-grained soils.

Let's look at the USCS for four special cases on the basis of *percentage of fines*: 0 to 5%, 5 to 12%, 12 to 50% and 50 to 100%:

- 0–5% fines: A coarse-grained soil with negligible fines. Classify as GW, GP, SW, or SP.
- 12–50% fines: A coarse-grained soil with substantial fines that can have a significant influence on the soil behavior. Classify as GM, GC, SM, or SC.
- 50–100% fines: A fine-grained soil. Classify as ML, MH, CL, or CH. Coarse grains are ignored (even if significant in presence) in assigning the symbol. If the LL and PL values plot in the hatched area in Figure 3.10, the soil is given a dual symbol, CL-ML.

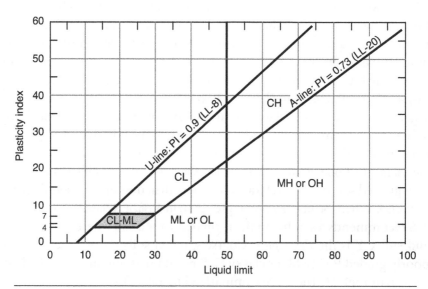

Figure 3.10 Casagrande's *PI-LL* chart

- 5–12% fines: A coarse-grained soil with some fines that can influence the soil behavior. Classify as XY-XZ, where *X* is the major coarse (*G* or *W*), *Y* defines the gradation (W or P), and Z is the major fines (M or C), with possible symbols of GW-GC, SP-SM, etc.

All possible symbols and the four groups of the USCS are summarized in Figure 3.11.

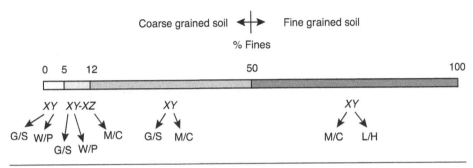

Figure 3.11 USCS summary

3.4.2 AASHTO Soil Classification System

The AASHTO soil classification system is used mainly for roadwork, and it groups soils into eight groups from A-1 to A-8. Groups A-1 to A-3 denote coarse-grained soils (defined as soils where % fines ≤ 35), and groups A-4 to A-7 denote fine-grained soils (defined as soils where % fines > 35). Group A-8 includes highly organic soils (e.g., peats). As with other classification systems, sieve analysis and Atterberg limits are used in assigning the above symbols. A-1 soils are well-graded gravels or sands with fines (≤ 25%) of little plasticity (≤ 6), and are further subdivided into A-1-a (% finer than 2 mm ≤ 50; % finer than 0.425 mm ≤ 30; and % fines ≤ 15) and A-1-b (% finer than 0.425 mm ≤ 50 and % fines ≤ 25). A-3 soils are clean, poorly graded fine sands with less than 10% nonplastic fines. A-2 group soils are coarse-grained with ≤ 35% fines and are differentiated on the basis of PI and LL using Figure 3.12. Depending on the quadrant they fall into, they are assigned symbols A-2-4, A-2-5, A-2-6, and A-2-7. When the fine fraction is greater than 35%, the soil is grouped as A-4, A-5, A-6, or A-7 on the basis of PI and LL values, as shown in Figure 3.12. Here, the horizontal line of PI = 10 separates clays from silts. When the PI and LL are both high, the soil is subdivided into A-7-5 and A-7-6 by a 45° line.

It is also necessary to assign a number known as *group index* (*GI*) within parentheses after the symbol. Group index is defined as:

$$GI = (F - 35)[0.2 + 0.005(LL - 40)] + 0.01(F - 15)(PI - 10) \qquad (3.7)$$

where *F* is the percentage of fines. In the case of A-2-6 and A-2-7, GI is calculated from:

$$GI = 0.01(F - 15)(PI - 10) \qquad (3.8)$$

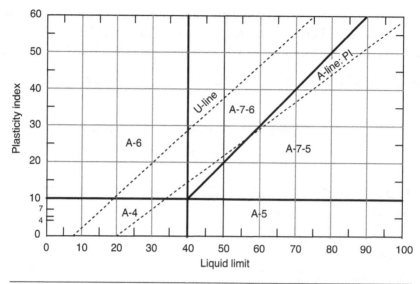

Figure 3.12 AASHTO classification of A-4, A-5, A-6, and A-7 subgroups

GI should be rounded off to the nearest integer and should be taken as zero when negative and for soil groups A-1-a, A-1-b, A-3, A-2-4, and A-2-5. The AASHTO symbol is assigned by a process of elimination, trying from group A-1 to A-8 (from low to high). The first group that fits the data gives the correct classification.

3.4.3 Visual Identification and Classification of Soils

A good geotechnical engineer must be able to identify and classify soils in the field simply by the feel. This is easier with coarse-grained soils where one can include qualitative information on *grain size* (fine, medium, or coarse), *grain shape, color, homogeneity, gradation, state of compaction* or *cementation, presence of fines*, etc. Based on these data and relative proportions, it is possible to assign the USCS symbol and a description (ASTM D2488). Fine-grained soils can be identified as clays or silts based on *dry strength* or *dilatancy*. A moist pat of clay feels *sticky* between the fingers, and silts feel *gritty*. Dry strength is a qualitative measure of how easy it is to crush a dry lump of fine-grained soil between the fingers. Clays have high dry strength, and silts have low dry strength. A dilatancy test involves placing a moist pat of soil in the palm and shaking it vigorously to see how quickly water rises to the surface. The standard terms used for describing dilatancy are quick, slow, none, etc. Silts show quick dilatancy and clays show slow to none.

Based on what we have discussed up to now, a comparison of clays and nonclays (i.e., silts, sands, and gravels) is made in Table 3.2.

Table 3.2 Clays vs. non-clays

Clays	Non-clays (silts, sands and gravels)
Grains are 1 (needle) or 2 (plate) dimensional	Grains are equidimensional
Grains < 2 μm	Grains > 2 μm
Negatively charged grains	Inert—no charge imbalance
Cohesive and hence sticky	Non-cohesive and gritty
Plastic (i.e., $PI > 0$)	Non-plastic ($PI \approx 0$)
High specific surface	Low specific surface
Colloidal (surface forces are significant)	Non-colloidal

Reminder

- ❖ 0.075 mm (75 μm) separates coarse- and fine-grained soils.
- ❖ Uniformly graded soils are poorly graded.
- ❖ Grain size distributions are mainly for coarse-grained soils; Atterberg limits are for fines.
- ❖ Clay particles are negatively charged flakes with a high surface area and are smaller than 2 μm in size; they are plastic and sticky (cohesive). Silts are nonplastic (PI ≈ 0).
- ❖ A fine-grained soil is classified as clay or silt based on Atterberg limits—not on relative proportions.
- ❖ The first thing one should know when classifying a soil is the % of fines. This determines how the symbol is assigned and how the soil is described.
- ❖ In AASHTO, the general rating as a subgrade decreases from left to right, A-1 being the best and A-8 being the worst.

WORKED EXAMPLES

1. The grain size distribution data for three soils are given below. The fines in Soil *A* showed low dry strength and the LL and PL of Soil *C* are 45 and 23 respectively. Classify the three soils.

Sieve size	Percentage passing		
(mm)	Soil A	Soil B	Soil C
19.0	100.0		99.0
9.5	69.0		83.0
4.75	48.8	100	61.5
2.36	34.4	95.0	36.0
1.18	24.3	36.0	32.0
0.600	17.3	4.0	31.0
0.300	12.2	0.0	30.0
0.150	8.7		26.5
0.075	6.1		9.0

Solution: The D_{10}, D_{30}, D_{60}, C_u, and C_c values, and the percentages of gravels, sands, and fines within the three soils are summarized:

	Soil A	Soil B	Soil C
D_{10} (mm)	0.2	0.73	0.08
D_{30} (mm)	1.8	1.1	0.3
D_{60} (mm)	7.3	1.5	4.5
C_u	36.5	2.1	56.3
C_c	2.2	1.1	0.25
% gravel	51.2	0	38.5
% sands	42.7	100	52.5
% fines	6.1	0	9

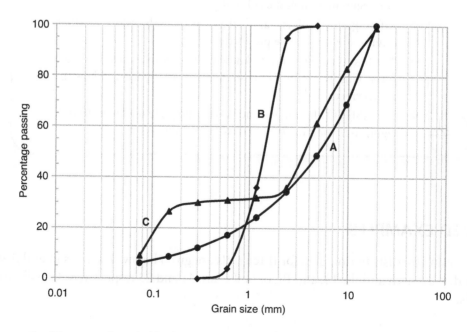

With 6.1% fines, Soil A would be classified as a coarse-grained soil with dual symbols. Since the fines have low dry strength, they are silty. It can be classified as *well-graded, silty, sandy gravel* with a symbol of GW-GM.

Soil B is uniformly graded sand, with all grains in the range of 0.5–3.0 mm. It can be classified as *uniformly graded sand* with a symbol of SP.

Soil C is a gap-graded soil that has no grains present in the size range of 0.5–2.0 mm. With 9% fines, it requires a dual symbol. PI and LL values plot above the A-line in Casagrande's PI-LL chart, implying that the fines are clayey. Therefore, the soil can be classified as *gap-graded, clayey gravelly sand* with a symbol of SP-SC.

2. The grain size distribution curve of a soil is described as:

$$p = \sqrt{\frac{D}{D_{max}}} \times 100$$

where p = percentage passing, D = grain size, and D_{max} = maximum grain size within the soil.

   a. Is the soil well graded or poorly graded?
   b. Assuming the largest grain within the soil is 50 mm, describe the soil with the USCS symbol.

Solution:

   a. At 10%, 30% and 60% passing:

$$10 = \sqrt{\frac{D_{10}}{D_{max}}} \times 100; \quad 30 = \sqrt{\frac{D_{30}}{D_{max}}} \times 100; \quad \text{and } 60 = \sqrt{\frac{D_{60}}{D_{max}}} \times 100$$

From the above three equations, it is a fairly straightforward exercise to show that:

$$C_u = \frac{D_{60}}{D_{10}} = 36, \text{ and } C_c = \frac{D_{30}^2}{D_{10} D_{60}} = 2.25 \rightarrow \text{A well graded soil}$$

Note: This equation was proposed by Fuller and Thompson (1907) for mix design of aggregates in selecting the right mix for a well-graded soil.

   b. Substituting D_{max} = 50 mm in the equation used in 2a:

$$p_{0.075} = 3.9\% \text{ and } p_{4.75} = 30.8\%$$
$$\therefore \% \text{ gravels} = 69.2, \% \text{ sands} = 26.9, \text{ and } \% \text{ fines} = 3.9$$

The soil can be classified as well-graded sandy gravel with negligible fines, with a USCS symbol of GW.

3. Classify the following soils using the given grain size distribution and Atterberg limits data.

    a. 68% retained on 4.75 mm sieve; 11% passed 0.075 mm sieve; fines showed quick dilatancy; $C_u = 34$ and $C_c = 0.83$

    b. 77% passed 4.75 mm sieve; 20% passed 0.075 mm sieve; fines have high dry strength

    c. 42% passed 4.75 mm sieve; 4% passed 0.075 mm sieve; $C_u = 18$, $C_c = 2.1$

    d. 14% retained on 4.75 mm sieve; 60% passed 0.075 mm sieve; LL = 65, PL = 35

Solution:

    a. % gravel = 68; % sands = 21; % fines = 11

        Fines showing quick dilatancy → silty fines
        $C_u = 34$ and $C_c = 0.83$ → Poorly graded
        ∴ GP-GM: Poorly graded, silty sandy gravel

    b. % gravel = 23; % sands = 57; % fines = 20

        Fines have high dry strength → clayey fines
        ∴ SC: Clayey gravelly sands

    c. % gravel = 58; % sands = 38; % fines = 4

        $C_u = 18$, $C_c = 2.1$ → well graded
        ∴ GW: Well-graded sandy gravel

    d. % gravel = 14; % sands = 26; % fines = 60

        LL= 65, PI = 30 → lies below A-line and hence silt
        ∴ *MH*: Gravelly, sandy high-plastic silt

REVIEW EXERCISES

1. State whether the following are true or false.
    a. The coefficient of uniformity has to be greater than unity
    b. The density of the soil-water suspension in a hydrometer test increases with time
    c. The plastic limit is always greater than the plasticity index
    d. The shrinkage limit is always less than the plastic limit
    e. Soils with larger grains have larger specific surfaces
    f. A 10 mm cube and 10 mm diameter sphere have the same specific areas

2. List 10 different sieve numbers and the corresponding aperture diameters.

3. How are the density-time measurements in a hydrometer translated into grain size percentage-passing data?

4. Write a 300-word essay on clay mineralogy covering *cation exchange capacity*, *isomorphous substitution*, and *diffuse double layer* in relation to what was discussed in 3.3.1 Clay Mineralogy.

5. Two coarse-grained Soils A and B have grain size distribution curves that are approximately parallel. A is coarser than B. Compare their D_{10}, D_{50}, e_{max}, and e_{min} values, stating which is larger. Give your reasons.

6. Calculate the specific surface of 1 mm, 0.1 mm, and 0.01 mm diameter soil grains assuming specific gravity of 2.70. See how the specific surface increases with the reduction in grain size. Compare these values to those of the flakey clay minerals such as kaolinite, illite, and montmorillonite.

   Answer: 2.2×10^{-3} m^2/g, 2.2×10^{-2} m^2/g, 0.22 m^2/g

7. The maximum and minimum void ratios of a granular soil are 1.00 and 0.50 respectively. What would be the void ratio at 40% relative density? What are the porosities at maximum and minimum void ratios? Assuming $G_s = 2.65$, determine the maximum and minimum dry densities.

   Answer: 0.80; 50%, 33.3%; 1.77 t/m^3, 1.33 t/m^3

8. The grain size distribution curves of four Soils A, B, C, and D are shown below and their LL and PL are: Soil C = 40, 16; Soil D = 62, 34. Classify the soils, giving their USCS symbols and descriptions.

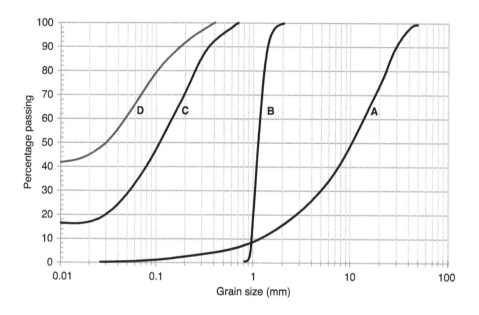

9. List all USCS symbols and align them with the corresponding and *most likely* AASHTO symbols. In some cases, there may be more than one. Once you have finished, go from AASHTO to USCS. This exercise will reinforce your understanding of AASHTO.

Quiz 2: Phase relations and soil classification

Duration: 30 minutes

1. Which of the following can exceed 100%?

   (a) Relative density   (b) Degree of saturation   (c) Water content   (d) Percentage passing   (e) Porosity

   (½ point)

2. Which of the following can exceed 1?

   (a) Void ratio   (b) Liquidity index   (c) Activity   (d) Coefficient of curvature

   (½ point)

3. Which of the following values is likely for the mass of a 1 m^3 rock?

   (a) 29 kg   (b) 290 kg   (c) 2900 kg   (d) 29 ton

   (½ point)

4. Which are the three most abundant elements found in the earth's crust?

   (a) O, Si, Al   (b) O, Si, N   (c) O, Si, Fe   (d) Si, Al, Mg

   (½ point)

5. Which of the following terms is not used with fine-grained soils?

   (a) Relative density   (b) Activity   (c) Liquidity index   (d) Plasticity

   (½ point)

6. Which of the following is not a valid USCS symbol?

   (a) GP-GM   (b) SW-WC   (c) SP   (d) CL

   (½ point)

7. The sieve analysis data of a soil are given below. The fines showed very low dry strength. Without plotting the grain size distribution curve, describe the soil, giving it the USCS symbol.

Sieve size (mm)	9.5	4.75	2.38	0.85	0.075
% passing	100	60	40	30	10

   (3 points)

8. Two samples of crushed mine tailings *A* and *B* are mixed in equal proportions by weight. Sample *A* contains 20% fines and has a specific gravity of 2.80. Sample *B* contains 30% fines and a specific gravity of 3.70. Find the percentage of fines and the average specific gravity of the grains in the mix.

   (4 points)

Compaction

4

4.1 INTRODUCTION

Natural ground is not always suitable in its present state for the proposed construction work. For example, the granular soils at a proposed site for a high-rise building may be in a looser state than desired, suggesting potential future *stability* problems or *settlement* problems, or both. The landfill clay liner that lies at the bottom of a landfill may allow more leachate than desired to flow through, polluting the groundwater. The simplest remedy in both circumstances is to compact the soils to ensure they have adequate strength and stiffness to limit any postconstruction settlement and stability problems, and to limit the quantity of seepage through the soils. Compaction is one of the most popular *ground improvement techniques* carried out in earthworks associated with roads, embankments, landfills, buildings, and backfills behind retaining walls. Generally, the main objective is to *increase* the strength and stiffness of the soil and *reduce* the permeability of the soil, all of which are achieved through a reduction in the void ratio. Some common machinery used in earthmoving is shown in Figures 4.1a through 4.1e. The soil excavated from the borrow area is transported to the site, where it is sprinkled with a specific quantity of water and compacted to the appropriate density. Acting like a lubricant, water sticks to the soil grains and facilitates the compaction process, thus densifying the soil.

Reduction in void ratio is a measure of the effectiveness of compaction. Since void ratio is never measured directly, it is indirectly quantified through the dry density of the compacted earthwork. It can be seen intuitively and in Equation 2.9 that lower void ratios equate to larger dry densities.

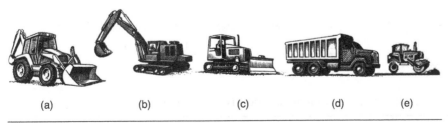

Figure 4.1 Some earthmoving machinery: (a) excavator (b) backhoe (c) spreader (d) dump truck (e) roller

4.2 VARIABLES IN COMPACTION

Water content and *compactive effort* are the two major variables that influence the degree of compaction and the engineering behavior of the compacted soil. This is illustrated through Example 4.1.

Example 4.1: A soil is compacted in a cylindrical mold with a volume of 1000 cm³ at six different water contents, using the same compactive effort (Test 1). After compaction, the samples were extruded and weighed. The same test was repeated, but with a larger compactive effort (Test 2). The water contents and wet masses of the samples from the two tests are given.

Water content (%)	Wet mass (g)	
	Test 1	Test 2
11	1867	1937
13	1956	2034
15	2044	2108
17	2106	2118
19	2090	2097
21	2036	2055

Compute the dry densities and plot them against the water content for both tests.

Solution: The volume of the compacted sample is 1000 cm³. The dry density can be determined using the equation $\rho_m = \rho_d(1 + w)$ from Chapter 2 under Worked Example 1. The computed values are shown.

w (%)	Wet mass (g)		Dry density (t/m³)	
	Test 1	Test 2	Test 1	Test 2
11	1867	1937	1.682	1.745
13	1956	2034	1.731	1.800
15	2044	2108	1.777	1.833
17	2106	2118	1.800	1.810
19	2090	2097	1.756	1.762
21	2036	2055	1.683	1.698

The dry density vs. water content variation is shown on page 51.

Continues

Example 4.1: *Continued*

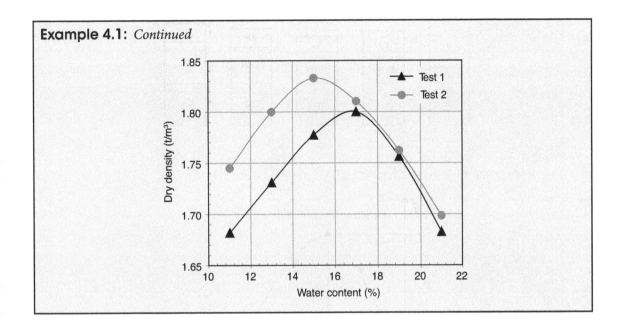

From both tests in Example 4.1, it can be seen that the dry density increases with the water content up to a certain value, where the dry density is known as the maximum dry density $\rho_{d,\,max}$ and the corresponding water content is known as *optimum water content*. A further increase in water content results in a reduction in the dry density. Increasing the compactive effort (see Example 4.1) leads to a reduction in the optimum water content and an increase in the maximum dry density. The optimum water content and the maximum dry density of the two tests are:

Test 1: optimum water content = 17.0%; $\rho_{d,\,max} = 1.80$ t/m^3
Test 2: optimum water content = 15.0%; $\rho_{d,\,max} = 1.83$ t/m^3

A curve drawn through the peaks of all compaction curves with different compactive efforts on the same soil is known as the *line of optimum*. The compacted earthwork will have very good geotechnical characteristics (i.e., strength, stiffness, permeability, etc.) when it is compacted near the optimum water content. Particularly in clayey soils, the behavior of the compacted earthwork is quite sensitive to the water content in the vicinity of the optimum water content. Therefore, it is necessary to know the optimum water content and the maximum dry density of a soil under a specific compactive effort in order to specify the right values for the field work. Terms such as *dry of optimum* or *wet of optimum* are used depending on if the compaction is carried out at a water content less or greater than the optimum water content.

The phase diagrams of the compacted soil at different water contents are shown in Figure 4.2a. The variations of dry density and void ratio against the water content are shown in Figures 4.2b and 4.2c respectively.

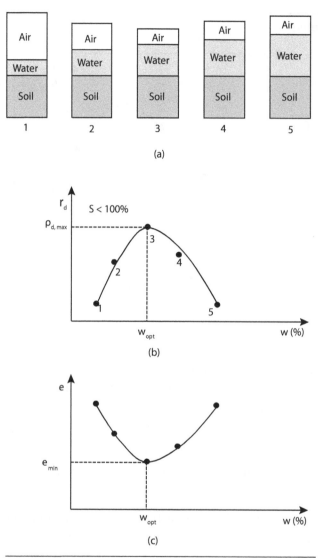

Figure 4.2 Compaction: (a) phase diagrams (b) ρ_d vs. w plot (c) e vs. w plot

4.3 LABORATORY TESTS

In the field, soil is compacted in 150–500 mm thick layers (known as *lifts*) using a wide range of rollers. Laboratory compaction tests were developed by R. R. Proctor in the 1930s, replicating the field compaction process in a cylindrical compaction mold with a volume of about 1 liter. The *standard Proctor compaction test* (ASTM D698; AS1289.5.1.1) and the *modified Proctor compaction test* (ASTM D1557; AS 1289.5.2.1) are the two popular compaction tests carried out for developing the compaction curve, and hence derive the optimum water content and maximum dry density. Here, a hammer of specific mass falling through a specific height is used for compacting the soil in a few layers of equal thickness, as shown in Figure 4.3. The test details are summarized in Table 4.1.

4.3.1 Zero Air Void Curve

From Equations 2.6 and 2.9, it can be shown that:

$$\rho_d = \frac{G_s \rho_w}{1 + \dfrac{wG_s}{S}} \tag{4.1}$$

Example 4.2: Show that the compactive effort imparted to the soil in a standard Proctor compaction test is 552 kJ/m³.

Solution: Work done per blow = 2.5 × 9.81 × 0.3 Nm = 7.36 Nm (or Joules)

When compacted in three layers with 25 blows per layer, the total energy imparted to the soil is:

$$3 \times 25 \times 7.36 = 552 \text{ J}$$

Volume of the compacted soil = 1000 cm³ = 10^{-3} m³

∴ Compactive effort = 552 kJ/m³ (See Table 4.1)

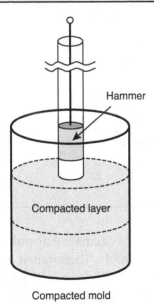

Hammer

Compacted layer

Compacted mold

Figure 4.3 Compaction mold and hammer

Table 4.1 Standard and modified Proctor compaction test details

	Standard Proctor	Modified Proctor
Mass of hammer (kg)	2.5	4.5
Hammer drop (mm)	300	450
Number of layers	3	5
Blows per layer	25	25
Compactive effort (kJ/m³)	552	2483

Therefore, in any soil (i.e., for a known value of G_s), the value of S is fixed for a specific pair of values of w and ρ_d. In other words, every point in the ρ_d-w space (see figure in Example 4.1) has a specific value of S. Thus, Equation 4.1 can be used to draw contours of S in a ρ_d-w space.

Example 4.3: Draw the contours of $S = 100\%$, $S = 90\%$, and $S = 70\%$ in the plot shown in Example 4.1, assuming $G_s = 2.72$.

Solution: Let's substitute $S = 70\%$ and $G_s = 2.72$ in Equation 4.1, which gives ρ_d as a function of w. This can be repeated for $S = 90$ and 100%. The calculated values are shown.

w (%)	ρ_d (t/m³) for S-contours		
	S = 70	S = 90	S = 100
11	1.906	2.041	2.094
13	1.807	1.953	2.009
15	1.718	1.872	1.932
17	1.638	1.797	1.860
19	1.565	1.728	1.793
21	1.498	1.664	1.731

These are plotted as shown.

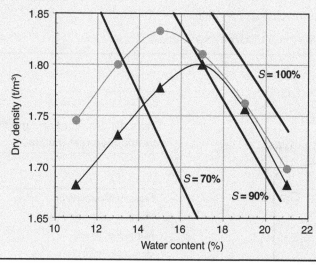

The contour of $S = 100\%$ in the ρ_d-w space is known as the *zero air void curve*. Any point to the right of the zero air void curve implies $S > 100\%$, which is not possible. Therefore, it is necessary that any compaction test point must lie to the left of the zero air void curve, which is a good check. It is quite common to show the zero air void curve along with the compaction curves. The S-contours in Example 4.3 give an idea of the degree of saturation of all test points. Sometimes, they are replaced by air content a contours where air content is defined as the ratio of the air void V_a volume to the total volume V_t. In terms of a, Equation 4.1 becomes:

$$\rho_d = \frac{G_s(1-a)\rho_w}{1+wG_s} \qquad (4.2)$$

Similar to $S = 70\%$, 90%, and 100%, one can draw $a = 30\%$, 10%, and 0% using Equation 4.2. *They are not the same.*

4.4 FIELD COMPACTION, SPECIFICATION, AND CONTROL

There is a wide range of rollers that are being used for compacting soils in the field. The compactive effort can be in the form of static pressures (e.g., smooth-wheeled roller), kneading (e.g., sheepsfoot roller), vibration (e.g., vibratory plates), or impact (e.g., impact roller), or any of these combined. While clays can be compacted effectively by a kneading action, vibratory compaction is the most effective in granular soils. Figure 4.4a shows an impact roller. Figure 4.4b shows a water truck sprinkling water to the soil in preparation for the compaction.

In *clayey soils* in particular, the behavior of the compacted earthwork can be very sensitive to the water content. A comparison is given in Table 4.2.

Compacting dry or wet of optimum has its own advantages and disadvantages. Depending on the expected performance of the compacted earthwork in service, one would select the appropriate water content. For example, a landfill liner should have low permeability and ductility to minimize future cracking. Therefore, it is better to compact it wet of optimum. On the other hand, a foundation base requires higher strength and stiffness, and hence it is better to compact it dry of optimum.

There are two ways of specifying compaction of earthworks, namely, *method specification* and *end-product specification*. In method specification, the engineer representing the client takes responsibility for the finished product and specifies every detail including type of roller, number of passes, lift thickness, water content, etc. In end-product specification, the contractor is required to select the variables and take responsibility for meeting the requirements of the end product. The specified requirements generally include a narrow range of water content and dry density of the compacted earthwork. The dry density is often specified as a certain percentage of the laboratory maximum dry density (e.g., 95% of $\rho_{d,\,max}$ from the modified Proctor compaction test in the laboratory). This is expressed through a variable known as relative compaction R, defined as:

$$R = \frac{\rho_{d,\text{field}}}{\rho_{d,\text{max\_lab}}} \times 100\% \qquad (4.3)$$

where $\rho_{d,\,field}$ is the dry density of the compacted earthwork, and $\rho_{d,\,max\_lab}$ is the maximum dry density determined by the laboratory compaction test. R can exceed 100% due to a larger

(a)

(b)

(c)

Figure 4.4 Field compaction: (a) impact roller (b) water truck sprinkling water (c) nuclear densometer

Table 4.2 Effects of compacting dry vs. wet of optimum in clays

	Dry of optimum	Wet of optimum
Clay fabric	Flocculated	Dispersed
Strength	High	Low
Stiffness	High (brittle)	Low (ductile)
Permeability	High	Low
Swelling potential	High	Low
Shrinkage potential	Low	High

compactive effort in the field. Dry density of the compacted earthwork and the water content are determined by a *sand cone/replacement test* (ASTM D1556; AS1289.5.3.1) or *nuclear densometer* (ASTM D2922; AS 1289.5.8.1, see Figure 4.4c). Sand cone tests are *destructive* (i.e., requires that a hole be dug into the compacted ground) and nuclear densometer tests are *nondestructive* and faster, hence more popular. These control tests are carried out on the compacted earthwork at a specified frequency (e.g., one test per 500 m³) to ensure the specifications are met. When discussing coarse-grained soils, it is possible to specify the density in terms of relative density than relative compaction. Lee and Singh (1971) suggested that they are related by:

$$R = 80 + 0.2D_r \qquad (4.4)$$

Example 4.4: Standard Proctor compaction was carried out on a clayey sand, and the compaction curve is shown in the figure. The specific gravity of the soil grains is 2.71.

a. What are the maximum dry density and the optimum water content?
b. Find the void ratio and degree of saturation at the optimum water content.

The compaction specifications require that the relative compaction be at least 98% and the water content to be within -2 to $-\frac{1}{2}$% from the optimum water content. A sand replacement test was carried out on the compacted earthwork where a 1240 cm³ hole was dug into the ground. The mass of the soil removed from the hole was 2748 g, which became 2443 g on drying. Does the compaction meet the specifications?

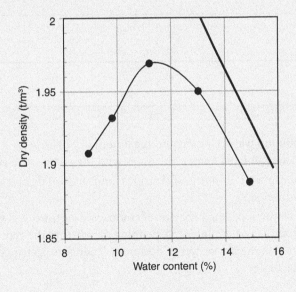

Continues

Example 4.4: *Continued*

Solution:

a. From the figure,

optimum water content (owc) = 11.5% and $\rho_{d,\max}$ = 1.97 t/m³

b. At optimum,

$$\rho_d = \frac{G_s \rho_w}{1+e} \rightarrow e = \frac{2.71}{1.97} - 1 = 0.376$$

$$w = \frac{Se}{G_s} \rightarrow S = \frac{0.115 \times 2.71}{0.376} = 82.9\%$$

Specifications:

1. $9.5\% \leq w_{field} \leq 11.0\%$
2. $\rho_{d,field} \geq 0.98 \times 1.97 (= 1.93)$ t/m³

Sand replacement test:

$$V_t = 1240 \text{ cm}^3, M_t = 2748 \text{ g}, M_s = 2443 \text{ g}$$
$$w_{field} = 12.5\% \text{ and } \rho_{d,field} = 1.97 \text{ t/m}^3$$

The compaction does not meet the specifications; it satisfies dry density but not the water content.

Reminder

❖ Optimum water content and maximum dry density for a specific soil are not fixed values; they vary with the compactive effort.

❖ You can work in terms of densities (and masses) or unit weights (and weights).

❖ In clayey soils, the behavior of compacted earthwork is very sensitive to the water content, depending on whether the clay is compacted to the dry or wet of optimum. Therefore, a stringent control is necessary.

WORKED EXAMPLES

1. A standard Proctor compaction test is carried out on the soil sample (G_s = 2.74) collected from an earthwork and the compaction curve is shown in the figure. Draw the zero air void curve to see if it intersects the compaction curve.

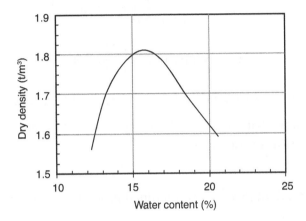

The compaction specifications require that the earthwork be compacted to a relative compaction of at least 95% with respect to the standard Proctor compaction test, and that the water content be within ± 1½% of the optimum water content. A field density test was later carried out to check the quality of compaction. A hole was dug in the compacted earthwork and 957 g soil was removed. The volume of the hole, as measured through a sand cone test, was 450 cm³. A 26.3 g soil sample that was removed from the hole was then dried in the oven and had a mass of 22.8 g. Does the compaction meet the specifications?

Solution: Let's use Equation 4.1 with G_s = 2.74 to locate a few points for the zero air void curve. This gives:

$$\rho_d = 2.74/(1 + 2.74w)$$

on the zero air void curve. Substituting w (%) = 17, 18, 19, and 20 in this equation gives ρ_d (t/m³) = 1.87, 1.84, 1.80, and 1.77. Plotting these four points on the above plot shows that the compaction curve fully lies to the left of the zero air void curve.

From the laboratory:

$$owc_{lab} = 16.0\% \text{ and } \rho_{d, max} = 1.81 \text{ g/cm}^3 \text{ or t/m}^3 \text{ (see figure)}$$

$$w_{field} = \frac{26.3 - 22.8}{22.8} \times 100 = 15.35\%$$

and

$$\rho_{m,\text{field}} = \frac{957}{450} = 2.13 \text{ g/cm}^3$$

$$\therefore \rho_{d,\text{field}} = \frac{2.13}{1+0.1535} = 1.84 \text{ g/cm}^3$$

$$\therefore \text{Relative compaction} = \frac{1.84}{1.81} \times 100 = 102\%$$

Specifications: (a) $R \geq 95\%$ (standard Proctor) and (b) $14.5\% \leq w_{\text{field}} \leq 17.5\%$

The control test shows that the compaction meets the specifications with respect to both water content and relative compaction.

2. The data from a standard Proctor and modified Proctor compaction test on a soil ($G_s = 2.64$) are given:

Standard Proctor:

Water content (%)	9.3	11.8	14.3	17.6	20.8	23.0
Dry density (t/m³)	1.691	1.715	1.755	1.747	1.685	1.619

Modified Proctor:

Water content (%)	9.3	12.8	15.5	18.7	21.1
Dry density (t/m³)	1.873	1.910	1.803	1.699	1.641

a. Plot the compaction curves along with the zero air void curve and find the optimum water content and the maximum dry density for each test.
b. Compaction control tests were carried out at four different field locations, and the results are as follows:

Control test no.	Volume of soil (cm³)	Mass of wet soil (g)	Mass of dry soil (g)
1	946	1822	1703
2	980	2083	1882
3	957	1960	1675
4	978	2152	1858

Compute the dry density, bulk density, and the water content for each test and plot the points in the above graph along with the compaction curves.

c. The compaction specification requires that the in situ dry density be greater than or equal to 95% of the maximum dry density from the modified Proctor compaction test and for the water contents to be within ± 2% of the modified Proctor optimum water content. Determine which of the four control tests meet the specifications, and give reasons why the specifications were not met for the tests that failed.

Solution: The computed values are shown in the plot.

$$\text{owc}_{\text{modified Proctor}} = 12.5\%, \text{ and}$$
$$\rho_{d,\,\text{max\_modified Proctor}} = 1.91 \text{ t/m}^3$$

Specifications require that: (a) $\rho_{d,\,\text{field}} \geq 1.81$ t/m^3 and (b) $10.5\% \leq w_{\text{field}} \leq 14.5\%$.

∴ Only the control tests falling within the shaded region would meet both water content and relative compaction criteria.

Control test 1: Too dry and low dry density

Control test 2: Meets the specifications (falls within the shaded region)

Control test 3: Too wet and low dry density

Control test 4: Control test itself is invalid—lies to the right of zero air void curve

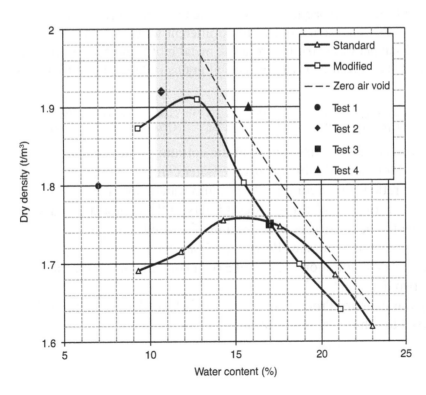

REVIEW EXERCISES

1. Write a 500-word essay on the different types of rollers used in compaction, clearly stating where each is suitable. Include pictures wherever possible.

2. Discuss the ground improvement techniques *dynamic compaction* and *vibroflotation*. Include pictures.

3. From phase relations (Chapter 2), show that the air content a is given by:

$$a = \frac{e(1-S)}{(1+e)}$$

and use this relation to show that:

$$\rho_d = \frac{G_s(1-a)\rho_w}{1+wG_s}$$

4. A standard Proctor compaction test was carried out on a silty clay, using a 1L compaction mold. The tests were carried out with six different water contents. Every time, the entire compacted sample was extruded from the metal mold, and the wet and dry masses were determined. The specific gravity of the soil grains is 2.69. The test data are summarized below.

Mass of wet sample (g)	1751	1907	2054	2052	2009	1976
Mass of dry sample (g)	1516	1634	1735	1700	1639	1590

a. Plot the compaction curve and find the optimum water content and maximum dry density. Plot the void ratio against the water content in the same plot to show that the void ratio is the minimum at optimum water content.
b. Draw the zero air void curve. Does it intersect the compaction curve?
c. What would be the degree of saturation of a sample compacted at the optimum water content in a standard Proctor compaction test?
d. Draw the 95% saturation curve and 5% air content curve in the above plot. Why are these two different?
e. Using the standard Proctor compactive effort, at what water content would you compact to achieve 80% saturation?

Answer: 19%, 1.75 t/m³, 94%, 17.5%

5. A compacted fill was made to the following specifications:

- Relative compaction to be at least 95% with respect to the standard Proctor compaction test, and
- Water content to be within the range of optimum − ½% to optimum + 2%

The dry density vs. water content plot from a standard Proctor compaction test is shown in the figure below. A sand cone test was done as part of the control measure. Here, an 840 cm³ hole was dug into the ground, from which 1746 g soil was removed. An 85 g sample of this soil was dried in an oven to 70.4 g. The specific gravity of the soil grains is 2.71.

a. Determine if the compacted earthwork meets the specifications
b. Find the degree of saturation and the air content at the optimum water content

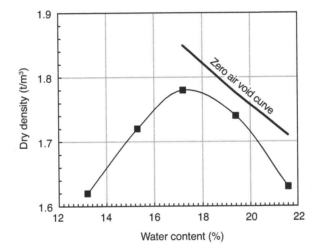

Answer: Does not meet the specs, 88.4%, 4.0%

Effective Stress, Total Stress, and Pore Water Pressure

5

5.1 INTRODUCTION

Soils are *particulate media*. They are made of an assemblage of soil grains of different sizes and shapes. They contain three phases: namely, air, water, and soil grains. In geotechnical engineering analyses, the soil mass is often assumed to be a *continuous medium* for convenience, where the presence of three phases is neglected and the entire soil mass is assumed to behave as a *homogeneous* and *isotropic* elastic body. This is far from reality, but it enables us to solve the problem.

In a particulate medium where the voids are filled with air and water, the normal stresses σ are shared by the soil grains, water, and air. In this chapter, you will learn how to compute the normal stresses acting separately on soil grains and water in a saturated soil. We will not worry about partially saturated soils where some of the normal stresses are carried by the air within the voids, which are too complex for now.

5.2 EFFECTIVE STRESS PRINCIPLE

In a *saturated* soil, the *total normal stress* σ at any point, in *any direction*, is shared by the soil grains and the water within the voids (known as *pore water*). The component of normal stress acting on the soil grains is known as the *effective stress* or *intergranular stress* σ'. The remainder of the normal stress carried by the water within the voids is known as *pore water pressure* or *neutral stress u*. Therefore, the total stress at any point, in any direction, can be written as:

$$\sigma = \sigma' + u \tag{5.1}$$

From now on, we will denote vertical normal stress and horizontal normal stress as σ_v and σ_h respectively. Therefore,

$$\sigma_v = \sigma'_v + u \text{ and} \tag{5.2}$$

$$\sigma_h = \sigma'_h + u \tag{5.3}$$

Note that pore water pressure, being hydrostatic, is the same in any direction. In this chapter, we will only deal with the vertical stresses, both effective and total.

5.3 VERTICAL NORMAL STRESSES DUE TO OVERBURDEN

In a dry soil mass having unit weight of γ (see Figure 5.1a), the vertical normal stress σ_v at point X, depth h below the ground level is simply given by $\sigma_v = \gamma h$. This is often called *overburden pressure*. If a uniform surcharge pressure of q is placed at the ground level, then $\sigma_v = \gamma h + q$. If there are three different soil layers as shown in Figure 5.1b, the vertical normal stress at X is given by $\sigma_v = \gamma_1 h_1 + \gamma_2 h_2 + \gamma_3 h_3$.

Now, let's see what happens when water is present. Let the saturated unit weight and submerged unit weight be γ_{sat} and γ' respectively. The total vertical stress at point X in Figure 5.1c is given by:

$$\sigma_v = \gamma_{sat} h \tag{5.4}$$

The pores are all interconnected, and hence the hydrostatic pore water pressure at this point is:

$$u = \gamma_w h \tag{5.5}$$

where γ_w is the unit weight of water. Therefore, the effective vertical normal stress becomes:

$$\sigma'_v = \sigma_v - u = (\gamma_{sat} - \gamma_w)h = \gamma' h \tag{5.6}$$

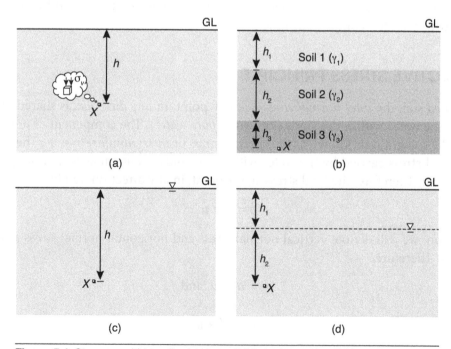

Figure 5.1 Stresses within soils: (a) dry soil (b) dry layered soil (c) saturated soil with water table at ground level (d) saturated soil with water table below ground level

When the water table is at some depth below the ground level as shown in Figure 5.1d, σ_v, u, and σ_v' can be written as:

$$\sigma_v = \gamma_m h_1 + \gamma_{sat}\, h_2 \tag{5.7}$$

$$u = \gamma_w h_2 \tag{5.8}$$

$$\sigma_v' = \gamma_m h_1 + \gamma' h_2 \tag{5.9}$$

Example 5.1: In a sandy terrain, the water table lies at a depth of 3 m below ground level. Bulk and saturated unit weights of the sand are 17.0 kN/m³ and 18.5 kN/m³ respectively. What is the effective vertical stress at 10 m depth?

Solution:

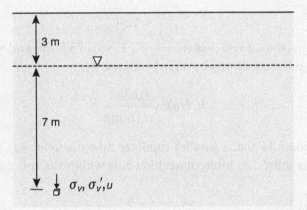

At 10 m depth, applying Equation 5.9,

$$\sigma_v' = 3 \times 17.0 + 7 \times (18.5 - 9.81) = 111.8 \text{ kPa}$$

Alternatively, σ_v (from Equation 5.7) and u (from Equation 5.8) can be determined and σ_v' can be obtained by subtracting u from σ_v. (That is a slightly longer way.)

When the soil is partially saturated, the situation is more complex. Here, the normal stresses on the soil elements are shared by the soil grains, pore water, and the pore air. Thus Equation 5.1 becomes:

$$\sigma = \sigma' + \chi u_w + (1-\chi)u_a \tag{5.10}$$

where u_w and u_a are the pore water pressure and pore air pressure respectively, and χ is a constant between 0 and 1 that can be determined from a triaxial test. In dry soils, $\chi = 0$. In saturated soils, $\chi = 1$.

5.4 CAPILLARY EFFECTS IN SOILS

Let's review some simple physics of *capillary*. When a glass capillary tube of inner diameter d is placed in a dish containing water as shown in Figure 5.2a, there is an immediate rise of water within the tube to a height of h_c with a meniscus at the top. Here, the capillary effect is caused by surface tension T between the interfaces of the glass tube, water, and air. The water column of height h_c that appears to be hanging from the inner walls is in equilibrium under two forces: the surface tension T around the perimeter at the top, and the self-weight of the water column. Therefore, for equilibrium:

$$T\cos\alpha \times \pi d = \frac{\pi d^2}{4} \times h_c \times \gamma_w$$

$$\therefore h_c = \frac{4T\cos\alpha}{\gamma_w d}$$

For a clean glass tube in contact with water, $\alpha = 0°$; $T = 0.073$ N/m; and $\gamma_w = 9810$ N/m^3. Substituting these values in the equation above, h_c becomes:

$$h_c\ (\text{m}) = \frac{0.0298}{d\ (\text{mm})} \tag{5.11}$$

It is clear from Equation 5.11 that a smaller capillary tube diameter has a larger capillary rise. How does this relate to soils? The interconnected voids within the soil skeleton act as capillary

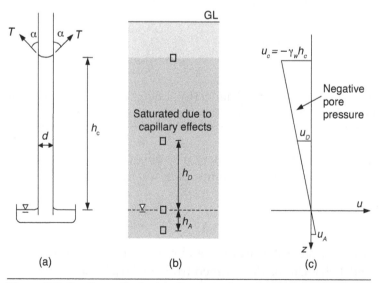

Figure 5.2 Capillary effects: (a) glass tube in water (b) field situation (c) pore water pressure variation with depth

tubes (not straight though), enabling water to rise to significant heights above the water table. We can assume that the effective pore size is about 1/5 of D_{10}. Therefore, the capillary rise h_c in soils can be written as:

$$h_c \text{ (m)} \approx \frac{0.15}{D_{10} \text{ (mm)}}$$ (5.12)

Example 5.2: Estimate the capillary rise in a sandy silt where $D_{10} = 30\ \mu m$.

Solution: Substituting $D_{10} = 0.030$ mm in Equation 5.12:

$$h_c = 5.0 \text{ m}$$

Capillary rise can vary from a few mm in gravels to several meters in clays. Capillary pressures are similar to suction and hence the resulting pore water pressures are negative (i.e., tensile). The capillary effects are present when there is no change in total stress. Therefore, the net effect is an increase in effective stress (remember, $\sigma = \sigma' + u$). Due to the high capillary pressures in clays, the effective stresses near the ground level can be significantly higher than we would expect.

Figure 5.2b shows a soil profile with a capillary rise of h_c above the water table, where the soil can be assumed to be saturated but *not submerged*. In other words, water rises into the voids, almost filling them but not having any buoyancy effect. Below the water table, the soil is saturated and submerged. The pore water pressures at A, B, C, and D are given by: $u_A = \gamma_w h_A$, $u_B = 0$, $u_C = -\gamma_w h_c$, and $u_D = -\gamma_w h_D$. Variation of pore water pressure with depth is still linear from C to A, being negative above the water table and positive below it (see Figure 5.2c).

Reminder

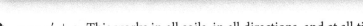

❖ $\sigma = \sigma' + u$. This works in all soils, in all directions, and at all times.
❖ When computing effective stresses, use γ_m above the water table and γ' below the water table.
❖ A smaller grain size means a larger capillary rise. It is insignificant in coarse-grained soils.
❖ Capillary pressures are *negative*. They increase the effective stresses.
❖ Capillary zone can be assumed to be saturated (i.e., use γ_{sat} in calculating σ_v), but not submerged.

WORKED EXAMPLES

1. Plot the variations of total and effective vertical stresses and pore water pressure with depth for the soil profile shown.

    Solution: The values of σ_v, u, and σ_v' computed at the layer interfaces are shown. Within a layer, the unit weights being constants σ_v, u, and σ_v' increase linearly.

z (m)	σ_v (kPa)	u (kPa)	σ_v' (kPa)
0	0.0	0.0	0.0
4	71.2	0.0	71.2
6	108.2	19.6	88.6
10	186.2	58.9	127.3
15	281.2	107.9	173.3

    The plots are shown:

    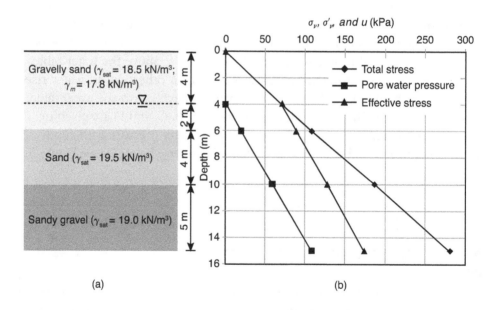

    (a)   (b)

2. The water table in an 8 m thick silty sand deposit lies at a depth 3 m below the ground level. The entire soil above the water table is saturated by capillary water and the saturated unit weight is 18.8 kN/m³. Plot the variation of total and effective vertical stresses and pore water pressure with depth.

    Solution: The values σ_v, u, and σ_v' computed at the layer interfaces are shown in the table. Note the negative capillary pressure and the effective stress of 29.4 kPa at the ground level.

z (m)	σ_v (kPa)	u (kPa)	σ_v' (kPa)
0	0	−29.4	29.4
3	56.4	0.0	56.4
8	150.4	49.1	101.4

The soil profile and the plots generated using the values given in the table are shown in the following figures.

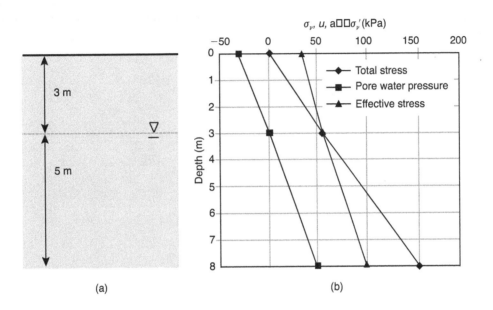

(a) (b)

REVIEW EXERCISES

1. A soil profile at a site consists of a 5 m of gravelly sand (γ_{sat} = 18.5 kN/m³; γ_m = 17.0 kN/m³) layer underlain by 4 m of sandy gravel (γ_{sat} = 18.0 kN/m³). The water table is 4 m below the ground level. Plot the variation of σ_v, σ_v', and u with depth. Neglect the capillary effects.

2. A river is 3 m deep with the riverbed consisting of a thick bed of sand having a saturated unit weight of 19.0 kN/m³. What would be the effective vertical stress at 4 m below the riverbed? If the water level *rises* by 2 m, what would be the new effective vertical stress at 4 m below the riverbed? If the water level *drops* by 2 m, what would be the new effective vertical stress at 4 m below the riverbed?

Answer: 36.8 kPa, 36.8 kPa, 36.8 kPa

3. The Pacific Ocean is 200 m deep at some locations. The seabed consists of a sandy deposit with a saturated unit weight of 20.0 kN/m³. Find the total and effective vertical stresses and pore water pressure at 5 m depth below the seabed.

   Answer: 2062 kPa, 50.9 kPa, 2011.1 kPa

4. In a clayey sandy silt deposit, the water table is 3.5 m below the surface, but the sand to a height of 1.5 m above the water table is saturated by capillary water. The top 2 m of sand can be assumed to be dry. The saturated and dry unit weights of the soil are 19.5 kN/m³ and 18.0 kN/m³ respectively. Calculate the effective vertical stress at 8 m below the surface.

   Answer: 108.9 kPa

Permeability and Seepage

6

6.1 INTRODUCTION

Permeability, as the name implies (the ability to permeate), is a measure of how easily a fluid can flow through a *porous medium*. In the context of geotechnical engineering, the porous medium is soils, and the fluid is water at ambient temperature. A petroleum engineer may be interested in the flow of oil through rocks. An environmental engineer may be looking at the flow of leachate through the compacted clay liner at the bottom of the landfill. Generally, coarser soil grains means larger voids and higher permeability. Therefore, gravels are more permeable than silts. *Hydraulic conductivity* is another term used for permeability, especially in environmental engineering literature.

The flow of water through soils is called *seepage*, which takes place when there is a difference in water levels on two sides (*upstream* and *downstream*) of a structure such as a dam (Figure 6.1a) or sheet pile (Figure 6.1b). Sheet piles are watertight walls made of interlocking sections of steel, timber, or concrete that are driven into the ground.

6.2 BERNOULLI'S EQUATION

Bernoulli's equation in fluid mechanics states that for steady, nonviscous, and incompressible flow, the *total head* at a point (*P* in Figure 6.2a) can be expressed as the summation of the three independent components *elevation head*, *pressure head*, and *velocity head* as shown in Equation 6.1 below:

$$\text{Total head} = \text{Elevation head} + \text{Pressure head} + \text{Velocity head}$$

$$= z + \frac{p}{\gamma_w} + \frac{v^2}{2g} \tag{6.1}$$

where p is the pressure and v is the velocity at P. The heads in Equation 6.1 are forms of energy that are expressed in the unit of length. The elevation head z is simply the height of the point above a *datum* (a reference level), which can be selected at any height. When the point of interest lies below the datum, the elevation head is negative. At point P in Figure 6.2a, the pressure is $\gamma_w h$, and hence the pressure head is h.

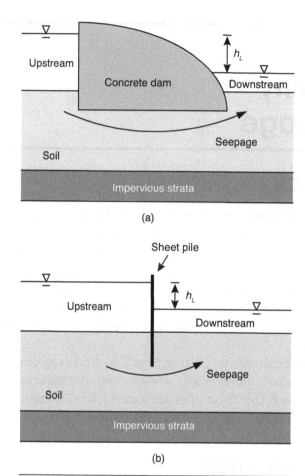

Figure 6.1 Seepage through soils: (a) beneath a concrete dam (b) beneath a sheet pile

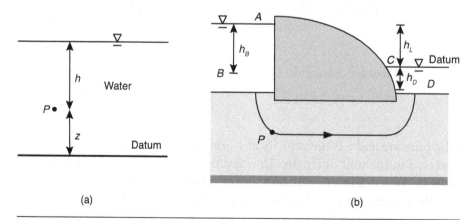

Figure 6.2 Bernoulli's energy principle: (a) a fluid particle in motion (b) seepage beneath a dam

Example 6.1: Seepage takes place beneath a concrete dam as shown in Figure 6.2b, where P is a point on a flow path (known as *streamline*). The pore water pressure is 42 kPa at P, 6 m below the datum that is taken at the downstream water level. The velocity of flow at P is 1 mm/s. Find the total head at P.

Solution: Elevation head $= -6.0$ m

Pressure head $= 42/9.81 = 4.28$ m

Velocity head $= \dfrac{(0.001)^2}{2 \times 9.81} = 5.1 \times 10^{-8}$ m (Negligible)

$\therefore$ Total head $= -6.0 + 4.28 + 5.1 \times 10^{-8} = -1.72$ m

When water flows through soils, the seepage velocity is very small. It gets smaller when squared, and the velocity head becomes negligible compared to the elevation and pressure heads, as seen in Example 6.1.

Example 6.2: In Example 6.1 (Figure 6.2b), points A and C are at the top of the upstream and downstream reservoirs. Points B and D are at depths of h_B and h_D respectively. Find the elevation, pressure, and total heads at A, B, C, and D.

Solution: A: Pressure head $= 0$, Elevation head $= h_L \rightarrow$ Total head $= h_L$

B: Pressure head $= h_B$, Elevation head $= h_L - h_B \rightarrow$ Total head $= h_L$

C: Pressure head $= 0$, Elevation head $= 0 \rightarrow$ Total head $= 0$

D: Pressure head $= h_D$, Elevation head $= -h_D \rightarrow$ Total head $= 0$

It can be seen from Example 6.2 that the total head remains the same within both reservoirs (h_L upstream and 0 downstream with respect to the selected datum at the downstream water level). Streamline is the path of a water molecule. Along a streamline, the total head gradually decreases from h_L upstream to 0 downstream. Here, water molecules expend energy in overcoming the frictional resistance provided by the soil skeleton in their travel from upstream to downstream. The *total head loss* across the dam is h_L, which is simply the difference in water level from upstream to downstream.

Flow takes place from higher total head to lower total head. If seepage takes place from A to B (i.e., $\text{TH}_A > \text{TH}_B$), the *average hydraulic gradient* between these two points is defined as the ratio of the total head difference between the two points to the length of the flow path between the points. Hydraulic gradient i is the head-loss per unit length and is dimensionless. It is a constant in a *homogeneous* soil, and can vary from point to point in a *heterogeneous* soil.

Example 6.3: Water flows from top to bottom through a 900 mm soil sample placed in a cylindrical tube as shown and the water levels are maintained at the levels shown. Find the pore water pressure at A, assuming the soil is homogeneous.

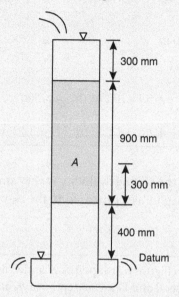

Solution: Let's select the *tail* (bottom) *water level* as the datum. Being at atmospheric pressure, the total head at the datum is 0. This must be the same within the entire water beneath the soil sample. The total head at the *head* (top) *water level* is 1600 mm, which is the same within the entire water above the soil sample. Therefore, the total head loss across the soil sample is 1600 mm, which occurs across a length of 900 mm. Therefore, the hydraulic gradient is 1600/900 = 1.78.

$$\text{Total head at the top of the sample} = 1600 \text{ mm}$$

$$\therefore \text{ Total head at } A = 1600 - 1.78 \times 600 = 532 \text{ mm}$$

$$\text{Elevation head at } A = 700 \text{ mm}$$

$$\therefore \text{ Pressure head at } A = 532 - 700 = -168 \text{ mm}$$

$$\therefore \text{ Pore water pressure at } A = -0.168 \times 9.81 = -1.65 \text{ kPa}$$

6.3 DARCY'S LAW

In 1856, French engineer Henry Darcy proposed that when the flow through a soil is laminar, the discharge velocity v is proportional to the hydraulic gradient i:

$$v \propto i$$
$$v = ki \tag{6.2}$$

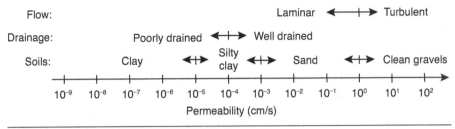

Figure 6.3 Typical values of permeability

Here, the constant k is known as the *hydraulic conductivity*, the *coefficient of permeability*, or simply *permeability*. It has the unit of velocity, and is commonly expressed in cm/s or m/s. Some approximate values of permeability for the major soil groups are shown in Figure 6.3.

In clean uniform sands, Hazen (1930) suggested that k can be related to D_{10} by:

$$k \text{ (cm/s)} \approx D_{10}^2 \text{ (mm)} \tag{6.3}$$

Here, D_{10} is also known as *effective grain size*, which regulates the flow of water through soils. It is also possible for k to be related to a function of void ratio such as e^2, $e^2/(1 + e)$, and $e^3/(1 + e)$. One can intuitively see that larger void ratios have larger void volumes, and hence a larger permeability.

Reynolds number is defined as:

$$R = \frac{vD\rho_w}{\mu_w} \tag{6.4}$$

where D = average diameter of the soil grains, ρ_w = density of water (1000 kg/m³), and μ_w = dynamic viscosity of water (approximately 10^{-3} kg/ms). Provided the Reynolds number R is less than 1, it is reasonable to assume that the flow is laminar.

6.4 LABORATORY AND FIELD PERMEABILITY TESTS

In the laboratory, permeability can be determined by a *constant head permeability test* (ASTM D2434; AS 1289.6.7.1) in a coarse-grained soil, and a *falling head permeability test* (ASTM D5856; AS 1289.6.7.2) in a fine-grained soil. The samples can either be undisturbed samples collected from the field or reconstituted samples prepared in the laboratory. In granular soils where it is difficult to get undisturbed samples, it is common to use reconstituted samples where the granular soil grains are packed to a specific density, replicating the field condition. Schematic diagrams for these two tests are shown in Figure 6.4.

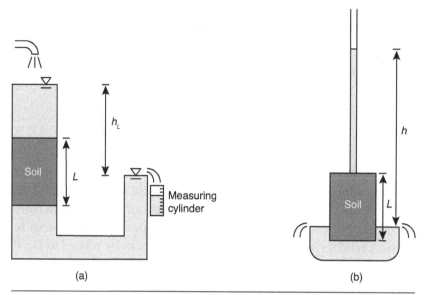

Figure 6.4 Laboratory permeability tests: (a) constant head (b) falling head

6.4.1 Constant Head Permeability Test

In a constant head permeability test, water flows through a cylindrical soil sample of a cross-sectional area A and length L, under a constant total head h_L, as shown in Figure 6.4a. From the water collected in a measuring cylinder or a bucket in time t, the flow rate Q is calculated. The hydraulic gradient across the soil sample is h_L/L. Applying Darcy's law:

$$\frac{Q}{A} = k\frac{h_L}{L} \qquad (6.5)$$

Therefore, k is given by:

$$k = \frac{QL}{Ah_L} \qquad (6.6)$$

In fine-grained soils, it just takes too long to collect an appreciable quantity of water in the measuring cylinder to get a reliable value of the flow rate.

6.4.2 Falling Head Permeability Test

In the falling head permeability test shown in Figure 6.4b, the tail water level is maintained at a constant level, and the water from the standpipe is allowed to flow through the saturated soil sample, with h decreasing with time. Let's equate the flow rate within the standpipe and the soil sample.

Standpipe: Soil sample:

Flow rate $Q = -a\dfrac{dh}{dt}$ Flow rate $Q = vA = k\dfrac{h}{L}A$

$$\therefore -a\dfrac{dh}{dt} = k\dfrac{h}{L}A$$

If h has fallen from h_1 at the start of the test to h_2 after time t, then:

$$-a\int_{h_1}^{h_2}\dfrac{dh}{h} = \dfrac{kA}{L}\int_0^t dt$$

$$k = \dfrac{aL}{At}ln\dfrac{h_1}{h_2} \qquad\qquad (6.7)$$

Example 6.4: In Example 6.3, if the diameter of the soil sample was 60 mm, and 800 ml of water was collected in 10 minutes, determine the permeability. If the average grain diameter is 0.5 mm, determine if the flow is laminar.

Solution: Cross-sectional area of the sample:

$A = \pi \times 3^2 = 28.3$ cm²; flow rate, $Q = 800/600 = 1.33$ cm³/s; $h_L = 160$ cm; $L = 90$ cm

Substituting these values in Equation 6.6:

$$k = \dfrac{1.33 \times 90}{28.3 \times 160} = 0.0264 \text{ cm/s}$$

$$v = \dfrac{Q}{A} = \dfrac{1.33}{28.3} = 0.047 \text{ cm/s}$$

Substituting Equation 6.4, Reynolds number R can be estimated as:

$$R = \dfrac{(0.00047\,\text{m/s})(0.5\times10^{-3}\,m)(1000\,\text{kg/m}^3)}{10^{-3}\,\text{kg/ms}} = 0.24 < 1 \rightarrow \text{Laminar flow}$$

Why can't we do falling head tests on coarse-grained soils? The flow rate is so high that the water level will drop from h_1 to h_2 rapidly, which will not provide enough time to take the proper measurements. Permeability tests can be carried out in the field by pumping water from wells. At steady state, permeability is related to the flow rate.

In the field, a pumping out test can be carried out to determine the permeability of the soil in situ. Here, a 300–450 mm diameter casing is driven into the bedrock as shown in Figure 6.5. The casing is perforated to allow the free flow of water into the well. Two observation wells of 50 mm diameter are also bored into the soil to a depth well below the current water table. The test consists of pumping out water until the flow rate Q and the water levels within the observation wells (h_1 and h_2) remain constant—a steady state.

At steady state, let's consider a cylindrical zone of radius r and height h above the impervious stratum. The hydraulic gradient at the perimeter of the cylinder is $\frac{dh}{dr}$. Therefore, the flow rate into the cylinder is the same as the flow rate out of the well, which is given by:

$$Q = k\frac{dh}{dr}2\pi rh$$

$$\int_{r_1}^{r_2} \frac{dr}{r} = \frac{2\pi k}{Q}\int_{h_1}^{h_2} h\,dh$$

$$\therefore k = \frac{Q}{\pi\left(h_2^2 - h_1^2\right)}\ln\frac{r_2}{r_1} \tag{6.8}$$

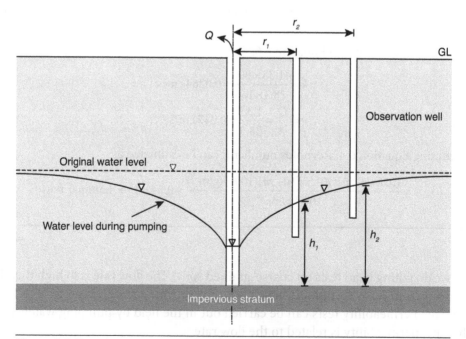

Figure 6.5 Pumping out test to determine permeability in situ

Example 6.5: A 10 m-thick sandy, silt deposit overlies an impermeable stratum. The water table is at a depth 3 m below the ground level. During a pumping-out test, at steady state, water is being pumped out of a 450 mm diameter well at the rate of 5140 liters/min. At the observation wells, at radial distances of 3.5 m and 25.0 m, the water levels dropped by 2.5 m and 1.2 m respectively. Determine the permeability of the soil. What would be the height of the water in the pumping well?

Solution: $Q = 5140 \times 10^{-3}/60 = 0.08567$ m³/s; $r_1 = 3.5$ m; $r_2 = 25.0$ m;

$$h_1 = 10.0 - 3.0 - 2.5 = 4.5 \text{ m}; h_2 = 10.0 - 3.0 - 1.2 = 5.8 \text{ m}$$

$$k = \frac{Q}{\pi(h_2^2 - h_1^2)} \ln\frac{r_2}{r_1} = \frac{0.08567}{\pi(5.8^2 - 4.5^2)} \ln\frac{25}{3.5} = 4\times10^{-3} \text{ m/s} = 4\times10^{-5} \text{ cm/s}$$

with $k = 4 \times 10^{-5}$ cm/s, $r_2 = 25.0$ m, $h_2 = 5.8$ m, $r_0 = 0.225$ m $\rightarrow h_0 = 1.24$ m

The height of water in the pumping well would be 1.24 m.

6.5 STRESSES IN SOILS DUE TO FLOW

Three different scenarios of three identical soil samples that have been subjected to different flow conditions are shown in Figure 6.6. Let's compute the effective vertical stress and pore water pressure at X for all three cases. In Figure 6.6a, there is no flow and the water is static; hence, the computations are straightforward. In the next two cases, flow takes place due to the total head difference of h_L with a hydraulic gradient of h_L/L, and is *upward* through the sample

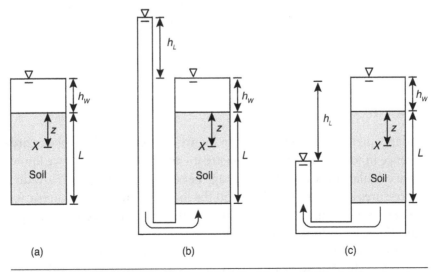

(a) (b) (c)

Figure 6.6 Three different scenarios: (a) static (b) upward flow (c) downward flow

in Figure 6.6b and *downward* through the sample in Figure 6.6c. The total vertical stress at X is the same in all three situations. The pore water pressures can be computed as in Example 6.3, and are summarized below along with the effective vertical stresses:

(a) Static:

$\sigma_v = \gamma_w h_w + \gamma_{sat} z$

$u = \gamma_w (h_w + z)$

$\sigma_v' = \gamma' z$

(b) Upward flow:

$\sigma_v = \gamma_w h_w + \gamma_{sat} z$

$u = \gamma_w (h_w + z) + i z \gamma_w$

$\sigma_v' = \gamma' z - i z \gamma_w$

(c) Downward flow:

$\sigma_v = \gamma_w h_w + \gamma_{sat} z$

$u = \gamma_w (h_w + z) - i z \gamma_w$

$\sigma_v' = \gamma' z + i z \gamma_w$

It is clear from the above that when the flow is upward, the pore water pressure increases by $i z \gamma_w$ and the effective vertical stress decreases by $i z \gamma_w$. When the flow is downward, the pore water pressure decreases by $i z \gamma_w$ and the effective stress increases by $i z \gamma_w$. Larger hydraulic gradients correspond to larger changes in u and σ_v'.

Now, let's have a closer look at the upward flow situation in a *granular soil*. The effective vertical stress is positive as long as $i z \gamma_w$ is less than $\gamma' z$. If the hydraulic gradient is large enough, $i z \gamma_w$ can exceed $\gamma' z$, and the effective vertical stress can become negative. This implies that there is no intergranular stress, and that the grains are no longer in contact. When this occurs (i.e., $i z \gamma_w = \gamma' z$), the granular soil is said to be in *quick condition*. The hydraulic gradient in this situation is known as *critical hydraulic gradient* i_c, given by:

$$i_C = \frac{\gamma'}{\gamma_w} = \frac{G_s - 1}{1 + e} \tag{6.9}$$

This is what creates the *quicksand* you may have seen in movies, and the *liquefaction* of granular soils that are subjected to vibratory loads such as pile driving. While total stress remains the same, a sudden rise in pore water pressure reduces the effective stress and soil strength to zero, causing failure. You will see in Chapter 9 that the strength of a granular soil is proportional to the effective stress.

6.6 SEEPAGE

In the concrete dam and the sheet pile shown in Figure 6.1, seepage takes place through the soil due to the difference in total heads between upstream and downstream. If we know the permeability, how do we calculate the quantity of seepage per day (i.e., flow rate)? How do we calculate the pore water pressures at various locations and the loadings on the structures caused by seepage? In the case of granular soil, is there a problem with hydraulic gradients being too high? To answer these questions, let us look at some fundamentals in flow through soils.

In Figure 6.7 the concrete dam is impervious and there is an impervious stratum underlying the soil. Let's select the downstream water level as the datum, which makes the total heads within the downstream and upstream water 0 and h_L respectively. A *streamline* or *flow line* is the path of a water molecule in the flow region; it originates from upstream and finishes at

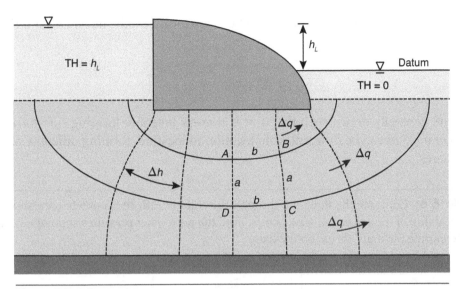

Figure 6.7 A flow net

downstream, and the total head loss is h_L along each of them. The passage of water between two adjacent streamlines is a *flow channel*. There are thousands of streamlines and flow channels in any flow region. Joining the points having the same total head in the flow region gives an *equipotential line*, which is simply a contour of the total head. There are thousands of equipotential lines within the flow region. Total head h at any point in a two-dimensional flow region with respect to the Cartesian coordinate system can be expressed as (see Worked Example 6.5):

$$\frac{\partial^2 h}{\partial x^2} + \frac{\partial^2 h}{\partial y^2} = 0 \tag{6.10}$$

The above is Laplace's equation, and it can be shown that the streamlines and equipotential lines intersect at 90° (see Worked Example 6 for proof).

Only a few selected streamlines and equipotential lines (dashed) are shown in Figure 6.7. A *flow net*, such as the one shown in Figure 6.7, is a network of these *selected* streamlines and equipotential lines. Let's select the equipotential lines such that the total head difference between two adjacent lines is the same ($= \Delta h = h_L/N_d$). We will select the streamlines such that the flow rate Δq is the same in all flow channels. Let's say there are N_f flow channels and N_d equipotential drops as shown in the figure ($N_f = 3$ and $N_d = 6$ in this particular example).

Let's consider the zone ABCD. The velocity of flow from AD to BC is $v_{AD-BC} = k\frac{\Delta h}{b}$. Considering a unit thickness (perpendicular to the plane), the flow rate is $\Delta q = k\frac{\Delta h}{b}a$. Since there are N_f flow channels, the total flow rate Q becomes:

$$Q = kh_L \frac{N_f}{N_d} \frac{a}{b}$$

If the streamlines are selected such that $a = b$ (at every location), the flow rate *per unit thickness* is given by:

$$Q = kh_L \frac{N_f}{N_d} \qquad (6.11)$$

Flow nets are generally drawn such that $a = b$ at every location, forming *curvilinear squares.* The values of a ($= b$) can be different from location to location, forming different sizes of curvilinear squares.

Example 6.6: Compute the flow rate through the soil beneath the concrete dam shown if the permeability of the soil is 3.2×10^{-4} cm/s. Find the pore water pressures at points A, B, and C, and compute the uplift thrust on the dam.

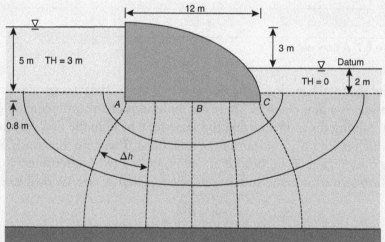

Solution: Substituting $N_f = 3$, $N_d = 6$, $h_L = 3$ m in Equation 6.11:

$$Q = \left(3.2 \times \frac{10^{-6} \text{ m}}{s}\right) \times (3 \text{ m}) \times \frac{3}{6} \times 24 \times 3600 = 0.41 \text{ m}^3/\text{day per m}$$

The change in the total head between two equipotential lines $\Delta h = 3/6 = 0.5$ m. Therefore, total heads at A, B, and C are 2.5 m, 1.5 m, and 0.5 m respectively. Elevation head is -2.8 m at all three points. Therefore, the pressure heads are 5.3 m, 4.3 m, and 3.3 m respectively, and the corresponding pore water pressures are:

$$u_A = 5.3 \times 9.81 = 52.0 \text{ kPa}, u_B = 42.2 \text{ kPa, and } u_C = 32.4 \text{ kPa}$$

Plotting the three values at the bottom of the dam, the uplift force can be computed as the area within the plot. This becomes:

$$\text{Uplift thrust} = \left(\frac{52.0 + 42.2}{2}\right) \times 6 + \left(\frac{42.2 + 32.4}{2}\right) \times 6 = 506 \text{ kN per m}$$

As seen in Example 6.6, once the flow net is drawn, it is a straightforward exercise to determine the pore water pressure at a point within the flow region.

6.6.1 Piping in Granular Soils

The hydraulic gradient at the exit i_{exit} decreases with the distance from the dam or sheet pile, and is the maximum right next to the structure. In granular soils, if this maximum exit hydraulic gradient $i_{exit, max}$ exceeds the critical hydraulic gradient defined in Equation 6.9, the effective stress at the downstream side near the structure becomes zero and the soil grains get washed away. The situation is even worse now with the flow path getting shorter! This mechanism can progressively work its way from downstream to upstream, eroding away the soil and forming a sort of pipe beneath the structure, which would provide free passage to the water and eventually flood the downstream. Therefore, it is necessary to ensure that $i_{exit, max}$ is well below i_c. The safety factor with respect to piping is defined as:

$$F_{piping} = \frac{i_c}{i_{exit,max}} \tag{6.12}$$

Several dams across the globe (e.g., Baldwin Hills Reservoir Dam, Los Angeles, 1963; Teton Dam, Idaho, 1976; Val di Stava Dam, Italy, 1985) have suffered catastrophic failures due to piping—often with short notice. Piping failures are often catastrophic and can cause severe human and economic losses at the downstream side. As a result, large safety factors (as high as 5) are commonly used against possible failures by piping. For temporary structures such as cofferdams, this can be lower. *The Canadian Foundation Engineering Manual (2006)* recommends a safety factor of 2–3. Some examples of dams that have failed possibly due to piping are shown in Figure 6.8.

6.6.2 Flow Net Construction

Graphical construction of a quality flow net is a trial and error process that is carried out with a pencil, eraser, and paper. An experienced engineer should be able to sketch one within 20 minutes. First, identify the boundary conditions and take advantage of any symmetry. For example, the vertical line of symmetry in Figure 6.7 is an equipotential line. The flow net has to be symmetrical about this line; thus half the flow net is adequate. Sketch the streamlines and equipotential lines so that they intersect at 90° at all locations (no exceptions!), and make sure they form approximate squares. A good check is to see that you can fit it in a circle touching all four sides within each of these curvilinear squares. The size of the curvilinear squares can vary from location to location.

6.6.3 Flow Net in Anisotropic Soils

When the soils are *anisotropic* with horizontal permeability k_h, which are generally larger than the vertical permeability (k_v), streamlines and equipotential lines don't intersect at 90° (see

(a)

(b)

(c)

Figure 6.8 Some piping failures: (a) upper clear Boggy Dam upstream, USA (b) Tunbridge Dam, Australia (c) Ouches Breche Dam, France (Courtesy of Professor Robyn Fell)

Worked Example 6). This makes it difficult to sketch the flow net. Here, we will use a transformed section where the entire flow region is redrawn with horizontal dimensions multiplied by $\sqrt{k_v/k_h}$, without change in the vertical scale. It can be shown mathematically that the streamlines and equipotential lines in the transformed section intersect at 90°, and hence the flow net can be sketched and used as before. The flow rate can be computed using Equation 6.10 with $k = \sqrt{k_v \times k_h}$.

6.7 DESIGN OF GRANULAR FILTERS

Filters, known as *protective filters*, are commonly used in earth dams, within the backfills behind retaining walls, etc., where seepage takes place. Traditionally, they are made of granular soils; but today, geofabrics are becoming popular. The purpose of a filter is to protect the upstream soils such that the fines are not washed away. Here, the pore channels must be small enough to prevent the migration of fines. This is known as *retention criterion*. On the other hand, the pore channels must be large enough to allow the free flow of water, thus preventing any buildup of excess pore water pressure. This is known as the *permeability criterion*. In addition, it is a common practice to select the filter material such that the grain size distribution curves of the filter grains and the soil being protected have the same shape. These criteria can be summarized as:

Retention criterion: $D_{15, \text{filter}} < 5\,D_{85, \text{soil}}$
Permeability criterion: $D_{15, \text{filter}} > 4\,D_{15, \text{soil}}$
Grain size distribution: Approximately parallel to grain size distribution of the soil

Here, D_{15} is taken as the average pore size of the filter. The permeability and retention criteria define the lower and upper bounds for the grain size distribution curve of the filter. The U.S. Navy (1971) suggests two additional conditions to reinforce the retention criterion, as shown below:

$$D_{15, \text{filter}} \leq 20\,D_{15, \text{soil}} \text{ and } D_{50, \text{filter}} \leq 25\,D_{50, \text{soil}}.$$

6.8 EQUIVALENT PERMEABILITIES FOR ONE-DIMENSIONAL FLOW

When the flow is horizontal or vertical, and if the soil profile consists of more than one layer of soil with different permeabilities $k_1, k_2, \ldots k_n$ as shown in Figure 6.9, it can be represented by an equivalent homogeneous soil profile of the same thickness. Such situations arise in sedimentary deposits that are comprised of layers of different permeabilities. The permeability of this homogeneous soil mass will vary depending on whether the flow is horizontal or vertical.

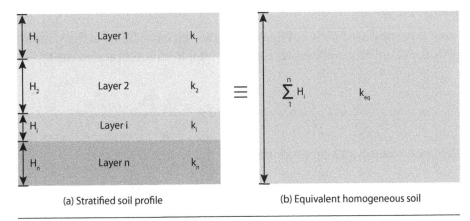

(a) Stratified soil profile   (b) Equivalent homogeneous soil

Figure 6.9 Equivalent permeabilities

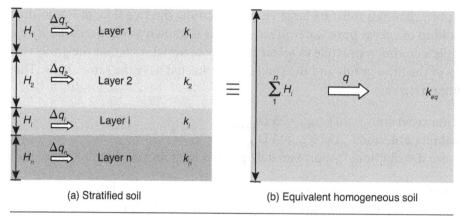

(a) Stratified soil   (b) Equivalent homogeneous soil

Figure 6.10 Horizontal flow through stratified soil

6.8.1 Horizontal Flow

When the flow is horizontal as shown in Figure 6.10, k_{eq} is estimated such that $q = \Delta q_1 + \Delta q_2 + \cdots + \Delta q_n$. Assuming that the hydraulic gradient i is the same across each layer as well as the equivalent soil profile:

$$\Delta q_1 = k_1 i H_1,\ \Delta q_2 = k_2 i H_2,\ \Delta q_n = k_n i H_n,\ \text{and}\ q = k_{eq}\, i \sum_1^n H_i$$

$$\therefore k_1 H_1 + k_2 H_2 + \cdots + k_n H_n = k_{eq}(H_1 + H_2 + \cdots + H_n)$$

Therefore:

$$k_{eq} = \frac{k_1 H_1 + k_2 H_2 + \cdots + k_n H_n}{H_1 + H_2 + \cdots + H_n} \tag{6.13}$$

6.8.2 Vertical Flow

When the flow is vertical as shown in Figure 6.11, the velocity of flow is the same within each layer as well as the equivalent soil profile. Here, the total head losses across the layers are $h_1, h_2, \ldots h_n$:

$$v = k_1 \frac{h_1}{H_1} = k_2 \frac{h_2}{H_2} = \cdots = k_n \frac{h_n}{H_n} = k_{eq} \frac{(h_1 + h_2 + \cdots + h_n)}{(H_1 + H_2 + \cdots + H\ )}$$

The equivalent permeability can be obtained from:

$$\frac{H_1 + H_2 + \cdots + H_n}{k_{eq}} = \frac{H_1}{k_1} + \frac{H_2}{k_2} + \cdots + \frac{H_n}{k_n} \tag{6.14}$$

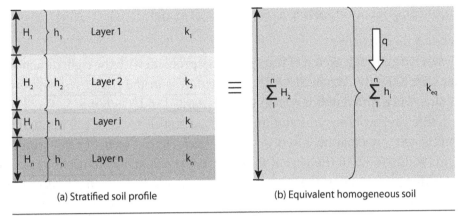

(a) Stratified soil profile   (b) Equivalent homogeneous soil

Figure 6.11 Vertical flow through stratified soil

6.9 SEEPAGE ANALYSIS USING *SEEP/W*

A DVD containing the *Student Edition* of GeoStudio 2007 is included with this book. One of the eight different programs that come up when you click the *GeoStudio 2007* icon is *SEEP/W*, a versatile finite element software that can be used to draw flow nets and compute pore water pressures and flow rates. The *Student Edition* of *SEEP/W* has a few limitations that make it suitable mainly for learning and evaluating. It can handle up to 500 elements, 10 different regions, and three different materials. It can model saturated and unsaturated flow problems. This section describes how to use *SEEP/W* to solve seepage problems.

The full version has several advanced features, and it has no limits on the number of elements, regions, and materials. It is available from *GEO-SLOPE* International, Canada (http://www.geo-slope.com).

6.9.1 Getting Started with *SEEP/W*

When running *GeoStudio*, select *Student License* from the start page. All *GeoStudio* project files are saved with the extension *.gsz* so that they can be called by any of the applications (e.g., *SIGMA/W, SLOPE/W*) within the suite.

Familiarize yourself with the different toolbars that can be made visible through the $\boxed{View/Toolbar...}$ menu. Moving the cursor over an icon displays its function. In the $\boxed{Analysis}$ toolbar, you will see three icons: $\boxed{DEFINE}$, $\boxed{SOLVE}$, and $\boxed{CONTOUR}$ next to each other. $\boxed{DEFINE}$ and $\boxed{CONTOUR}$ are two separate windows and you can switch between them. The problem is fully defined in the $\boxed{DEFINE}$ window and is saved. Clicking the $\boxed{SOLVE}$ icon solves the problem as specified. Clicking the $\boxed{CONTOUR}$ icon displays the results in the $\boxed{CONTOUR}$ window. The input data can be changed by switching to the $\boxed{DEFINE}$ window and then $\boxed{SOLVE}$d again.

The major components in solving a seepage problem are:

1. *Defining the geometry:*
   Always have a rough sketch of your geometry problem with the right dimensions before you start *SEEP/W*. When *SEEP/W* is started, it is in the DEFINE window. The Set menu has two different but related entries: Page... and Units and Scales... can be used to define your working area and units. A good start is to use a 260 mm (width) × 200 mm (height) area that fits nicely on an A4 sheet. Here, a scale of 1:200 would represent 52 m (width) × 40 m (height) of the geometry problem. Try to use the same scale in x and y directions so that the geometry is not distorted. Units and Scales... can also be used for defining the problem as two-dimensional (plane strain) or axisymmetric. All problems discussed in this chapter are two-dimensional. Grid... will allow you to select the grid spacing. Make it visible and snap to the grid points. Axes... will allow you to draw the axes and label them. Sketch/Axes... may be a better way to draw the axes and label them. Use View/Preferences... to change the way the geometry and fonts are displayed and to change the way the flow net is graphically presented.

   Use Sketch/Lines to sketch the geometry using *free* lines. Use Modify/Objects... to delete or move these. Sketch is different from Draw. Use Draw/Regions... on the sketched outlines to create the real geometry and to define the zones of different materials. Alternatively, one may omit Sketch and start from Draw instead. While *Sketch*ing, *Draw*ing, or *Modify*ing, right clicking the mouse ends the action. The Sketch menu is useful for drawing dimension lines with arrowheads and for labeling the dimensions and objects.

2. *Defining soil properties and assigning them to regions:*
   Use Draw/Materials... to assign the soil permeabilities and apply them to the regions by dragging. The *Student Edition* can accommodate up to three different materials. Write 3.5×10^{-5} m/s as 3.5e-5 m/s. In anisotropic soils, specify the value of horizontal permeability k_x as saturated conductivity and give k_y/k_x as the conductivity ratio, which is generally less than 1.

3. *Defining the boundary conditions:*
   Assign the boundary conditions through Draw/Boundary Conditions... . Here, specify the equipotential lines at the upstream and downstream boundaries and give the values of total heads. Use the horizontal axis as the datum. Use a separate name tag for each boundary since the total-head value specified is different. Once a boundary condition is created, it can be applied to a point, line, or a region. Apply the boundary conditions by dragging them to the relevant location.

4. *Defining the finite element mesh:*

   This step gives us some taste of finite element modeling. The *Student Edition* of *SEEP/W* limits the number of elements to 500. The default mesh would be adequate for all our work here. The mesh can be seen through $\boxed{Draw/Mesh\ Properties...}$. The mesh size can be varied by adjusting the global element size; the larger the element size is, the coarser the mesh will be.

   In seepage problems, you must often compute the flow rate. *SEEP/W* computes this by calculating the flux crossing a specific section. Turn off *Grid/Snap* to precisely define the section. Select $\boxed{Draw/Flux\ Sections...}$, and a dialog box with a section number will appear. Select $\boxed{OK}$ and the cursor will change into a crosshair. Draw the flux line, which appears as a blue dashed line with an arrowhead.

5. *Solving the problem:*

   Once the problem is fully defined through steps 1–4, it can be $\boxed{SOLVE}$d, and the results can be viewed in a $\boxed{CONTOUR}$ window. You can switch between the $\boxed{DEFINE}$ and $\boxed{CONTOUR}$ windows while experimenting with the output. This can be very effective for a parametric study. $\boxed{Tools/Verify}$ can be used for checking the problem definition before solving.

6. *Displaying the results:*

   By default, the $\boxed{CONTOUR}$ window will show the total head contours, which are the equipotential lines. From $\boxed{Draw/Contours...}$ the intervals and colors can be varied. By clicking the $\boxed{Draw/Contour\ Labels}$, the cursor changes into a crosshair. By placing the crosshair on a contour line and clicking the mouse, the contour value is labeled. By clicking the $\boxed{Draw/Flow\ Paths}$, the cursor changes into a crosshair. By placing the crosshair on any point within the flow region and clicking mouse, the flow line is drawn. By clicking on it a second time, the flow line is removed.

   $\boxed{View/Result\ Information...}$ gives the full information about any point in the flow region, including the pore water pressure. To display the flow rate through the flux section defined below, select $\boxed{Draw/Flux\ Labels}$, and the cursor changes into a crosshair. Place the cursor at any point on the flux section and click to place a label showing the flow rate.

Example 6.7: Use *SEEP/W* to draw the flow net for the sheet pile arrangement shown. The permeability of the soil is 2.5×10^{-5} cm/s. The soil is underlain by an impervious stratum. Label the equipotential lines and show the flow rate. Show the finite element mesh used in the analysis.

Continues

Example 6.7: *Continued*

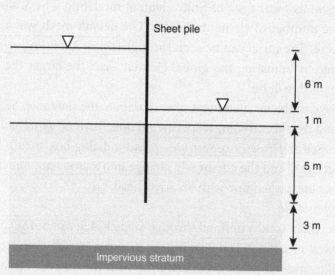

Solution: The finite element mesh with 363 elements and 423 nodes used in the analysis is shown.

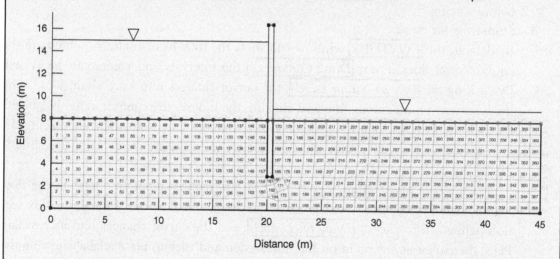

The flow rate is displayed as 5.831×10^{-7} m³/s. The horizontal axis (bottom of the soil layer) is the datum.

The flow rate can be computed using the flow net and Equation 6.10 as:

$$Q = 2.5 \times 10^{-7} \times 6 \times \frac{5}{12} = 6.3 \times 10^{-7} \text{ m}^3/\text{s,}$$

Continues

Example 6.7: *Continued*

which is in agreement with the value (5.831 × 10^{-7} m^3/s) calculated by *SEEP/W*. Note that N_f/N_d is only approximately 5/12.

❖ Elevation, pressure, and total heads are forms of energy expressed as length.

❖ Velocity head in soils is negligible.

❖ Elevation and total heads depend on the datum; pressure head is independent of the datum.

❖ Always show the datum when solving seepage problems.

❖ Pore water pressure = Pressure head × γ_w.

❖ Constant head permeability tests are for coarse-grained soils and falling head tests are for fine-grained soils.

❖ Streamlines and equipotential lines are orthogonal only when the soil permeability is isotropic.

❖ The *Student Edition* of *SEEP/W* can be used for drawing flow nets and computing flow rates, pore water pressures, etc.

Reminder

WORKED EXAMPLES

1. Water flows through a 100 mm diameter granular soil specimen as shown. The water levels on both sides are maintained constant during the test, and the void ratio of the soil is 0.82, and $G_s = 2.68$.

   a. What is the maximum possible value for h such that the soil does not reach quick condition?

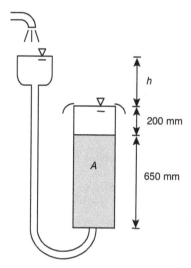

   b. For $h = 150$ mm, 175 ml of water was collected in 15 minutes. Find the permeability of the soil and the effective vertical stress at A 220 mm below the top of the sample.

Solution:

   a. Let's take tail water level as the datum. Head loss across the sample is 650 mm:

$$i_c = \frac{G_s - 1}{1 + e} = \frac{2.68 - 1}{1 + 0.82} = 0.923 \geq \frac{h}{650} \rightarrow h \leq 600 \text{ mm}$$

   b. For $h = 150$ mm, $i = 150/650 = 0.231$:

$$\text{Flow rate} = \frac{175}{900} = 0.1944 \text{ cm}^3/\text{s}$$

$$\text{cross-sectional area} = 78.5 \text{ cm}^2$$

$$\therefore \text{velocity} = \frac{0.1944}{78.5} = 2.48 \times 10^{-3} \text{cm/s}$$

$$k = \frac{2.48 \times 10^{-3}}{0.231} = 1.08 \times 10^{-2} \text{ cm/s}$$

Total heads at head water and tail water are 150 mm and 0 respectively:

$$\therefore (TH)_A = (0.231)(220) = 50.8 \text{ mm}; (EH)_A = -420.0 \text{ mm}$$
$$\therefore (PH)_A = 470.8 \text{ mm} \rightarrow u_A = 0.471 \times 9.81 = 4.09 \text{ kPa}$$

$$\gamma_{sat} = \frac{2.68 + 0.82}{1 + 0.82} \times 9.81 = 18.87 \text{ kN/m}^3$$

$$\sigma_v = 0.2 \times 9.81 + 0.22 \times 18.87 = 6.11 \text{ kPa}$$

$$\therefore \sigma'_v = 6.11 - 4.09 = 2.02 \text{ kPa}$$

2. A sheet pile is driven into sandy silt and seepage takes place under the head difference of 9.0 m as shown. The permeability of the soil is 1.6×10^{-4} cm/s and the water content of the soil is 33%. The specific gravity of the soil grains is 2.66. Using the flow net shown in the figure, compute the following:

   a. Flow rate in m³/day per meter run
   b. Pore water pressure at A
   c. Safety factor with respect to piping

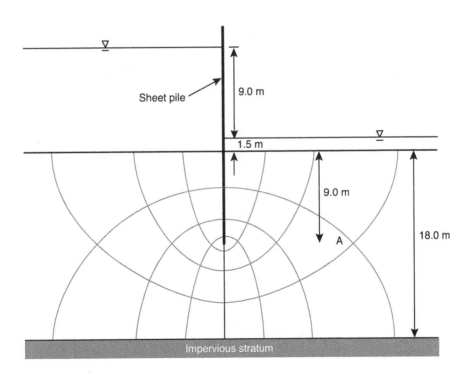

Solution:

a.

$N_f = 3, N_d = 8, h_L = 9.0$ m

$\therefore Q = (1.6 \times 10^{-6}) \times 9.0 \times \dfrac{3}{8} \times 24 \times 3600 = 0.47$ m³/day per m run

b. $\Delta h = 9.0/8 = 1.125$ m per equipotential drop

Let's take the downstream water level as the datum. Then, total head at A is 1.125 m. Elevation head at A is −10.5 m.

$\therefore$ Pressure head at A is 11.625 m.
$\therefore$ Pore water pressure at A $= 11.625 \times 9.81 = 114.0$ kPa.

c. In the curvilinear square to the right of the sheet pile at exit, the distance along the sheet pile is measured as 3.5 m:

$$\therefore i_{exit,\,max} = 1.125/3.5 = 0.32$$

Assuming $S = 1$ (below water table)

$e = 0.33 \times 2.66 = 0.88$

$$\therefore i_c = \frac{2.66 - 1}{1 + 0.88} = 0.88 \rightarrow F_{piping} = \frac{0.88}{0.32} = 2.75$$

3. A small area is protected from flooding by sheet piles as shown. The original water level was at the top of the clay layer. Later, the water level is expected to rise by 4 m outside the area

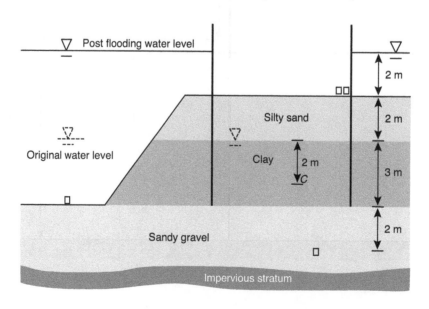

protected from flooding. This is expected to cause some upward seepage through the clay layer between the two sheet piles. A piezometer measurement shows that the pore water pressure at B is 88.2 kPa. The silty sand can be assumed to be saturated due to capillary effects. The bulk unit weights of silty sand, clay, and sandy gravel are 18.0, 17.5, and 18.5 kN/m^3 respectively.

    a. Calculate the total heads at A and B, taking the top of the sandy gravel layer as the datum. Show that there is hardly any head loss due to the flow through the gravel.

    b. Calculate the vertical total stress and vertical effective stress at B.

    c. Find the hydraulic gradient for the upward flow between the sheet piles in the clay layer.

    d. Find the total and pressure head and pore water pressure at C.

Solution:

a.
$$(EH)_A = 0.0 \text{ m}, (PH)_A = 7.0 \text{ m} \rightarrow (TH)_A = 7.0 \text{ m}$$
$$(EH)_B = -2.0 \text{ m}, (PH)_B = 88.2/9.81 = 8.99 \text{ m} \rightarrow (TH)_B = 6.99 \text{ m}$$

∴ Total head loss from A to B is 0.01 m, which is negligible; the total head loss within the gravel layer is negligible.

b. At B, $\sigma_v = 18.0 \times 2 + 17.5 \times 3 + 18.5 \times 2 = 125.5$ kPa and $u = 88.2$ kPa
∴ $\sigma_v' = 125.5 - 88.2 = 37.3$ kPa

c. At top of the clay layer, $EH = 3$ m, $PH = 0 \rightarrow TH = 3.0$ m. In the sandy gravel layer, $TH = 7.0$ m.

∴ Total head loss across the clay layer $= 7 - 3 = 4.0$ m
∴ Hydraulic gradient within the clay layer $= 4.0/3.0 = 1.33$

d.
$$(TH)_C = 7.00 - 1.33 \times 1.0 = 5.67 \text{ m}$$
$$(EH)_C = 1.0 \text{ m} \rightarrow (PH)_C = 4.67 \text{ m}$$
$$\therefore u_C = 4.67 \times 9.81 = 45.8 \text{ kPa}$$

4. An unlined irrigation canal runs parallel to a river and the cross section is shown on page 98. The soils in the region are generally stiff clays that are assumed to be impervious. There is a 200 mm-thick sand seam connecting the canal and river as shown, which continues to a length of 3.0 km along the river. Assuming that the permeability of the sand is 2.3×10^{-2} cm/s, compute the quantity of water lost from the irrigation canal per day.

Solution: Let's take the water level in the river as the datum.

    ∴ Total heads at the canal and the river is 20.0 m and 0 respectively, with the head loss across the sand seam being 20.0 m.
    ∴ Hydraulic gradient $= 20/250 = 0.080$.

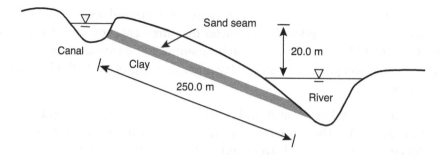

By Darcy's law, velocity of flow $= (2.3 \times 10^{-2}) \times (0.080) = 0.184 \times 10^{-2}$ cm/s.

Cross-sectional area of flow $= 3000 \text{ m} \times 0.2 \text{ m} = 600 \text{ m}^2$.

$\therefore$ Flow rate $= (0.184 \times 10^{-4} \text{ m/s}) \times (600 \text{ m}^2) \times (24 \times 3600 \text{ s/day}) = 954 \text{ m}^3/\text{day}$.

5. In a two-dimensional seepage problem (see the illustration on the next page), show that the equation of flow is given by:

$$k_x \frac{\partial^2 h}{\partial x^2} + ky \frac{\partial^2 h}{\partial y^2} = 0$$

where $h(x,y)$ is the total head at a point in the flow region.

Solution: The horizontal and vertical dimensions of the element shown in the figure are dx and dy respectively.

The net flow into the element being zero, and considering a unit width normal to the plane,

$$v_x dy + v_y dx = \left(v_x + \frac{\partial v_x}{\partial x} dx \right) dy + \left(vy + \frac{\partial v_y}{\partial y} dy \right) dx$$

$$\therefore \frac{\partial v_x}{\partial x} + \frac{\partial v_y}{\partial y} = 0$$

From Darcy's law,

$$v_x = -k_x \frac{\partial h}{\partial x} \quad \text{and} \quad v_y = -k_y \frac{\partial h}{\partial y}$$

Substituting these in the above equation,

$$k_x \frac{\partial^2 h}{\partial x^2} + k_y \frac{\partial^2 h}{\partial y^2} = 0$$

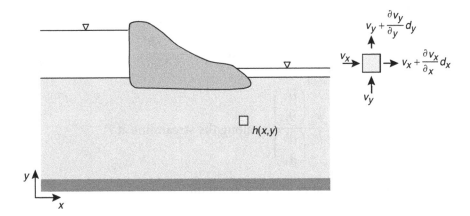

In three dimensions, the equation becomes:

$$k_x \frac{\partial^2 h}{\partial x^2} + k_y \frac{\partial^2 h}{\partial y^2} + k_z \frac{\partial^2 h}{\partial z^2} = 0$$

6. A streamline and an equipotential line are shown in the illustration. From the first principles, show that they are perpendicular to each other.

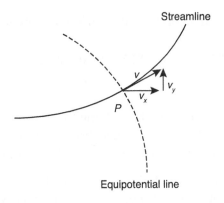

Solution: Let the velocity of the fluid particle at P be v, with horizontal and vertical components of v_x and v_y respectively. In time dt, point P moves a distance of dx and dy respectively, which are given by $dx = v_x\, dt$, and $dy = v_y\, dt$. Therefore:

$$\frac{dy}{dx} = \frac{v_y}{v_x}$$

From Darcy's law:

$$v_x = -k_x \frac{\partial h}{\partial x} \quad \text{and} \quad v_y = -k_y \frac{\partial h}{\partial y}$$

$$\therefore \frac{dy}{dx} = \frac{k_y}{k_x} \frac{\left(\dfrac{\partial h}{\partial y} \right)}{\left(\dfrac{\partial h}{\partial x} \right)} \quad \dots \text{along the streamline at } P$$

Along the equipotential line at P, $h(x,y) = $ constant:

$$\therefore dh = \frac{\partial h}{\partial x} dx + \frac{\partial h}{\partial y} dy = 0$$

$$\therefore \frac{dy}{dx} = -\frac{\left(\dfrac{\partial h}{\partial x} \right)}{\left(\dfrac{\partial h}{\partial y} \right)} \quad \dots \text{along the equipotential line at } P$$

For the two to intersect at 90°, the product of the gradients must be -1. This is true if the soil is isotropic and hence, $k_y / k_x = 1$.

7. The grain size distribution data of the soil in an embankment are given:

Size (mm)	0.02	0.04	0.075	0.15	0.30	0.425	1.18	2.36	4.75
% finer	1	6	23	47	70	80	98	100	100

It is required to design a granular filter satisfying the four criteria given in Section 6.7, Design of Granular Filters. Plot the grain size distribution curve for the soil and mark the upper and lower bounds for the possible grain size distribution curve of the filter.

In the contract specifications, the geotechnical consultants have proposed the following upper and lower bounds as the criteria for filter grains. Does this meet your expectations?

Grain size (mm)	9.5	6.7	4.75	2.36	1.18	0.425	0.3	0.15	0.075
% finer (lower bound)	100	100	100	80	57	25	15	5	3
% finer (upper bound)	100	90	80	60	37	5	1	0	0

Solution: $D_{15, soil} = 0.06$ mm, $D_{50, soil} = 0.16$ mm, $D_{85, soil} = 0.55$ mm

Permeability Criteria 1: $D_{15, filter} > 4\, D_{15, soil} \rightarrow D_{15, filter} > 0.24$ mm
Retention Criteria 2: $D_{15, filter} < 5\, D_{85, soil} \qquad \rightarrow D_{15, filter} < 2.75$ mm
Retention Criteria 3: $D_{15, filter} \leq 20\, D_{15, soil} \quad \rightarrow D_{15, filter} \leq 1.2$ mm
Retention Criteria 4: $D_{50, filter} \leq 25\, D_{50, soil} \quad \rightarrow D_{50, filter} \leq 4.0$ mm

These four values are shown in the grain size distribution plot.

The grain size distributions of the soil and the upper and lower bounds for the filter grains as specified by the consultant are shown in the figure. The four criteria from Section 6.7 are calculated here and are also shown in the figure below. The band suggested by the consultant fully lies within the bounds specified by the four criteria.

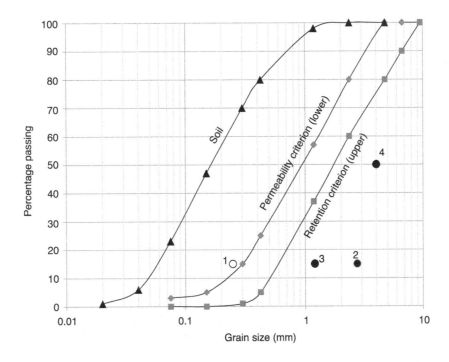

8. Seepage takes place beneath a concrete dam with an upstream blanket and a sheet pile cutoff wall as shown in the top figure on the next page. The permeability of the soil is 7.5×10^{-6} cm/s. Using *SEEP/W*, draw the flow net and determine the following:

   a. Flow rate
   b. Pore water pressures at *A* and *B*

   Repeat steps a. and b. for $k_x = 7.5 \times 10^{-6}$ cm/s and $k_y = 1.5 \times 10^{-6}$ cm/s.

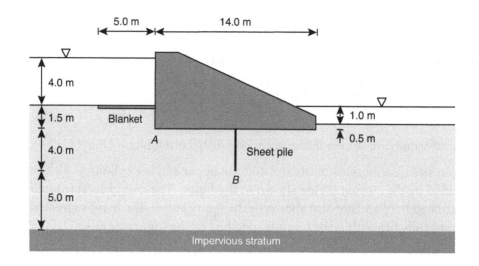

Solution: The flow net obtained from *SEEP/W* is shown. The flow rate is 7.26×10^{-8} m³/s per m width:

$$u_A = 43.6 \text{ kPa}$$
$$u_B = 69.2 \text{ kPa}$$

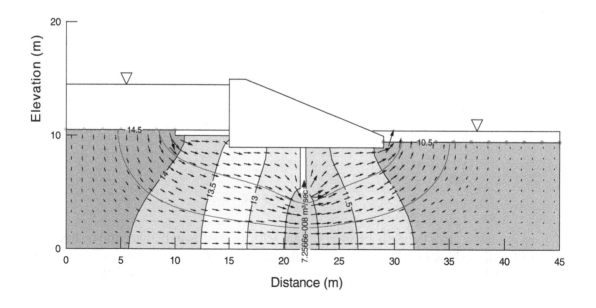

If the soil is anisotropic with:

$$k_x = 7.5 \times 10^{-6} \text{ cm/s and } k_y = 1.5 \times 10^{-6} \text{ cm/s}$$

the flow rate is:

$$4.11 \times 10^{-8} \text{ m}^3\text{/s per m}$$
$$u_A = 43.8 \text{ kPa}; \; u_B = 69.4 \text{ kPa}$$

The flow net is shown below. The equipotential lines and streamlines do not intersect at 90° (see the following illustration).

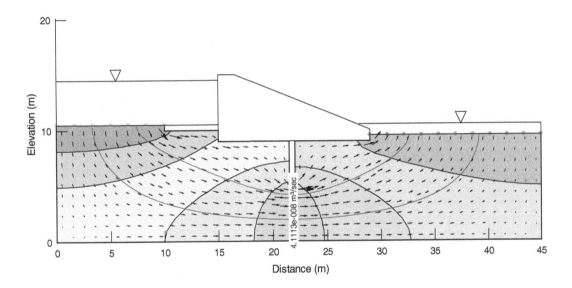

REVIEW EXERCISES

1. Three cylindrical granular soil samples of the same length and diameter are subjected to a constant head flow as shown in the figure on the following page. If the permeability of the sand, silty sand, and gravelly sand is 2×10^{-2}, 6×10^{-3}, and 4×10^{-2} cm/s respectively, find h_1 and h_2.

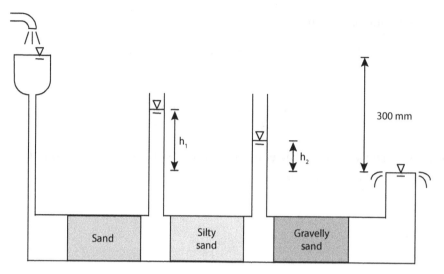

Answer: 238 mm, 31 mm.

2. A 50 mm diameter and 90 mm-long silty clay sample was subjected to a falling head permeameter test using a setup similar to the one shown in Figure 6.4b, where the inner diameter of the standpipe was 3.0 mm. The head dropped from 870 mm to 450 mm in 5 minutes. What is the permeability of the sample?

Answer: 7.1×10^{-5} cm/s

3. Write a 500-word essay on *liquefaction in granular soils*.

4. Write a 500-word essay on *piping problems and quicksand*, giving examples of dams that have had failures attributed to the dam's piping.

5. List the empirical correlations on permeability of granular soils and list their limitations.

6. Discuss the methods of determining permeability in the field.

7. Water flows under constant head through the two soil samples 1 and 2, as shown in the figure. The cross-sectional area of the sample is 2000 mm². In five minutes, 650 ml water flows through the samples.
   a. Find the permeability of the samples, and
   b. In sample 2, find the pore water pressure at a point 40 mm above the bottom.

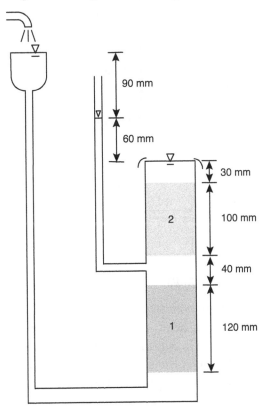

Answer: 0.145 cm/s, 0.181 cm/s, 1.23 kPa

8. Water flows through the constant head setup in the laboratory, as shown in the figure, where two identical dense sand samples *A* and *B* are placed—one horizontally and the other vertically. The samples are 50 mm diameter and 100 mm in length. The water levels in the left and right sides are maintained at the levels shown, ensuring constant head throughout the test. The void ratio of the sand is 0.92 and the specific gravity of the grains is 2.69. If 165 g of water was collected in the bucket within 15 minutes, what is the permeability of the soil?

   What are the pore water pressure and vertical effective stress at the mid height of sample *B*?

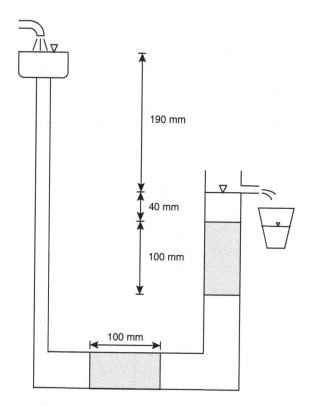

Answer 0.98 × 10⁻² cm/s; 1349 Pa, 78 Pa

9. An experimental setup in the laboratory is shown. Two 50 mm diameter soil samples A and B are placed under constant head. (All dimensions are in mm.) The permeability of sample A is twice that of sample B. Assuming the tail water level as the datum, what is the total head at the interface between the soil samples? If 200 ml of water flows through the sample in 5 minutes, determine the permeabilities of the two soil samples.

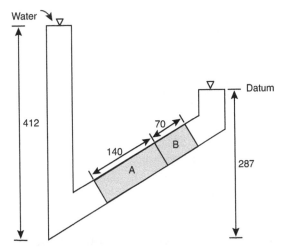

Answer: 62.5 mm, 0.076 cm/s, 0.038 cm/s

10. A soil sample within a sampling tube is connected to an experimental setup (as shown in the figure) to carry out a constant head permeability test. The cross-sectional area of the tube is 75 mm and the length of the sample is 250 mm. If 875 ml of water flows through the sample in 5 minutes, find the permeability of the soil.

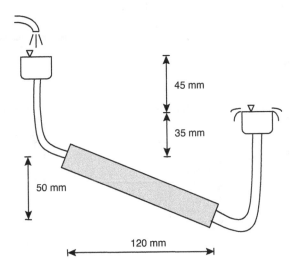

Answer: 0.19 cm/s

11. A 500 m-long levee made of compacted clay impounds water as shown in the figure on page 108. There is a 1 m-thick sand seam along the entire length of the levee at a 15° horizontal inclination that connects the reservoir to the ditch. The permeability of the sand is 3×10^{-3} cm/s. Determine the quantity of water that flows into the ditch in m^3/day.

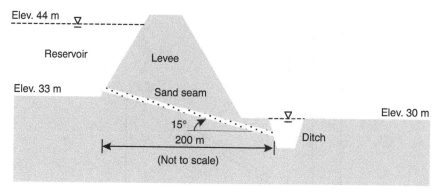

Answer: 87.6 m³/day

12. Seepage takes place beneath the concrete dam shown in the figure, where a sheet pile is present at the downstream end. Permeability of the fine, sandy, silty soil beneath the dam is 3.6×10^{-4} cm/s. Find the following:
    a. The flow rate in m³/day per m run;
    b. The safety factor with respect to piping, assuming that the void ratio is 0.8 and the specific gravity of the soil grains is 2.66; and
    c. The uplift force on the bottom of the dam.

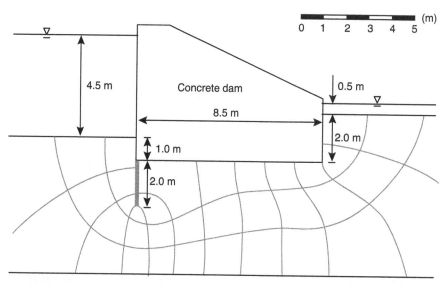

Answer: 0.23 m³/day per m, 4.5, approximately 300 kN per m width

13. Everything else being the same, which of the two dams in the figure will experience larger seepage through the underlying soil? Why?
Which of the two will have a larger exit hydraulic gradient? Why?
Which of the two dams will have a larger uplift? Why?

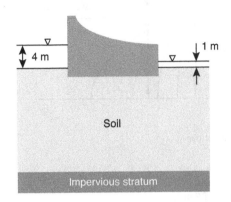
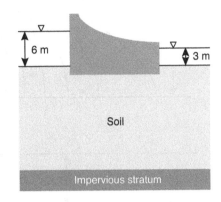

14. The equipotential lines are shown on page 110 for three seepage problems: (a) seepage beneath a concrete dam, (b) seepage beneath a sheet pile, and (c) seepage near a cofferdam. Draw the streamlines and complete the flow nets. Assuming that the permeability of the underlying clayey sand is 2×10^{-5} cm/s, compute the flow rates.

Answer: 4.7×10^{-7} m³/s per m, 2.4×10^{-7} m³/s per m, 3.1×10^{-7} m³/s per m

15. A 10 m-wide and 20 m-high mine stope has a 4 m-high and 2 m-wide drain as shown in the figure on page 111. The stope is backfilled with saturated hydraulic fill that is essentially a silty sand material. Once filled, the permeability of the hydraulic fill is 5.6×10^{-4} cm/s. Draw the flow net. Estimate the flow rate and the location and magnitude of the maximum pore water pressure within the stope.

Answer: 1.0 m³/day per m, 111 kPa at bottom corner

16. A sheet pile is driven into the ground in a waterfront area during some temporary construction work, as shown in the figure on page 110. The silty sand has a permeability of 4.2×10^{-3} cm/s and a water content of 28%. The specific gravity of the soil grains is 2.65. Draw the flow net and calculate the flow rate and safety factor with respect to piping.

Answer: 3.9 m³/day per m, 7.9

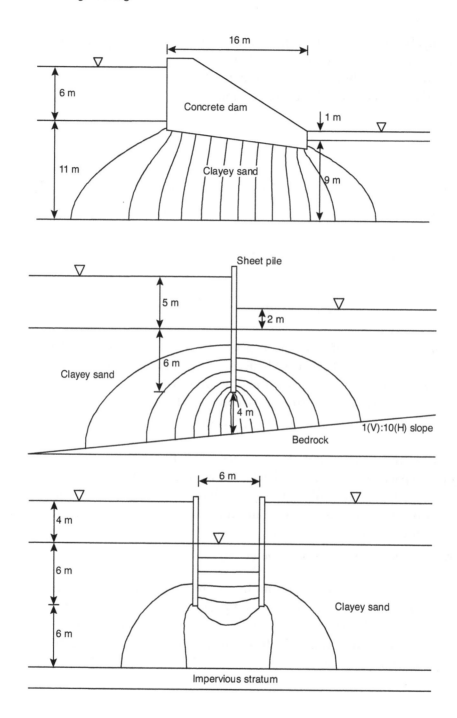

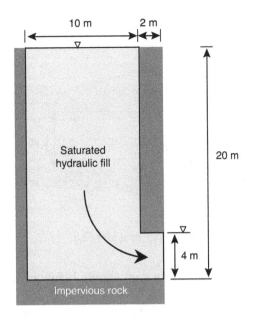

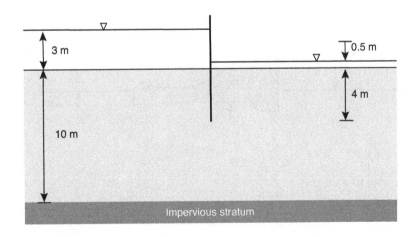

17. A concrete dam shown on page 112 rests on a fine, sandy silt having a permeability of 5 × 10⁻⁴ cm/s, which is underlain by an impervious clay stratum. The saturated unit weight of the sandy silt is 18.5 kN/m³. Draw a flow net. Compute the flow rate beneath the dam in m³/day per meter width and the uplift force on the base of the dam per meter width. What is the safety factor of the dam with respect to piping?

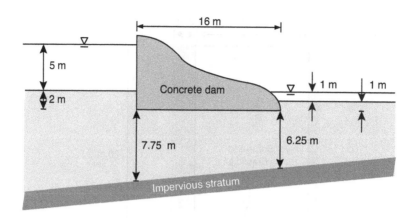

Answer: 0.6 m³/day per m width, 700 kN per m width, 2.8

18. Seepage takes place beneath a concrete dam shown below resting on a fine, sandy silty soil having a permeability of 5×10^{-4} cm/s and a saturated unit weight of 19 kN/m³. A sheet pile is also provided at the upstream end of the dam in an attempt to reduce the seepage. Determine the quantity of seepage in m³/day per meter width, the safety factor with respect to piping, and the uplift thrust on the dam.

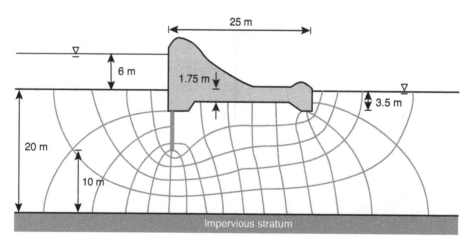

Answer: 0.81 m³/day per m width, 1200 kN per m width, 5.2

19. A long porous drain is placed at a depth 3 m below the ground level as shown in the figure on page 113 to collect the water percolating through the soil above. The permeability of the soil is 2.0×10^{-5} cm/s. There is an impervious stratum at the depth of 6 m. Assuming atmospheric conditions within the drain and at ground level, draw the flow net and estimate the flow rate. [Hint: The perimeter of the drain is an equipotential line; make use of symmetry and draw only one half of the flow net.]

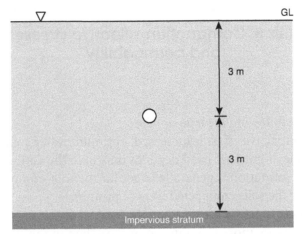

Answer: 0.07 m³/day per m run

20. In a layered soil system (see Section 6.8 Equivalent Permeabilities for One-Dimensional Flow) where the flow is one-dimensional and is either horizontal or vertical, is the horizontal permeability always greater than the vertical permeability? Discuss.
    [Hint: Select a three- to four-layer soil profile and use a spreadsheet to compute the equivalent permeabilities for a wide range of values.]

21. Try Review Exercises 15, 16, 17, and 19 using *SEEP/W*.

22. Seepage takes place beneath a concrete dam underlain by a two-layer soil profile shown below. Use *SEEP/W* to draw the flow net and compute the flow rate, pore water pressures at *A* and *B*, uplift on the dam, and the exit hydraulic gradient.

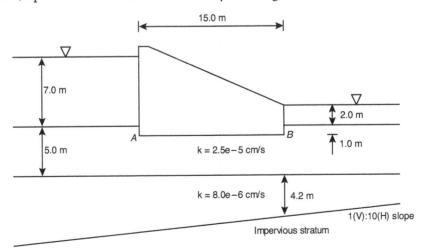

Answer: 3.1 × 10⁻⁷ m³/s; 74 kPa, 34 kPa; 810 kN per m; 0.45

Quiz 3: Compaction, effective stresses, and permeability

Duration: 15 minutes

1. State whether the following are true or false:
   a. When the compactive effort is increased, optimum water content increases
   b. The magnitude of pressure head depends on where the datum is selected
   c. The degree of saturation is generally larger for the soils compacted dry of optimum than for the soils compacted wet of optimum
   d. Effective stress cannot be greater than the total stress
   e. Capillary effects are more pronounced in clays than in sands

   (2.5 points)

2. A 3 m-thick sand layer is underlain by a deep bed of clay where the water content is 29%. The water table is at the bottom of the sand layer. The unit weight of sand is 18 kN/m^3, and G$_s$ for clay grains is 2.70. Find the effective vertical stress at a depth 10 m below the ground level.

   (2.5 points)

3. A sheet pile is driven into a 12 m deep clayey sand bed as shown. Without drawing the flow net, determine the pore water pressure at the bottom tip of the sheet pile.

   (5 points)

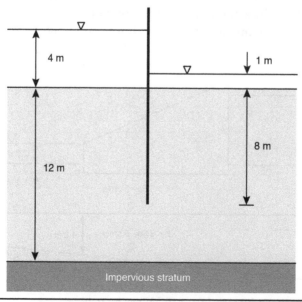

Vertical Stresses Beneath Loaded Areas

7

7.1 INTRODUCTION

In Chapter 5, we saw that the computation of the vertical normal stresses σ_v at any depth within a soil profile is fairly straightforward. It is simply the sum of the product of the layer thickness and the unit weight of the soil lying above the point of interest, which is written as:

$$\sigma_v = \Sigma_1^n \gamma_i H_i \qquad (7.1)$$

where H_i = thickness of the ith layer above and γ_i = unit weight of the soil in the ith layer. When a foundation or embankment is placed on the ground, the stresses within the underlying soil are increased. It is often necessary to compute the increase in the vertical stress $\Delta\sigma_v$ induced by these surface loads.

Soils are *particulate media* formed of granular skeletons made of soil grains. The load transfer mechanism can be very complex here. As a matter of simplicity, soils are treated as *continuous media* (or *continuum*) in stress calculations and in the designs of foundations, retaining walls, slope stability, etc. Here, soils are treated like any other engineering material (e.g., steel) that is a continuum.

Stress-strain diagrams of soils are often simplified as either linear elastic (Figure 7.1a), rigid perfectly plastic (Figure 7.1b), or elastic perfectly plastic (Figure 7.1c). In Figures 7.1b and 7.1c, the material yields when σ reaches the values of σ_y, known as the *yield stress*. Here, the material becomes plastic, undergoing very large deformation while there is no increase in σ. In reality, soils can be *strain hardening* or *strain softening* with the stress-strain plots as shown in Figure 7.1d. Nevertheless, at low stress levels, it is reasonable to assume that the stress-strain variation is linear. In this chapter, we will assume that soil is a *linear elastic continuum*.

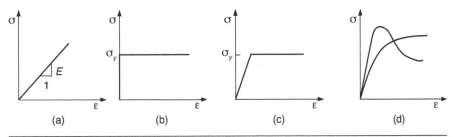

Figure 7.1 Stress-strain plots: (a) linear elastic (b) rigid perfectly plastic (c) elastic perfectly plastic (d) strain hardening and strain softening

7.2 STRESSES DUE TO POINT LOADS

Figure 7.2 shows a point load acting on an *elastic half space*. Here it is assumed that beneath the level at which the load is being applied, the material is elastic and extends to infinity in all directions. Boussinesq (1885) showed that the vertical normal stress increase $\Delta\sigma_v$ at a point within the elastic mass can be written as:

$$\Delta\sigma_v = \frac{3Q}{2\pi z^2 \left[1+\left(\dfrac{r}{z}\right)^2\right]^{5/2}} = \frac{Q}{z^2}I_B \tag{7.2}$$

where z = depth of the point below the horizontal surface and r = horizontal distance of the point from the vertical centerline.

Westergaard (1938) treated soil as an elastic material interspersed with a large number of infinitely thin, perfectly rigid sheets that allow only vertical deformations, and showed that $\Delta\sigma_v$ can be expressed as:

$$\Delta\sigma_v = \frac{Q}{2\pi z^2}\frac{\sqrt{(1-2v)/(2-2v)}}{\left\{\left(\dfrac{1-2v}{2-2v}\right)+\left(\dfrac{r}{z}\right)^2\right\}^{3/2}} = \frac{Q}{z^2}I_W \tag{7.3}$$

where v is the Poisson's ratio of the elastic medium, which can vary in the range of 0–0.5 for linear elastic materials. The Boussinesq and Westergaard influence factors (I_B and I_W) are compared in Figure 7.3 for Poisson's ratio values of 0, 0.1, and 0.2. The Boussinesq equation gives larger values of $\Delta\sigma_v$ when $r/z < 1.5$ (i.e., when the stress increase is significant). For larger r/z, the values of $\Delta\sigma_v$ are very small and are about the same for both methods. In geotechnical engineering practice, the Boussinesq equation is widely used for two reasons: It is simpler than the

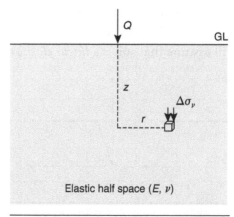

Figure 7.2 Point load on an elastic half space

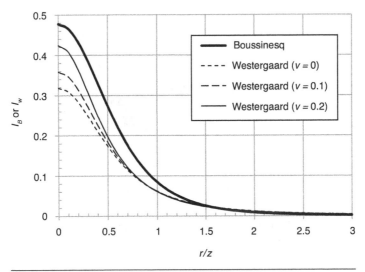

Figure 7.3 Comparison of Boussinesq and Westergaard values

Westergaard equation, and since the $\Delta\sigma_v$ estimates are greater from the Boussinesq equation, it can only overestimate the loadings within the soil, and hence be conservative—this is not a bad thing in geotechnical engineering. Using Equations 7.2, 7.3, or Figure 7.3, one can calculate the vertical normal stress increase at any point within the soil mass. From now on, we will limit our discussions to the Boussinesq equation.

Example 7.1: A 500 kN point load is applied on an elastic half space. Plot the variation of the normal stress increase $\Delta\sigma_v$ with depth (a) along the vertical centerline, (b) along a vertical line 1 m away from the load, and (c) along a vertical line 3 m away from the load.

Solution:

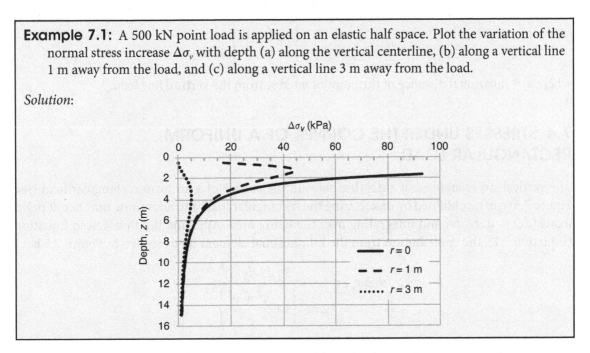

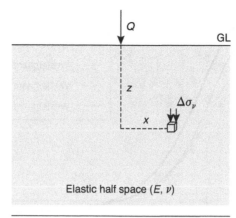

Figure 7.4 Line load on an elastic half space

7.3 STRESSES DUE TO LINE LOADS

When a long line load Q per unit length acts on an elastic half space as shown in Figure 7.4, the vertical stress increase $\Delta\sigma_v$ at a point can be obtained by discretizing the line load into several point loads, applying Equation 7.2, and integrating over the entire line length. This exercise gives:

$$\Delta\sigma_v = \frac{Q}{z}\frac{2}{\pi\left[1+\left(\dfrac{x}{z}\right)^2\right]^2} \tag{7.4}$$

where x = horizontal distance of the point of interest from the vertical line load.

7.4 STRESSES UNDER THE CORNER OF A UNIFORM RECTANGULAR LOAD

The vertical stress increase at a depth z beneath the corner of a uniform rectangular load (see Figure 7.5) can be obtained by discretizing the rectangular load into an infinite number of point loads ($dQ = q\,dx\,dy$) and integrating over the entire area. Applying the Boussinesq Equation (Equation 7.2), the contribution from the infinitesimal element $dx\,dy$ shown in Figure 7.5 is:

$$d(\Delta\sigma_v) = \frac{3q\,dx\,dy}{2\pi z^2\left\{1+\dfrac{x^2+y^2}{z^2}\right\}^{2.5}}$$

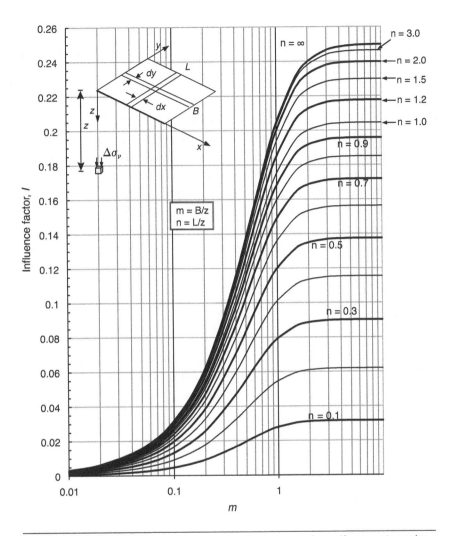

Figure 7.5 Influence factors for $\Delta\sigma_v$ under the corner of a uniform rectangular load

Therefore, the vertical stress increase $\Delta\sigma_v$ is given by:

$$\Delta\sigma_v = \frac{3q}{2\pi} \int\limits_{x=0}^{x=L} \int\limits_{y=0}^{y=B} \frac{1}{z^2\left\{1+\frac{x^2+y^2}{z^2}\right\}^{2.5}} \, dx \, dy$$

$$\Delta\sigma_v = Iq \qquad (7.5)$$

where q is the uniform applied pressure and the influence factor I is given by:

$$I = \frac{1}{4\pi} \left\{ \left(\frac{2mn\sqrt{m^2+n^2+1}}{m^2+n^2+m^2n^2+1} \right) \left(\frac{m^2+n^2+2}{m^2+n^2+1} \right) + \tan^{-1} \left(\frac{2mn\sqrt{m^2+n^2+1}}{m^2+n^2-m^2n^2+1} \right) \right\} \qquad (7.6)$$

where m = B/z and n = L/z. Here, B and L are the breadth and length respectively of the loaded area, and z is the depth of the point of interest under a corner. Variation of I with m and n is shown in Figure 7.5 where m and n are interchangeable. The influence factor obtained or Figure 7.5 can be used with Equation 7.5 to determine the vertical stress increase at any depth within the soil under the *corner* of a uniformly loaded rectangular footing. This can be extended to obtain $\Delta\sigma_v$ at any point within the soil mass—not necessarily under the corner. This will require breaking up the loaded area into four rectangles and applying the principle of superposition (see Example 7.2). This can be extended further to T-shaped or L-shaped areas, too (see Worked Example 7.2).

Example 7.2: A 3 m × 4 m rectangular pad footing applies a uniform pressure of 150 kPa to the underlying soil. Find the vertical normal stress increase at 2 m below points A, B, C, and D.

Solution: a. Under A: L = 4 m, B = 3 m, z = 2 m → m = 1.5, n = 2.0 → I = 0.224

$$\therefore \Delta\sigma_v = 0.224 \times 150 = 33.6 \text{ kPa}$$

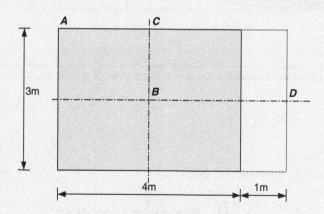

Continues

Example 7.2: *Continued*

b. Under *B*: Let's consider a quarter of the loaded area (and later multiply by 4) so that *B* becomes a *corner*, and we can apply Equation 7.5:

$$L = 2 \text{ m}, B = 1.5 \text{ m}, z = 2 \text{ m} \rightarrow m = 0.75, n = 1.0 \rightarrow I = 0.155$$
$$\therefore \Delta\sigma_v = 4 \times 0.155 \times 150 = 93.0 \text{ kPa}$$

c. Under *C*: Let's consider the left half of the loaded area (and later multiply by 2) where *C* is a corner, so that we can apply Equation 7.5:

$$L = 2 \text{ m}, B = 3 \text{ m}, z = 2 \text{ m} \rightarrow m = 1.5, n = 1.0 \rightarrow I = 0.194$$
$$\therefore \Delta\sigma_v = 2 \times 0.194 \times 150 = 58.2 \text{ kPa}$$

d. Under *D*: Let's consider the upper half above the centerline as shown, and later multiply by 2.

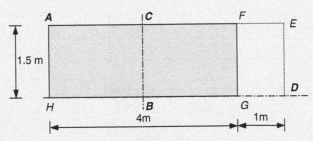

$$AFGH = DHAE - DGFE$$
$$\therefore I = I_{DHAE} - -I_{DGFE}$$
DHAE: $L = 5 \text{ m}, B = 1.5 \text{ m}, z = 2 \text{ m} \rightarrow m = 0.75, n = 2.5 \rightarrow I_{DHAE} = 0.177$
DGFE: $L = 1 \text{ m}, B = 1.5 \text{ m}, z = 2 \text{ m} \rightarrow m = 0.75, n = 0.5 \rightarrow I_{DGFE} = 0.107$
$$\therefore \Delta\sigma_v = 2 (0.177 - 0.107) \times 150 = 21.0 \text{ kPa}$$

We can see from Example 7.2 that $\Delta\sigma_v$ is the maximum under the center of the loaded area, as expected intuitively. Let's see how $\Delta\sigma_v$ varies laterally along a centerline with depth through Example 7.3.

Example 7.3: A 3 m × 3 m square footing carries a uniform pressure of 100 kPa. Plot the lateral variation of $\Delta\sigma_v$ along the horizontal centerline at depths of 1 m and 3 m.

Solution: Due to symmetry, we will only compute the values of $\Delta\sigma_v$ for the right half of the footing at points $A, B, C, \ldots H$, spaced at 0.5 m intervals as shown.

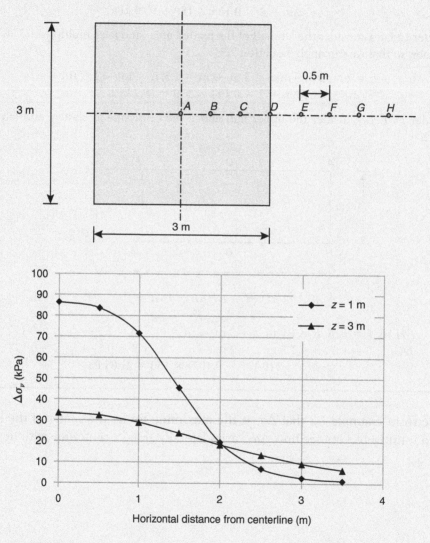

The values of $\Delta\sigma_v$ computed as in Example 7.2 are summarized on page 123 and are shown in the plot.

Continues

Example 7.3: *Continued*

Under	A	B	C	D	E	F	G	H
$\Delta\sigma_v$ (kPa) @ 1 m depth	86.4	83.6	71.6	45.6	19.6	7.0	2.6	1.2
$\Delta\sigma_v$ (kPa) @ 3 m depth	33.6	32.4	29.0	24.0	18.4	13.6	9.4	6.4

7.5 2:1 DISTRIBUTION METHOD

It can be seen from Example 7.3 that $\Delta\sigma_v$ is the maximum under the center of the loaded area and decays laterally and with depth. Very often, we want a quick estimate of $\Delta\sigma_v$ at a specific depth z without any consideration of the lateral variations. A simple but crude empirical method for estimating the vertical stress increase at a specific depth z is discussed here. As shown in Figure 7.6, it is assumed that the load Q applied on a rectangular footing with dimensions of $B \times L$ is spread in a 2 (vertical):1 (horizontal) manner in both directions.

Just below the footing, the pressure applied to the underlying soil is $q = Q/BL$. Since the load is acting over a larger area at depth z, the additional vertical normal stress $\Delta\sigma_v$ is significantly less and is given by:

$$\Delta\sigma_v = \frac{Q}{(B+z)(L+z)} \tag{7.7}$$

At shallow depths, the 2:1 approximation gives lower values for $\Delta\sigma_v$ when compared to the maximum value obtained under the center using the Boussinesq equation. At very large depths, the 2:1 approximation gives higher values. See the figure on page 133. In the case of a strip footing ($L = \infty$) carrying a line load (load per unit length), Equation 7.7 becomes:

$$\Delta\sigma_v = \frac{Q}{(B+z)} \tag{7.8}$$

where Q is in kN/m.

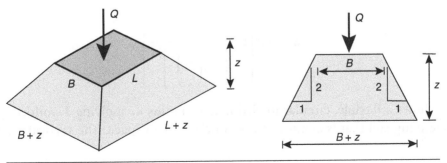

Figure 7.6 Estimating $\Delta\sigma_v$ by 2:1 distribution method

Example 7.4: Using Equation 7.7, estimate $\Delta\sigma_v$ at 1 m and 3 m depths below the 3 m square footing in Example 7.3 carrying 100 kPa. How do the values compare with those computed in Example 7.3?

Solution: $B = L = 3$ m; $Q = q\,B\,L = 900$ kN

At 1 m depth, $z = 1$ m $\rightarrow$ Using Eq. 7.6, $\Delta\sigma_v = 56.3$ kPa
At 3 m depth, $z = 3$ m $\rightarrow$ Using Eq. 7.6, $\Delta\sigma_v = 25.0$ kPa

These values are significantly less than the maximum values of $\Delta\sigma_v$ directly below the center observed in Example 7.3.

7.6 PRESSURE ISOBARS UNDER FLEXIBLE UNIFORM LOADS

Using the influence factors discussed in Section 7.4, it is possible to determine the vertical normal stress increase due to a uniform rectangular load at *any* point within the soil mass. Let's identify the points at which the value of $\Delta\sigma_v$ is $0.1q$, and then connect the points. This will give a *stress contour* or *isobar* for $0.1q$, which will be symmetrical about the vertical centerline of the footing. Such isobars can be drawn for any value of $\Delta\sigma_v$ for a square, rectangular, strip (very long in one direction), or circular footing. They are shown for a square and for strip footings of width B in Figure 7.7. Due to symmetry, only half is shown. It can be seen that the isobars extend significantly deeper for strip footings than for square ones. For example, the isobar of $0.1q$ extends to a depth of $2B$ for square footings and more than $5B$ for strip footings. At any depth, $\Delta\sigma_v$ would be greater under a strip footing than under a square one. These isobars can be used for a quick estimate of $\Delta\sigma_v$ beneath a square or strip footing.

7.7 NEWMARK'S CHART

By integration of the Boussinesq equation (Equation 7.2), it can be shown that the vertical normal stress increase $\Delta\sigma_v$ at a depth of z below the center of a flexible circular load of radius a is given by:

$$\Delta\sigma_v = q\left\{1 - \frac{1}{\left[1+\left(\dfrac{a}{z}\right)^2\right]^{3/2}}\right\} \tag{7.9}$$

Figure 7.8 shows a flexible, circular loaded area of radius a, applying a uniform pressure q to the underlying soil that is assumed to be an elastic half space. The vertical normal stress

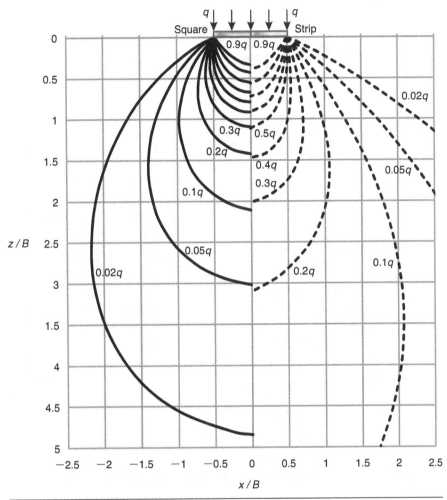

Figure 7.7 Pressure isobars for uniformly loaded flexible square and strip

increase $\Delta\sigma_v$ at point X at depth z below the center can be calculated using Equation 7.9. What would be the radius a in terms of z, such that $\Delta\sigma_v$ would be $0.1q$? From Equation 7.9, it can be calculated as $0.2698z$. Repeating this exercise for $\Delta\sigma_v$ of $0.2q$, $0.3q$, etc., the values are tabulated in Figure 7.8.

When $a = 0.9176z$, $\Delta\sigma_v$ at X is $0.6q$. When $a = 0.7664z$, $\Delta\sigma_v$ at X is $0.5q$. Therefore, when the annular zone between these two circular areas (see Figure 7.8) is subjected to a pressure of q, $\Delta\sigma_v$ at X would be $0.1q$. This is the underlying principle of Newmark's chart.

Newmark (1942) developed the influence chart shown in Figure 7.9, which consists of concentric circles of different radii, the values of which are given in Figure 7.8. In drawing the circles, the value of z was taken as the length of the line shown in Figure 7.9 as scale, which

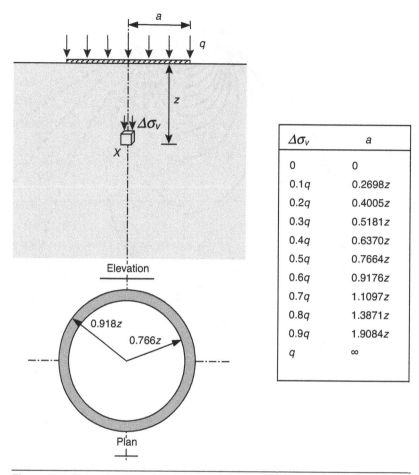

$\Delta\sigma_v$	a
0	0
$0.1q$	$0.2698z$
$0.2q$	$0.4005z$
$0.3q$	$0.5181z$
$0.4q$	$0.6370z$
$0.5q$	$0.7664z$
$0.6q$	$0.9176z$
$0.7q$	$1.1097z$
$0.8q$	$1.3871z$
$0.9q$	$1.9084z$
q	∞

Figure 7.8 Stress increase beneath the center of a flexible circular load

is simply the depth of the point of interest, X. If a pressure q is applied over the annular zone between any two adjacent circles, this would increase the vertical normal stress at X by $0.1q$. The radial lines divide the annular zones into 12 equal blocks. There are 120 equal blocks in Newmark's chart in Figure 7.9, and pressure q applied on any of them will lead to an increase in normal vertical stress of $1/120q$ at X.

How do we use Newmark's chart to find $\Delta\sigma_v$ under a point within a loaded area at a certain depth, z^*? Newmark's chart has a scale that is shown along with the figure, which is simply the depth z at which $\Delta\sigma_v$ would be computed. The radii of the circles were computed on the basis of this length z. Therefore, all that is required now is to redraw the loaded area to a new scale where the length shown in Newmark's chart equals the depth of interest, z^*. The point under which $\Delta\sigma_v$ is required is placed exactly on the center of the chart and the number of blocks that

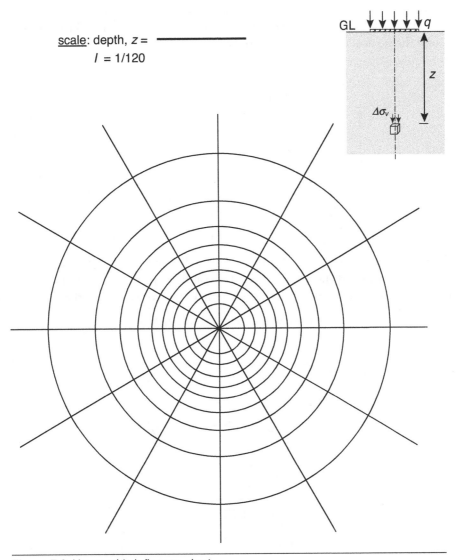

scale: depth, $z =$ ─────────
$I = 1/120$

Figure 7.9 Newmark's influence chart

are covered in Newmark's chart are counted. If n blocks are covered by the loaded area, $\Delta\sigma_v$ is given by:

$$\Delta\sigma_v = n\,I\,q \qquad\qquad (7.10)$$

where q is the applied pressure and I is the influence factor, which is simply the reciprocal of the number of blocks in Newmark's chart.

Example 7.5: The loaded area shown carries a uniform pressure of 60 kPa. Using Newmark's chart, find the vertical normal stress increase at 6 m below X.

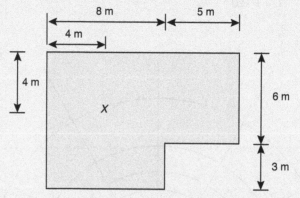

Solution: Let's redraw the area to a scale of 6 m = scale length shown in the chart, and overlay the rescaled area on the chart such that the point X is at the center. Counting the blocks covered by the loaded area, including those fractions when they are covered only partially, $n = 69.5$.

$$\therefore \Delta\sigma_v \text{ at 6 m below } X = 69.5 \times (1/120) \times 60 = 35 \text{ kPa}$$

<u>scale</u>: depth, $z =$ _____
$I = 1/120$

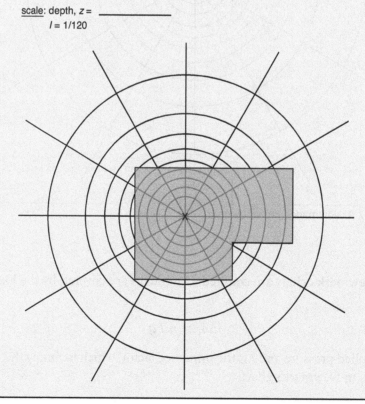

7.8 STRESS COMPUTATIONS USING *SIGMA/W*

A DVD containing the *Student Edition* of *GeoStudio 2007* is included with this book. One of the eight different programs that come up when you click the *GeoStudio 2007* icon is *SIGMA/W*, a versatile finite element software that can be used to compute and plot stresses under loaded areas that can be modeled as two-dimensional plane strain or axisymmetric problems. The *Student Edition* of *SIGMA/W* has a few limitations that make it suitable mainly for learning and evaluation. It can handle up to 500 elements, 10 different regions, and three different materials. It can model two-dimensional plane strain problems (e.g., stresses beneath a long embankment or strip footing) as well as axisymmetric problems (e.g., stresses beneath a circular footing). The *Student Edition* allows the soil to be modeled only as an infinitely linear elastic material. The full version has several advanced features and no limits to the number of elements, regions, and materials. It is available from *GEO-SLOPE* International, Canada (http://www.geo-slope.com).

7.8.1 Getting Started with *SIGMA/W*

When running *GeoStudio*, select *Student License* from the start page. All *GeoStudio* project files are saved with the extension *.gsz* so that they can be called by any of the applications (e.g., *SEEP/W*, *SLOPE/W*) within the suite.

Familiarize yourself with the different toolbars that can be made visible through the View/Toolbar... menu. Moving the cursor over an icon displays its function. In the Analysis toolbar, you will see three icons, DEFINE, SOLVE, and CONTOUR, next to each other. DEFINE and CONTOUR are two separate windows and you can switch between them. The problem is fully defined in the DEFINE window and saved. Clicking the SOLVE icon solves the problem as specified. Clicking the CONTOUR icon displays the results in the CONTOUR window. The input data can be changed by switching to the DEFINE window and SOLVEd again.

The major components in solving a stress computation problem are:

1. *Defining the geometry:*
   Always have a rough sketch of your geometry problem with the right dimensions before you start *SIGMA/W*. When *SIGMA/W* is started, it is in the DEFINE window. The Set menu has two different but related entries, Page... and Units and Scales..., which can be used to define your working area and units. A good start is to use a 260 mm (width) × 200 mm (height) area that fits nicely on an A4 sheet. Here, a scale of 1:200 would represent 52 m (width) × 40 m (height) of problem geometry. Try to use the same scale in the *x* and *y* directions so that the geometry is not distorted. Units and Scales... should be used for defining the problem as two-dimensional (plane strain) or axisymmetric. Grid... will allow you to select the grid spacing, make it visible, and snap to the grid points. Axes... will allow you to draw the axes and label them. Sketch/Axes... may be a

better way to draw the axes and label them. Use $\boxed{View/Preferences...}$ to change the way the geometry, fonts, and graphical outputs are displayed.

Use $\boxed{Sketch/Lines}$ to sketch the geometry using free lines. You can use $\boxed{Modify/Objects...}$ to delete or move the lines. $\boxed{Sketch}$ is different from $\boxed{Draw}$. Use $\boxed{Draw/Regions...}$ on the sketched outlines to create the real geometry and to define the zones of different materials. Alternatively, one may omit $\boxed{Sketch}$ and start from $\boxed{Draw}$ instead, especially in simple problems. While *Sketch*ing, *Draw*ing, or *Modify*ing, right clicking the mouse ends the action. The $\boxed{Sketch}$ menu is useful for drawing dimension lines with arrowheads and for labeling the dimensions and objects. It is a good practice to break the soil into regions so that the finite element mesh can be made finer in the regions of interest.

2. *Defining soil properties and assigning to regions:*
Use $\boxed{Draw/Materials...}$ to assign the soil properties (e.g., Young's modulus) and apply them to the regions by dragging. The *Student Edition* can accommodate up to three different materials that are placed in 10 different zones, all of which are assumed to be linear elastic. When we are interested in the *change* in stresses caused by the applied loadings, we may assume the soil unit weight to be zero to neglect the gravitational stresses.

3. *Defining the boundary conditions:*
Assign the boundary conditions through $\boxed{Draw/Boundary\ Conditions...}$. Here, specify the fixities (no displacements along the x/y directions) along the boundaries and create new boundary conditions to specify the known loadings or displacements at the boundaries. Use a separate name tag for each boundary condition. Once a boundary condition is created, it can be applied to a point, line, or a region. Apply the boundary conditions by dragging them to the relevant location. Take advantage of symmetry and analyze only one-half of the problem in two-dimensional plane strain problems. Remember, we have to use the 500 elements wisely! Avoid the boundary interference by selecting them as far as possible. When we assume that there is no displacement in the x and y directions, the assumption must be realistic.

4. *Defining the finite element mesh:*
This step gives us a taste of finite element modeling. The *Student Edition* of *SIGMA/W* limits the number of elements to 500. The default mesh would be adequate for most of our work here. The mesh can be seen through $\boxed{Draw/Mesh\ Properties...}$. The mesh size can be varied by adjusting the global element size; as the element size increases, the coarser the mesh becomes.

The area of interest can be divided into a few regions (up to 10 in the *Student Edition*) and the mesh density can be varied within the regions—provided the total number of elements does not exceed 500.

5. *Solving the problem:*

   Once the problem is fully defined through steps 1–4, it can be $\boxed{SOLVE}$d, and the results can be viewed in a $\boxed{CONTOUR}$ window. You can switch between the $\boxed{DEFINE}$ and $\boxed{CONTOUR}$ windows while experimenting with the output. This can be very effective for a parametric study. $\boxed{Tools/Verify}$ can be used for checking the problem definition before solving.

6. *Displaying the results:*

   $\boxed{CONTOUR}$ can be used to display the stress contours and displacement contours. From $\boxed{Draw/Contour...}$, the intervals and colors can be varied. By clicking the $\boxed{Draw/Contour\ Labels}$, the cursor changes into a crosshair. By placing the crosshair on a contour line and clicking the mouse, the contour value is labeled. $\boxed{Draw/Mohr\ Circles}$ can be used to draw the Mohr circle representing the state of stress at any point, along with the elements showing the normal and shear stresses. $\boxed{Draw/Graph...}$ can be used to generate various plots of stress vs. depth, displacements vs. distance, etc. More than one graph can be selected by clicking the first one, holding the shift key, and then clicking the last one on the list. $\boxed{View/Result\ Information...}$ provides full information about the stresses and displacements at any point in a separate window.

Example 7.6: A 10 m diameter silo applies a uniform pressure of 200 kPa to the underlying soil. Assuming the soil to be linear elastic with $E = 10$ MPa and $v = 0.2$, estimate the settlement below the centerline using *SIGMA/W*. Show the vertical stress increase contours with 20 kPa intervals and the boundary conditions. Use the default finite element mesh. How many elements and nodes are there?

Solution: In *SIGMA/W*, let's take gravity as 0 and avoid the gravitational initial stresses. This is an axisymmetric problem and we will model along a radial plane.

Continues

Example 7.6: *Continued*

The default mesh has 450 elements and 496 nodes. Settlement beneath the center = 159 mm.

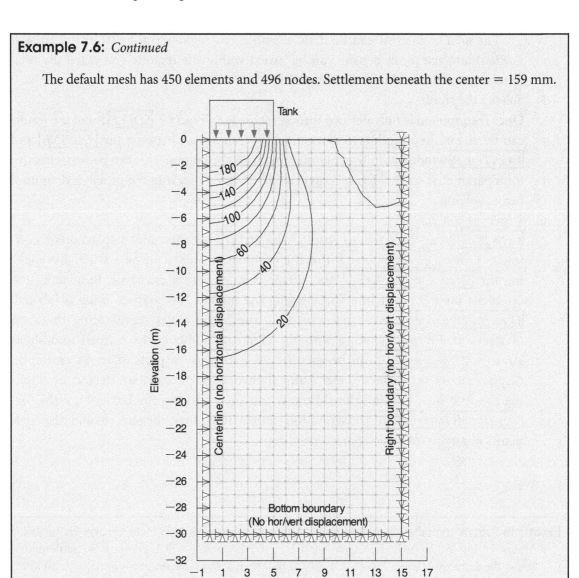

Reminder

❖ Soil is treated as an elastic continuum in this chapter.

❖ Boussinesq analysis is preferred over Westergaard's due to its simplicity and conservativeness.

❖ Equation 7.5 and Figure 7.5 can be applied *only under a corner* of a uniform rectangular load.

❖ Newmark's chart can be applied on any irregularly shaped, uniformly loaded area.

❖ *SIGMA/W* can be used to compute stresses and deformations in plane strain and axisymmetric problems.

WORKED EXAMPLES

1. A square, flexible footing of width *B* applies a uniform pressure of 150 kPa to the underlying soil. Compute the normal, vertical stress increase along the vertical centerline using the m-n coefficients (Equation 7.5) and the 2:1 distribution (Equation 7.7) at different depths and plot them.

Solution:

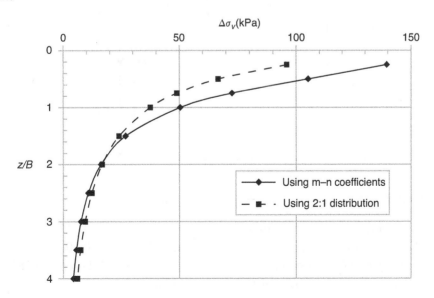

2. The area shown in the figure carries a uniform pressure of 200 kPa. Find the vertical normal stress increase at 5 m below A, B, and C.

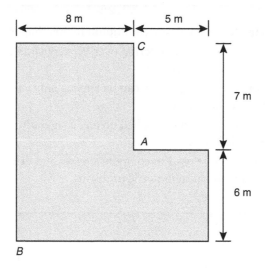

Solution: Let's use the m-n coefficients, remembering that they can only be used for $\Delta\sigma_v$ under corners. We will also add a few dashed lines and points as shown.

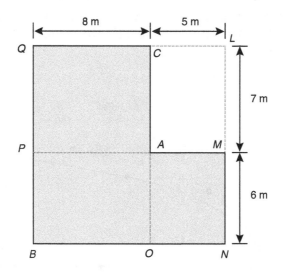

Under Point A: $\Delta\sigma_v = q\,I = q\,(I_{APQC} + I_{AOBP} + I_{AMNO})$

$APQC$: $L = 8$ m, $B = 7$ m, $z = 5$ m $\rightarrow$ m $= 1.4$, n $= 1.6 \rightarrow I = 0.215$
$AOBP$: $L = 8$ m, $B = 6$ m, $z = 5$ m $\rightarrow$ m $= 1.2$, n $= 1.6 \rightarrow I = 0.207$
$\qquad$ $AMNO$: $L = 6$ m, $B = 5$ m, $z = 5$ m $\rightarrow$ m $= 1.0$, n $= 1.2 \rightarrow I = 0.185$
$\therefore \Delta\sigma_v = 200 \times (0.215 + 0.207 + 0.185) = 121.4$ kPa

Under Point B: $\Delta\sigma_v = q\,I = q\,(I_{BQCO} + I_{BPMN} - I_{BPAO})$

$BQCO$: $L = 13$ m, $B = 8$ m, $z = 5$ m → m = 1.6, n = 2.6 → $I = 0.230$
$BPMN$: $L = 13$ m, $B = 6$ m, $z = 5$ m → m = 1.2, n = 2.6 → $I = 0.215$
$\quad\quad BPAO$: $L = 8$ m, $B = 6$ m, $z = 5$ m → m = 1.2, n = 1.6 → $I = 0.207$
∴ $\Delta\sigma_v = 200 \times (0.230 + 0.215 - 0.207) = 47.6$ kPa

Under Point C: $\Delta\sigma_v = q\,I = q\,(I_{COBQ} + I_{CLNO} - I_{CLMA})$

$COBQ$: $L = 13$ m, $B = 8$ m, $z = 5$ m → m = 1.6, n = 2.6 → $I = 0.230$
$CLNO$: $L = 13$ m, $B = 5$ m, $z = 5$ m → m = 1.0, n = 2.6 → $I = 0.203$
$\quad\quad CLMA$: $L = 7$ m, $B = 5$ m, $z = 5$ m → m = 1.0, n = 1.4 → $I = 0.191$
∴ $\Delta\sigma_v = 200 \times (0.230 + 0.203 - 0.191) = 48.4$ kPa

3. A square footing of width B applies a uniform pressure q to the underlying soil. Using the 2:1 distribution, estimate the depth at which $\Delta\sigma_v$ is 20% of the applied pressure q. How does this estimate compare with the estimate from pressure isobars?

   Solution:

   $$\Delta\sigma_v = \frac{qBL}{(B+z)(L+z)} \rightarrow 0.2q = \frac{qB^2}{(B+z)^2} \rightarrow \left(\frac{B+z}{B}\right)^2 = 5$$

   $z/B = \sqrt{5} - 1 \rightarrow z = 1.24B$

   From the pressure isobars in Figure 7.7, $\Delta\sigma_v = 0.2\,q$ at approximately 1.4B.

4. A 3 m-wide and very long strip footing applies a uniform pressure of 120 kPa to the underlying soil. Find the vertical stress increase at a 2 m depth under the centerline using m-n coefficients.

   Solution: Let's divide the strip into four quarters and find $\Delta\sigma_v$ under the corner of one. For each quarter,

   $$L = \infty, B = 1.5 \text{ m}, z = 2 \text{ m} \rightarrow I = 0.179$$
   $$\therefore \Delta\sigma_v = 4 \times 0.179 \times 120 = 85.9 \text{ kPa}$$

5. A 10 m diameter silo applies a uniform pressure of 200 kPa to the underlying soil. Assuming the soil to be linear elastic with $E = 10$ MPa and $v = 0.2$, draw the vertical stress increase $\Delta\sigma_v$ contours in 20 kPa intervals using *SIGMA/W*. Use a finer mesh close to the loaded area. Plot the lateral stress variation of $\Delta\sigma_v$ at 2.5 m, 5.0 m, and 15.0 m depths.

   Solution: Let's divide the mesh into six regions and adjust the mesh density using Draw/Mesh Properties..., such that the mesh is finer at the top left and coarser at the bottom right. This will be solved as an axisymmetric problem, and we will consider a radial plane as shown.

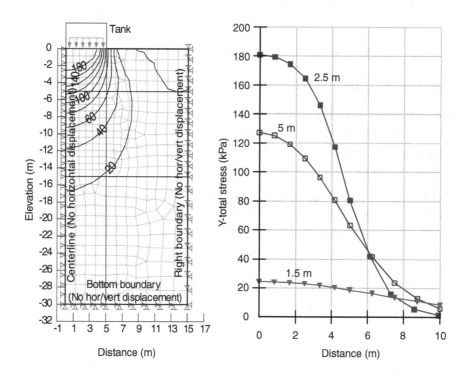

The boundary conditions are:

- No horizontal displacements along the left boundary (vertical centerline)
- Neither horizontal nor vertical displacements at the bottom and right boundaries
- 200 kPa applied from 0 to 5 m

The $\Delta\sigma_v$ versus distance plots at 2.5 m, 5.0 m, and 15.0 m depths created using $\boxed{Draw/Graph}$ are shown in the figure.

REVIEW EXERCISES

1. Try Worked Example 2 using Newmark's chart.

2. The loaded area shown on page 137 applies a uniform pressure of 80 kPa to the underlying soil. Find the normal vertical stress increase at 4 m below A, B, and C.

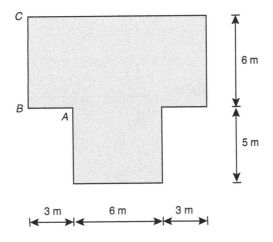

Answer: 48.2 kPa, 22.3 kPa, 18.8 kPa

3. A strip footing of width B applies a uniform pressure of q to the underlying soil. Using the 2:1 distribution, find the depth (in terms of B) at which $\Delta\sigma_v = 0.2q$. Compare this with the estimate from the pressure isobars shown in Figure 7.7.

Answer: 4B

4. A 2.5 m-wide strip footing applies a uniform pressure of 100 kPa to the underlying soil. The Young's modulus and Poisson's ratio of the soil are 14 MPa and 0.25 respectively. Using *SIGMA/W*, develop the pressure isobars and plot the variation in vertical stress increase $\Delta\sigma_v$ with depth along the vertical centerline.

5. Repeat the previous exercise substituting a 2.5 m diameter circular footing and compare the findings.

6. A 5 m-high embankment ($\gamma = 20$ kN/m³, $E = 16$ MPa, $\nu = 0.20$) is being constructed at a site where the top 6 m consists of Soil 1 (E_1, ν_1) underlain by a very large depth of Soil 2 (E_2, ν_2), as shown in the figure on page 138. The right half of the embankment with the soil profile is also shown. Model the embankment using *SIGMA/W*, neglecting the unit weights of Soil 1 and Soil 2. If $E_1 = 4$ MPa, $\nu_1 = 0.25$, $E_2 = 8$ MPa, and $\nu_2 = 0.30$, find the settlements at A and C and the vertical stress increase at B.

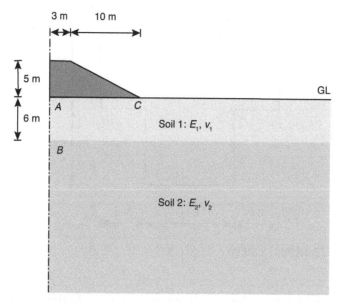

Answer: 325 mm, 175 mm, and 67 kPa

7. An embankment is being built with a berm as shown. The embankment soil properties are: $E = 18$ MPa, $\nu = 0.2$, $\gamma = 20$ kN/m³. The 5 m-thick foundation soil has $E = 8$ MPa, $\nu = 0.25$, $\gamma = 18$ kN/m³, and is underlain by bedrock. Use *SIGMA/W* to analyze the problem and report the settlement of the crest of the embankment at the centerline. Show the vertical stress increase $\Delta\sigma_v$ contours.

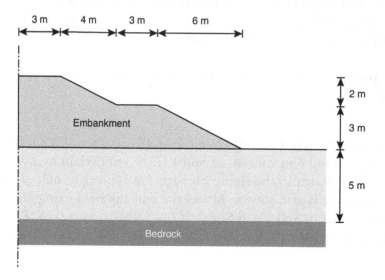

Answer: 83 mm

Consolidation

8

8.1 INTRODUCTION

When an embankment or a foundation is placed on soil, *settlement* takes place. The weaker the soil is, the greater the settlement will be. In the case of dry or saturated granular soils, the settlement is almost instantaneous, whereas in saturated clays, this occurs over a much longer time through the process of *consolidation*.

Consolidation is a process in saturated clays where the water is squeezed out by the applied external loads, thus gradually increasing settlement due to a reduction in void ratio. The settlement eventually stops, but only after a long time. We will assume this length of time to be infinity for now based on Terzaghi's consolidation theory, which we will discuss in Section 8.5. Let's consider a soil element X in Figure 8.1a where the initial values of total and effective vertical stresses and pore water pressures are $\sigma_{v0} = \gamma_{sat}h$, $\sigma'_{v0} = \gamma'h$, and $u_0 = \gamma_wh$, respectively. If a uniform surcharge of q kPa is applied at the ground level (see Figure 8.1b), the above values will increase by $\Delta\sigma(t)$, $\Delta\sigma'(t)$, and $\Delta u(t)$ respectively where (t) reflects the time dependence of these changes. Figure 8.1c shows the variation of the consolidation settlement with time where the final consolidation settlement s_c is reached only at a time of ∞. Figure 8.1d shows the variations of $\Delta\sigma(t)$, $\Delta\sigma'(t)$, and $\Delta u(t)$ with time.

$$\sigma = \sigma' + u$$

hence:

$$\Delta\sigma = \Delta\sigma' + \Delta u$$

At any time during consolidation:

$$\Delta\sigma(t) = q$$

hence:

$$\Delta\sigma'(t) + \Delta u(t) = q$$

Immediately after the surcharge is applied, water carries the entire load. Hence, at $t = 0^+$, $\Delta u = q$ and $\Delta\sigma' = 0$. With time, drainage takes place and the load is gradually transferred from the water to the soil skeleton (i.e., Δu decreases and $\Delta\sigma'$ increases). Finally, at $t = \infty$, the excess

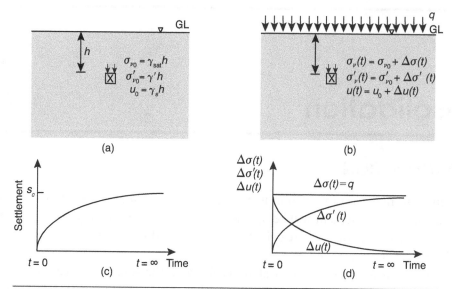

Figure 8.1 Consolidation fundamentals: (a) just before applying surcharge (b) during consolidation (c) settlement versus time (d) $\Delta\sigma$, $\Delta\sigma'$, and Δu versus time

pore water pressure induced by the applied surcharge is fully dissipated and the entire surcharge is carried by the soil skeleton, making $\Delta u = 0$ and $\Delta\sigma' = q$.

8.2 ONE-DIMENSIONAL CONSOLIDATION

The clay layer in Figure 8.2a is sandwiched between two free-draining granular soil layers. The surcharge $\Delta\sigma$ is spread over a very large area, so it is reasonable to assume that the *strains* and *drainage* within the clay are both vertical, and as such, in one-dimension, i.e., there is no water draining horizontally and there are no horizontal strains. Here, the clay is undergoing *one-dimensional consolidation*. In field situations, the consolidation is often three-dimensional with drainage and strains taking place in all directions, and it may be necessary to apply some corrections if we use one-dimensional consolidation theory.

One-dimensional consolidation is simulated in the laboratory in a 50–75 mm diameter metal *oedometer* ring, which restricts horizontal deformation and drainage. The undisturbed clay sample is placed in the oedometer ring, sandwiched between two porous stones that allow drainage (Figure 8.2b), thus simulating the field situation shown in Figure 8.2a.

8.2.1 Δe–ΔH Relation

The clay layer and the phase diagram (for V_s of unity) are shown in Figure 8.3a and b at both the beginning and end of the consolidation process. Due to consolidation, the thickness has

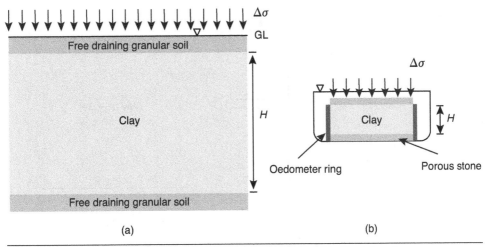

Figure 8.2 One-dimensional consolidation: (a) in the field (b) in the laboratory

decreased by ΔH (same as s_c in Figure 8.1c) from the initial value of H_0, and the void ratio has decreased by Δe from the initial value of e_0. Therefore, the average vertical strain within the clay is $\Delta H/H_0$.

From the phase diagrams, the average vertical strain can be computed as:

$$\frac{\Delta e}{(1+e_0)}$$

Therefore:

$$\frac{\Delta H}{H_0} = \frac{\Delta e}{1+e_0} \tag{8.1}$$

8.2.2 Coefficient of Volume Compressibility m_v

Coefficient of volume compressibility m_v is a measure of the compressibility of the clay. It is defined as the volumetric strain $\varepsilon_{\mathrm{vol}}$ per unit stress increase, and is expressed as:

$$m_v = \frac{\varepsilon_{\mathrm{vol}}}{\Delta\sigma'} = \frac{\Delta V/V_0}{\Delta\sigma'} \tag{8.2}$$

where V_0 = initial volume, ΔV = volume change, and $\Delta\sigma'$ = effective stress increase that causes the volume change ΔV. In one-dimensional consolidation where the horizontal cross-sectional area remains the same, $\Delta V/V_0 = \Delta H/H_0$. Therefore, Equation 8.2 can be written as:

$$s_c = \Delta H = m_v\Delta\sigma'H_0 \tag{8.3}$$

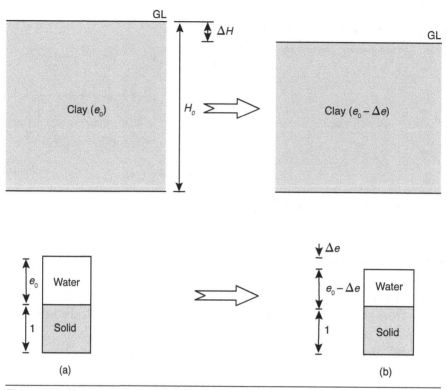

Figure 8.3 Changes in layer thickness and void ratio due to consolidation: (a) at $t = 0^+$ (b) at $t = \infty$

which is a simple and useful equation for estimating the final consolidation settlement, s_c. The coefficient of volume compressibility m_v is often expressed in MPa^{-1} or m^2/MN. It can be less than 0.05 MPa^{-1} for stiff clays and can exceed 1.5 MPa^{-1} for soft clays.

Example 8.1: A 5 m-thick clay layer is surcharged by a 3 m-high compacted fill with a bulk unit weight of 20.0 kN/m^3. The coefficient volume compressibility of the clay is 1.8 MPa^{-1}. Estimate the final consolidation settlement.

Solution:

$$\Delta\sigma' = 3 \times 20 = 60 \text{ kPa}$$

From Equation 8.3, $s_c = (1.8)\left(\dfrac{60}{1000}\right)(5000) = 540 \text{ mm}$

8.3 CONSOLIDATION TEST

The consolidation test (ASTM D2435; AS1289.6.6.1) is generally carried out in an oedometer in the laboratory (see Figure 8.2b) where a 50–75 mm diameter undisturbed clay sample is sandwiched between two porous stones and loaded in increments. Each pressure increment is applied for 24 hours, ensuring the sample is fully consolidated at the end of each increment. At the end of each increment with the consolidation completed, the vertical effective stress is known and the void ratio can be calculated from the measured settlement ΔH, using Equation 8.1. After reaching the required maximum vertical pressure, the sample is unloaded in a similar manner and the void ratios are computed. A typical variation of the void ratio against effective vertical stress (in logarithmic scale) for a good quality undisturbed clay sample is shown in Figure 8.4a. Here, the initial part of the plot from A to B is approximately a straight line with a slope of C_r (known as the *recompression index*), until the vertical stress reaches a critical value σ'_p, known as the *preconsolidation pressure*, which occurs at B. Once the preconsolidation pressure is exceeded and until unloading takes place, the variation from B to C is again linear, but with a significantly steeper slope C_c, known as the *compression index*. The variation is linear during unloading from C to D, again with a slope of C_r. Reloading takes place along the same path as unloading.

Figure 8.4b shows what happens in reality to a clay sample when loaded, unloaded, and reloaded along the path $ABCDCE$. It reaches the preconsolidation pressure at B and is loaded further along the path BC in increments. At C, the clay is unloaded to D. The loading and unloading paths do not exactly overlap as we idealized in Figure 8.4a, but it is reasonable to assume that they

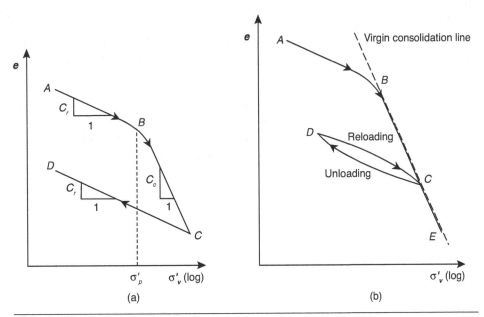

Figure 8.4 e vs. log σ'_v plot: (a) definitions (b) virgin consolidation line

do and that the path is a straight line with a slope of C_r. The dashed line shown in Figure 8.4b is the virgin consolidation line VCL, which has a slope of C_c. It can be seen that as soon as the reloading path meets the VCL near B and C, the slope changes from C_r to C_c and the clay sample follows the virgin consolidation line. Similarly, when unloading takes place from the VCL, the slope changes from C_c to C_r. Every time unloading takes place from the VCL (e.g., at B and C), a new preconsolidation pressure is established. The initial state of the clay at A had been attained by previous unloading from the VCL (near B) sometime in its history, which may have been hundreds of years ago. At no stage can the clay reach a state represented by a point lying to the right of the VCL. As in the case of the oedometer sample discussed above, the clays in nature also undergo similar loading cycles, and everything discussed above holds true for field situations as well. It should be noted that the preconsolidation pressure is the *maximum past pressure* that the clay has experienced *ever* in its history.

The ratio of the preconsolidation pressure to the current effective vertical stress on the clay is known as the overconsolidation ratio OCR. Thus:

$$OCR = \frac{\sigma'_p}{\sigma'_{v0}} \tag{8.4}$$

If $\sigma'_{v0} = \sigma'_p$ (i.e., the e_0 and σ'_{v0} values of the clay lie on the VCL), the clay is known as *normally consolidated* where the OCR = 1. If $\sigma'_{v0} < \sigma_p'$ (i.e., the e_0 and σ'_{v0} values of the clay plot to the left of the VCL), the clay is known to be *overconsolidated*. The OCR is larger the further the values get from the VCL. In Figure 8.4b, at A and D, the clay is overconsolidated and at B, C, and E, it is normally consolidated. The virgin consolidation line is unique for a clay, has a specific location and slope, and applies to all overconsolidated and normally consolidated states of that clay.

Typically, the compression index varies from 0.2 to 1.5, and is proportional to the natural water content w_n, initial void ratio e_0, or liquid limit LL. Skempton (1944) suggested that for undisturbed clays:

$$C_c = 0.009(LL - 10) \tag{8.5}$$

There are numerous correlations reported in the literature that relate C_c with e_0, w_n, and LL. The recompression index C_r, also known as the swelling index C_s, is typically 1/5 to 1/15 of C_c.

Example 8.2: A consolidation test was carried out on a 61.4 mm diameter and 25.4 mm-thick saturated and undisturbed soft clay sample with an initial water content of 105.7% and a G_s of 2.70. The dial gauge readings, which measure the change in thickness at the end of consolidation due to each pressure increment, are summarized. The sample was taken from a depth of 3 m at a soft clay site where the water table is at ground level.

Vertical stress (kPa)	0	5	10	20	40	80
Dial reading (mm)	12.700	12.352	12.294	12.131	11.224	9.053
Vertical stress (kPa)	160	320	640	160	40	5
Dial reading (mm)	6.665	4.272	2.548	2.951	3.533	4.350

a. Plot e vs. log σ'_v and determine σ'_p, C_c, C_r, and the overconsolidation ratio.
b. Plot m_v vs. log σ'_v.

Solution: The initial void ratio can be computed as $e_0 = 1.057 \times 2.70 = 2.854$. Initial height $H_0 = 25.4$ mm. With these, let's calculate the values of e and m_v at the end of consolidation due to the first pressure increment of 5 kPa:

$$\Delta H = 12.7 - 12.352 = 0.348 \text{ mm}$$

From Equation 8.1:

$$\Delta e = \left(\frac{0.348}{25.400} \right) \times (1 + 2.854) = 0.0528$$

$$\rightarrow e = 2.854 - 0.0528 = 2.801$$

$$\rightarrow m_v = \frac{\left(\dfrac{0.348}{25.4} \right)}{(5 \times 10^{-3})} = 2.74 \text{ MPa}^{-1}$$

For the next pressure increment where σ'_v increases from 5 kPa to 10 kPa:

$$H_0 = 25.4 - 0.348 = 25.052 \text{ mm}$$
$$e_0 = 2.801, \Delta\sigma' = 10 - 5 = 5 \text{ kPa, and } \Delta H = 12.352 - 12.294 = 0.058 \text{ mm}$$

From Equation 8.1, $\Delta e = 0.058/25.052 \times (1 + 2.801) = 0.0088$

Using these values, at the end of consolidation:

$$H = 24.994 \text{ mm}, e = 2.792, \text{ and } m_v = 0.46 \text{ MPa}^{-1}$$

Continues

Example 8.2: *Continued*

This can be repeated for all pressure increments during loading and then for unloading as well. The values computed are summarized in the following table:

σ'_v (kPa)	Dial reading (mm)	H_0 (mm)	ΔH (mm)	Δe	e	m_v (MPa^{-1})
0	12.7	25.4			2.854	
5	12.352	25.4	0.348	0.0528	2.801	2.74
10	12.294	25.052	0.058	0.0088	2.792	0.46
20	12.131	24.994	0.163	0.0247	2.768	0.65
40	11.224	24.831	0.907	0.1376	2.630	1.83
80	9.053	23.924	2.171	0.3294	2.301	2.27
160	6.665	21.753	2.388	0.3623	1.938	1.37
320	4.272	19.365	2.393	0.3631	1.575	0.77
640	2.548	16.972	1.724	0.2616	1.314	0.32
160	2.951	15.248	−0.403	−0.0611	1.375	
40	3.533	15.651	−0.582	−0.0883	1.463	
5	4.35	16.233	−0.817	−0.1240	1.587	

The plots of void ratio vs. effective stress and m_v vs. effective stress are shown on page 147.

The preconsolidation pressure σ'_p is approximately 35 kPa. Now, let's compute the values of C_r and C_c from the plot. The unloading path is relatively straight and we will use the values of e and σ'_v at the beginning and end of unloading to calculate C_r:

$$C_r = \frac{1.587 - 1.314}{\log 640 - \log 5} = \frac{1.587 - 1.314}{\log \dfrac{640}{5}} = 0.13$$

The average value of C_c can be computed from the slope of the VCL as 1.21:

$$G_s = 2.70, \; e_0 = 2.854 \rightarrow \gamma_{\text{sat}} = 14.14 \text{ kN/m}^3$$
$$\therefore \sigma'_{v0} = 3.0 \times (14.14 - 9.81) = 13 \text{ kPa}$$
$$\therefore \text{OCR} = 35/13 = 2.7$$

Note the stress-dependence of m_v; it is not a constant as are C_c and C_r.

Continues

Example 8.2: *Continued*

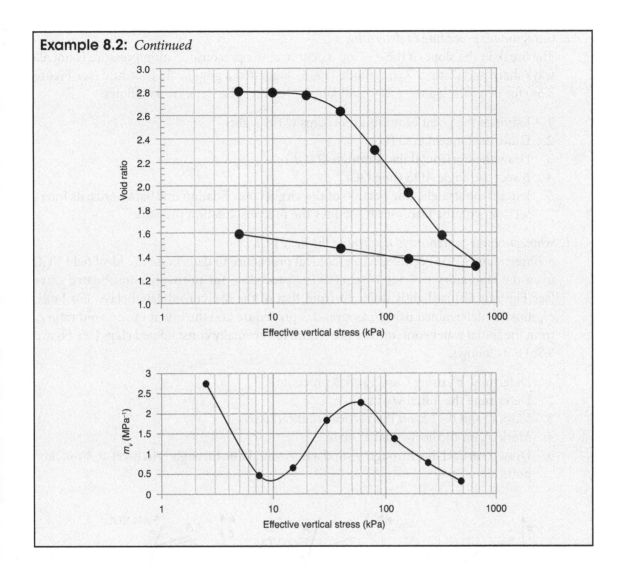

8.3.1 Field Corrections to the *e* Versus Log σ'_v Plot

When a clay sample is taken from the ground, it undergoes mechanical disturbance and stress relief, some of which are often inevitable. These disturbances can have a significant effect on the $e - \log \sigma'_v$ curve, making it difficult to arrive at realistic estimates of C_c, C_r, and σ'_p, which represent the field situation. What we really want are the values of the ideal undisturbed in situ clay element at the site. How does one use the somewhat disturbed laboratory sample to estimate the values of the in situ clay?

a. Casagrande's procedure to determine σ'_p

The break in the slope of the $e - \log \sigma'_v$ curve at the preconsolidation pressure is not always sharp and distinct. Casagrande (1936) suggested a graphical procedure (see Figure 8.5a) for determining the preconsolidation pressure. The steps are as follows:

1. Estimate the point of minimum radius O (by sight)
2. Draw the tangent at O (OA)
3. Draw the horizontal line through O (OB)
4. Bisect the angle AOB (line OC)
5. Extend the straight-line portion of the virgin consolidation line backwards; its intersection with the bisector OC defines the preconsolidation pressure

b. Schmertmann's procedure to determine the field VCL

Schmertmann (1955) developed a graphical procedure to determine the ideal field VCL from the laboratory $e - \log \sigma'_v$ curve. The procedure for normally consolidated clays (see Figure 8.5b) is slightly different from that of the overconsolidated clays. For both, σ'_p must be determined using Casagrande's procedure and the initial in situ void ratio e_0 from the initial water content. The procedure for normally consolidated clays (see Figure 8.5b) is as follows:

1. Determine σ'_p using Casagrande's procedure
2. Determine the initial void ratio e_0
3. Mark e_0 and $0.42\ e_0$ on the vertical void ratio axis
4. Mark σ'_p on the horizontal σ'_v axis
5. Draw a vertical line through σ'_p and a horizontal line through e_0 to meet at A (anchor point A)

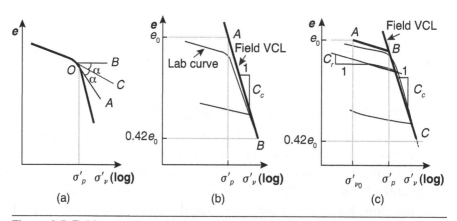

Figure 8.5 Field corrections: (a) determining σ'_p (b) field *VCL* of a normally consolidated clay sample (c) field *VCL* of an overconsolidated clay sample

6. Extend the straight-line part of the laboratory virgin consolidation line and draw a horizontal line through $0.42\ e_0$ to intersect at B (anchor point B)
7. Join the anchor points A and B, which is the field virgin consolidation line, the slope of which is the true C_c

In the case of overconsolidated clays, it is required to have an unload-reload cycle after the preconsolidation pressure to determine C_r. The procedure for overconsolidated clays (see Figure 8.5c) is as follows:

1. Determine σ'_p using Casagrande's procedure
2. Determine the initial void ratio e_0, and the initial in situ effective overburden pressure σ'_{v0}
3. Mark e_0 and $0.42\ e_0$ on the vertical void ratio axis
4. Mark σ'_p and σ'_{v0} on the horizontal σ'_v axis
5. Determine C_r from the unload-reload cycle
6. Draw the horizontal line through e_0 and the vertical line through σ'_{v0} to meet at anchor point A
7. Draw a line with a slope of C_r through A to intersect the vertical line through σ'_p at anchor point B
8. Extend the straight-line part of the laboratory virgin consolidation line to intersect the horizontal line through $0.42\ e_0$ at anchor point C
9. Join the anchor points A, B, and C to form the field $e - \log \sigma'_v$ plot (line BC is the field virgin consolidation line, the slope of which gives the field value of C_c, which should be used in the designs)

It can be shown from the first principles of the consolidation theory (discussed later in Section 8.5) that in normally consolidated clays, C_c and m_v are related by:

$$m_v = \frac{0.434\ C_c}{(1+e_0)\sigma'_{average}} \tag{8.6}$$

where $\sigma'_{average}$ is the average effective stress during consolidation. If the loading is entirely in the overconsolidated range, C_c can be replaced by C_r.

The consolidation test in an oedometer also generates stress-strain data. However, the vertical strains ($\Delta H/H$) take place under lateral constraints. Therefore, the coefficient of volume compressibility m_v, expressed as $(\Delta H/H)/\Delta\sigma'$, is the reciprocal of constrained modulus or oedometer modulus D, defined as $\Delta\sigma'/(\Delta H/H)$. Drained Young's modulus E and constrained modulus D are related by:

$$D = \frac{1}{m_v} = \frac{(1-v)}{(1+v)(1-2v)}E = K + \frac{4}{3}G \tag{8.7}$$

where ν = Poisson's ratio of the soil under drained conditions, K = bulk modulus of the soil, and G = shear modulus of the soil. With drained Poisson's ratio in the range of 0.10–0.33, D = 1–1.5 E. K and G are related to E by:

$$K = \frac{E}{3(1-2\nu)} \tag{8.8}$$

and

$$G = \frac{E}{2(1+\nu)} \tag{8.9}$$

8.4 COMPUTATION OF FINAL CONSOLIDATION SETTLEMENT

Final consolidation settlement s_c is the consolidation settlement after significant time ($t = \infty$) has elapsed, when all the excess pore water pressure is fully dissipated and the consolidation process is complete. The simplest way to compute s_c is to use Equation 8.3 as in Example 8.1, provided m_v is known. m_v is a stress-dependent parameter and is not a soil constant. To obtain a realistic estimate of s_c, it is necessary to know the value of m_v that corresponds to the stress level expected.

A more rational method of estimating s_c is to use Equation 8.1 and to express s_c as:

$$s_c = \frac{\Delta e}{1+e_0} H_0 \tag{8.10}$$

Since the clay is saturated, e_0 can be determined from Equation 2.6 as $e_0 = wG_s$. How do we find Δe? Here, we will look at three scenarios (see Figure 8.6). In each, the applied vertical stress increment $\Delta \sigma'$ causes the clay to consolidate from the initial void ratio of e_0 where the initial effective vertical stress is σ'_{v0}. The initial and final states are shown by points A and B respectively.

a. *In normally consolidated clays (Figure 8.6a):*
In normally consolidated clays, the initial state (point A) lies on the VCL. During consolidation, the point moves from A to B with a reduction in the void ratio of Δe, which can be computed as:

$$\Delta e = C_c \log \frac{\sigma'_{v0} + \Delta \sigma'}{\sigma'_{v0}} \tag{8.11}$$

b. *In overconsolidated clays where $\sigma'_{v0} + \Delta \sigma' \le \sigma'_p$ (Figure 8.6 b):*
In overconsolidated clays where the applied pressure is not large enough to take the clay past the preconsolidation pressure (i.e., $\sigma'_{v0} + \Delta \sigma' \le \sigma'_p$), the expression for Δe is similar to Equation 8.11 where C_c is replaced by C_r, and becomes:

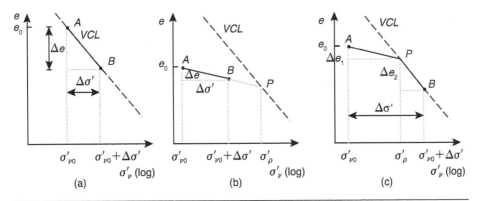

Figure 8.6 Three scenarios: (a) normally consolidated (b) overconsolidated where σ'_{v0} + $\Delta\sigma' \leq \sigma'_p$ (c) overconsolidated where σ'_{v0} + $\Delta\sigma' \geq \sigma'_p$

$$\Delta e = C_r \log\frac{\sigma'_{v0}+\Delta\sigma'}{\sigma'_{v0}} \tag{8.12}$$

c. *In overconsolidated clays where σ'_{v0} + $\Delta\sigma' > \sigma'_p$ (Figure 8.6c):*
In overconsolidated clays where the applied pressure is large enough to take the clay beyond the preconsolidation pressure (i.e., σ'_{v0} + $\Delta'\sigma > \sigma'_p$), the reduction in the void ratio ($\Delta e = \Delta e_1 + \Delta e_2$) is:

$$\Delta e = C_r \log\frac{\sigma'_p}{\sigma'_{v0}} + C_c \log\frac{\sigma'_{v0}+\Delta\sigma'}{\sigma'_p} \tag{8.13}$$

Depending on which case the situation falls into, the reduction in void ratio can be calculated using Equations 8.11, 8.12, or 8.13 and substituted in Equation 8.10 for determining the final consolidation settlement s_c.

Preloading is a very popular ground improvement technique that is carried out generally in normally consolidated clays where the expected consolidation settlements are too large. Here, a surcharge is applied over several months to consolidate the clay. On removal of the surcharge, the clay becomes overconsolidated. Later, when the load (e.g., building or embankment) is applied and thus the clay being overconsolidated, the settlement would be significantly less than what it would have been if it had been normally consolidated.

Example 8.3: The soil profile at a site consists of a 5 m-thick normally consolidated clay layer sandwiched between two sand layers as shown on page 152. The bulk and saturated unit weights of the sand are 17.0 kN/m³ and 18.5 kN/m³. An oedometer test carried out on an undisturbed clay sample obtained from the middle of the clay layer showed that the compression index and recompression index are 0.75 and 0.08 respectively. The natural water content of the clay is 42.5% and the specific gravity of the soil grains is 2.74. It is required to build a warehouse that would impose 30 kPa at the ground level.
Continues

Example 8.3: *Continued*

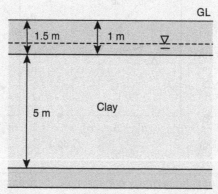

a. Estimate the final consolidation settlement of the warehouse, neglecting the settlements in sands.

In an attempt to reduce the post-construction consolidation settlements, a proposal has been made to carry out preloading at this site. A 40 kPa surcharge was applied over a large area, and the clay was allowed to consolidate. Once the consolidation was almost complete, the surcharge was removed.

b. What would be the net reduction in the ground level?

c. What would be the final consolidation settlement if the warehouse was built?

Solution: $e_0 = 0.425 \times 2.74 = 1.165 \rightarrow \gamma_{sat} = 17.69$ kN/m^3

a. At middle of the clay layer:

$$\sigma'_{v0} = (1 \times 17) + 0.5 \times (18.5 - 9.81) + 2.5 \times (17.69 - 9.81) = 41.0 \text{ kPa}$$
$$\Delta\sigma' = 30 \text{ kPa due to the proposed warehouse}$$

$$\Delta e = C_c \log \frac{\sigma'_{v0} + \Delta\sigma'}{\sigma'_{v0}} = 0.75 \log \frac{41.0 + 30.0}{41.0} = 0.1789$$

From Equation 8.10:

$$s_c = \frac{0.1789}{1 + 1.165} \times 5000 = 413 \text{ mm}$$

b. Due to 40 kPa surcharge:

$$\Delta e = 0.75 \log \frac{41.0 + 40.0}{41.0} = 0.2218$$

$$s_c = \frac{0.2218}{1 + 1.165} \times 5000 = 512 \text{ mm}$$

$$e = 1.165 - 0.2218 = 0.9432$$

$$H = 5000 - 512 = 4488 \text{ mm}$$

$$\sigma'_v = 81.0 \text{ kPa}$$

Continues

Example 8.3: *Continued*

For unloading:

$$\Delta e = 0.08 \log \frac{81.0}{41.0} = 0.0237$$

$$\Delta H = \frac{0.0237}{1+0.9432} \times 4488 = 55 \text{ mm}$$

Now:

$$e = 0.9432 + 0.0237 = 0.967, H = 4488 + 55 = 4543 \text{ mm}$$

Net reduction in ground level = 5000 − 4543 = 457 mm

c. If the warehouse is built now (H = 4543 mm and e = 0.967):

$$\Delta e = 0.08 \log \frac{41.0 + 30.0}{41.0} = 0.0191$$

$$s_c = \frac{0.0191}{1+0.967} \times 4543 = 44 \text{ mm}$$

(A significant reduction from the 413 mm originally expected.)

8.5 TIME RATE OF CONSOLIDATION

We now have the tools to compute the final consolidation settlement s_c that takes place after a very long time ($t = \infty$). We can use Equation 8.3 (see Example 8.1) or Equation 8.10 (see Example 8.3). Using Equation 8.3 is simpler but requires the correct value of m_v, which is a stress-dependent variable and hence a value appropriate to the stress level must be selected. The second approach using Equation 8.10 gives a more realistic estimate of s_c based on the values of C_c, C_r, and σ_p'.

Having computed s_c does not tell us anything about how long it takes to reach a 25 mm settlement or the magnitude of consolidation settlement in two years. In practice, when an embankment or footing is placed on a clayey soil, it is necessary to know how long it takes the settlement to reach a certain magnitude or how much settlement will take place after a certain time. Let's have a look at Terzaghi's one-dimensional consolidation theory, which assumes the following: (a) clay is homogeneous and saturated, (b) strains and drainage are both one-dimensional, (c) Darcy's law is valid, (d) strains are small and therefore k and m_v remain constants, and (e) soil grains and water are incompressible.

The clay layer shown in Figure 8.7a is sandwiched between two granular soil layers that are free draining, thus preventing the buildup of excess pore water pressures at the top and bottom boundaries of the clay layer. When the surcharge $\Delta\sigma$ is applied at the ground level, the entire load is immediately (at time $= 0^+$) carried by the pore water and there is an immediate increase in the pore water pressure at all depths by a value of Δu_0, which is equal to $\Delta\sigma$ (see Figure 8.7b). This excess pore water pressure dissipates with time due to the drainage from the top and bottom, gradually transferring the load to the soil skeleton in the form of an increase in effective stress $\Delta\sigma'$. At time $= t$, the variation of the *excess pore water pressure* Δu with depth z is shown in Figure 8.7c. At any time during the consolidation, $\Delta\sigma = \Delta\sigma'(z,t) + \Delta u(z,t)$ at any depth. Over time, $\Delta\sigma'$ increases and Δu decreases (at any depth) by the same amount. At the end of consolidation (time $= \infty$), the applied surcharge is transferred in its entirety to the soil skeleton; hence, $\Delta u = 0$ and $\Delta\sigma' = \Delta\sigma$ at all depths, as shown in Figure 8.7d. This is exactly what we hypothesized in Figure 8.1d, only qualitatively.

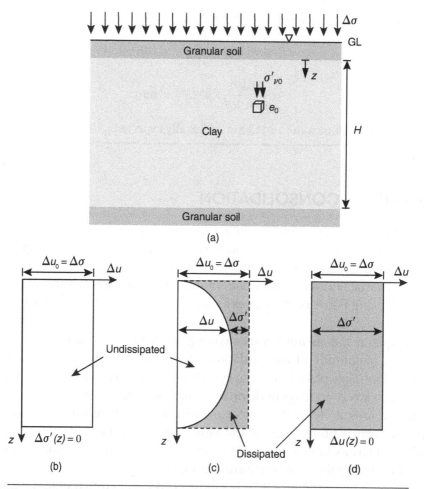

Figure 8.7 Dissipation of pore water pressure during consolidation: (a) doubly drained clay layer (b) at time $= 0^+$ (c) at time $= t$ (d) at time $= \infty$

Terzaghi (1925) showed that the governing differential equation for the excess pore water pressure can be written as:

$$\frac{\partial u}{\partial t} = c_v \frac{\partial^2 u}{\partial z^2} \tag{8.14}$$

where c_v is the *coefficient of consolidation*, defined as $c_v = \frac{k}{m_v \gamma_w}$, with a preferred unit of m^2/year. By solving the above differential equation with appropriate boundary conditions, it can be shown that the excess pore water pressure at depth z and time t can be expressed as:

$$\Delta u(z,t) = \Delta u_0 \sum_{m=0}^{m=\infty} \frac{2}{M} \sin(MZ) e^{-M^2 T} \tag{8.15}$$

where $M = (\pi/2)(2m + 1)$, and Z and T are a dimensionless *depth factor* and *time factor* defined as $Z = z/H_{dr}$ and $T = c_v t/H_{dr}^2$. H_{dr} is the *maximum length of the drainage path* within the clay layer. If the clay is drained from top and bottom as shown in Figure 8.7a, it is known as *doubly drained*, and $H_{dr} = H/2$. When the clay is underlain by an impervious stratum, drainage can only take place from the top. Therefore, $H_{dr} = H$. c_v can vary from less than 1 m^2/year for low permeability clays to as high as 1000 m^2/year for sandy clays of very high permeability. Figure 8.8 proposed by the U.S. Navy (1986) can be used as a rough guide for checking the c_v values determined by the laboratory.

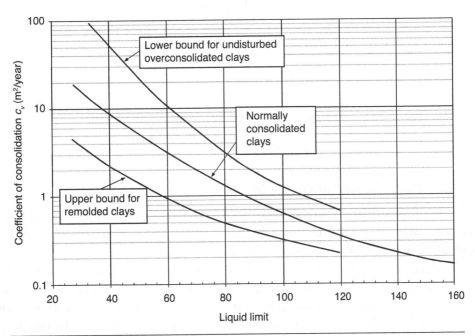

Figure 8.8 Approximate values of c_v (after U.S. Navy 1986)

8.5.1 Degree of Consolidation

The degree of consolidation at a depth z, at a specific time t, denoted by $U_z(t)$, is the fraction of the excess pore water pressure that has dissipated, expressed as a percentage. Therefore, it can be written as:

$$U_z(t) = \frac{\Delta u_0 - \Delta u(z,t)}{\Delta u_0} \times 100\%$$

$$= 1 - \sum_0^\infty \frac{2}{M} \sin(MZ) e^{-M^2 T} \tag{8.16}$$

The interrelationship among $U_z(t)$, T, and Z is shown graphically in Figure 8.9a. It can be seen that the degree of consolidation at any time is the minimum at the middle of the doubly drained clay layer, or at the impervious boundary of a singly drained clay layer.

At any time, the degree of consolidation varies with depth. How do we define an *average degree of consolidation* at a specific time for the entire thickness that we can also equate to the fraction of the consolidation settlement that has taken place at that time? The average degree of consolidation U_{avg} for the clay layer at a specific time is defined as the area of the dissipated excess pore water pressure distribution diagram in Figure 8.7c, divided by the initial excess pore water pressure distribution diagram in Figure 8.7c. It is given by:

$$U_{avg}(T) = 1 - \sum_0^\infty \frac{2}{M^2} e^{-M^2 T} \tag{8.17}$$

Equation 8.17 can be approximated as:

$$T = \frac{\pi}{4} U_{avg}^2 \quad \text{for} \quad U_{avg} \le 60\% \tag{8.18a}$$

$$T = 1.781 - 0.933 \log(100 - U_{avg}) \quad \text{for} \quad U_{avg} \ge 60\% \tag{8.18b}$$

The relationship between U_{avg} and T is also shown graphically in Figure 8.9b.

8.5.2 Laboratory Determination of c_v

The coefficient of consolidation can be determined from the time–settlement data obtained from a consolidation test during any pressure increment. A dial gauge is used to continuously measure the change in thickness of the clay sample during consolidation—usually over a period of 24 hours. As in the case of m_v, c_v is also a stress-dependent parameter. When overconsolidated (i.e., $\sigma_v' < \sigma_p'$), c_v is approximately an order of magnitude larger than when it is normally consolidated. It is a good practice to plot c_v against σ_v' (log) and use the value appropriate for the stress level.

In the laboratory consolidation tests, the sample in the oedometer is loaded in pressure increments, typically allowing 24 hours between two successive increments to ensure full consolidation. Two of the traditional empirical curve-fitting methods used for determining c_v are *Casagrande's log time method* and *Taylor's square root of time method*.

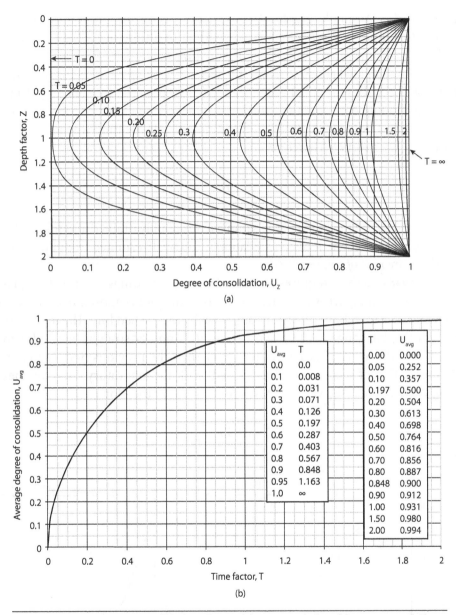

Figure 8.9 Degree of consolidation charts: (a) *U-Z-T* variation (b) $U_{avg} - T$ variation

a. Casagrande's log time method

Casagrande (1938) proposed this method where the dial gauge reading is plotted against the logarithm of time. The time-settlement plot shown in Figure 8.10a consists of a parabolic curve followed by two straight-line segments. The intersection of the two straight-line segments defines the 100% consolidation state, which is denoted by a dial gauge reading of d_{100}. A simple graphical construction using the properties of a parabola is required to define d_0, the reading corresponding to a time of 0^+ (i.e., just after loading, which cannot be measured). Mark an arbitrary time t and then $4t$ on the time axis, and note the corresponding dial gauge readings, the difference being x (see Figure 8.10a). Mark this offset distance x above the dial gauge reading corresponding to t, and this defines d_0. The dial gauge reading corresponding to $U_{avg} = 50\%$ is computed as $d_{50} = (d_0 + d_{100})/2$. The time t_{50} corresponding to d_{50} is read off the plot. This is the time when $U_{avg} = 50\%$. From Figure 8.9b, $T_{50} = 0.197$. Therefore:

$$T_{50} = 0.197 = \frac{c_v t_{50}}{H^2_{dr}} \tag{8.19}$$

where H_{dr} is half the thickness of the sample if it is doubly drained and full thickness if singly drained. The coefficient of consolidation c_v can be determined from Equation 8.19.

b. *Taylor's square root of time method*

Taylor's (1948) method requires plotting dial gauge readings against the square root of time, as shown in Figure 8.10b. The early part of the plot is approximately a straight line, which is extended in both directions as shown by the dashed line. The intersection of this line with the dial gauge reading axis defines d_0. Another straight line is drawn through d_0 such that the abscissa is 1.15 times larger than the previous line (see Figure 8.10b). The intersection of this second line (dotted) with the laboratory curve defines the 90% consolidation point. The value of $\sqrt{t_{90}}$ can be read off the plot:

$$T_{90} = 0.848 = \frac{c_v t_{90}}{H^2_{dr}} \tag{8.20}$$

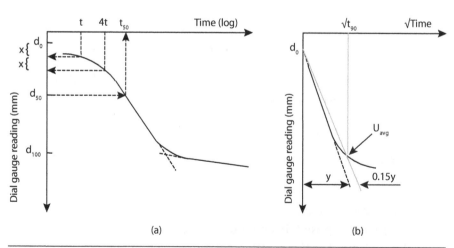

Figure 8.10 Laboratory determination of c_v: (a) Casagrande's log time method (b) Taylor's square root of time method

Generally, Taylor's method gives larger values than Casagrande's method. Nevertheless, both laboratory values are often significantly less than the c_v values that are back-calculated in the field. In other words, consolidation in the field takes place at a faster rate, and the laboratory methods underestimate the coefficient of consolidation. Shukla et al. (2008) reviewed the different methods reported in the literature for determining the coefficient of consolidation.

8.6 SECONDARY COMPRESSION

According to Terzaghi's consolidation theory, the consolidation process goes on forever. Remember, $U_z = 100\%$ and $U_{avg} = 100\%$ only when $T = \infty$. In reality, as we see in the consolidation tests in the laboratory, all clays fully consolidate after some time. This time, often denoted as t_p or t_{100}, is proportional to the square of the thickness. In the laboratory, this can be a few hours; in the field, this can be months or several years.

When consolidation is completed, the excess pore water pressure has fully dissipated at every point within the clay layer. Beyond this time, the clay continues to settle under constant effective stress—indefinitely—as seen in the laboratory consolidation test shown in Figure 8.10a. This process is known as *secondary compression* or *creep*, and occurs due to some changes in the microstructure of the clay fabric. This is more pronounced in organic clays. When the void ratio, settlement, or dial gauge reading is plotted against the logarithm of time, the variation is linear during secondary compression (e.g., Figure 8.10a). Here, the secondary compression index C_α is defined as the change in void ratio per log cycle of time, and is expressed as:

$$C_\alpha = \frac{\Delta e}{\Delta \log t} \qquad (8.21)$$

C_α can be determined from the tail end of the dial gauge reading versus the log time plot (Figure 8.11), which is used for determining c_v by Casagrande's method. Mesri and Godlewski (1977)

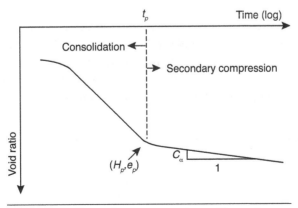

Figure 8.11 Secondary compression settlement

observed that C_α/C_c varies within the narrow range of 0.025–0.10 for all soils with an average value of 0.05. The upper end of this range applies to organic clays, peat, and muskeg, and the lower end applies to granular soils. The modified secondary compression index $C_{\alpha\varepsilon}$ is defined as:

$$C_{\alpha\varepsilon} = \frac{C_\alpha}{1+e_p} \qquad (8.22)$$

where e_p is the void ratio at the end of primary consolidation. For normally consolidated clays, $C_{\alpha\varepsilon}$ lies in the range of 0.005–0.02. For highly plastic clays or organic clays, $C_{\alpha\varepsilon}$ can be 0.03 or higher. For overconsolidated clays with OCR > 2, $C_{\alpha\varepsilon}$ is less than 0.001 (Lambe and Whitman 1979).

Between times t_p and t ($> t_p$), the reduction in the void ratio Δe and the secondary compression settlement s_s are related by (see Equation 8.1):

$$\Delta e = \frac{s_s}{H_p}(1+e_p) \qquad (8.23)$$

where H_p and e_p are the layer thickness and void ratio respectively at the end of primary consolidation (see Figure 8.11). From Equations 8.21 and 8.23, the secondary compression settlement s_s at time t ($> t_p$) can be expressed as:

$$s_s = C_\alpha \frac{H_p}{1+e_p} \log \frac{t}{t_p} \qquad (8.24)$$

In practice, it is quite difficult to arrive at a realistic estimate of H_p and e_p. On the other hand, H_0 and e_0, the values at the beginning of consolidation, are readily available, and therefore, $H_p/(1+e_p)$ in Equation 8.24 can be replaced by $H_0/(1+e_0)$.

Example 8.4: A 20 mm-thick clay sample at a void ratio of 1.71 is subjected to a consolidation test in an oedometer where the dial gauge reading is initially set to 0.0 mm. The vertical pressure on the sample was increased from 0 to 272.6 kPa in a few increments, each being applied for 24 hours. During the next increment when σ_v' was increased from 272.6 kPa to 543 kPa, the time-dial gauge readings were:

Continues

Example 8.4: *Continued*

Time	Dial gauge reading (mm)
0^-	3.590 (just before applying the pressure increment)
1.5 s	3.676
15 s	3.690
30 s	3.718
1 min	3.756
2 min	3.806
4 min	3.884
8 min	3.983
16 min	4.130
32 min	4.330
60 min	4.562
141 min	4.853
296 min	5.027
429 min	5.086
459 min	5.095
680 min	5.141
1445 min	5.204
1583 min	5.212

a. Determine the coefficient of consolidation during this pressure increment using Casagrande's log time and Taylor's square root of time methods
b. Determine the coefficient of volume compressibility during this increment
c. Determine the coefficient of secondary compression during this increment
d. Estimate the permeability during this increment

Solution: a. Casagrande's graphical construction for determining d_0, d_{100}, and t_{50} is shown in the figure on page 162 where $d_0 = 3.63$ mm:

$$d_{100} = 5.05 \text{ mm} \rightarrow d_{50} = 4.34 \text{ mm} \rightarrow t_{50} = 33 \text{ minutes}$$

The average thickness of the sample during consolidation (i.e., at t_{50}):

$$= 20 - 4.34 = 15.66 \text{ mm}$$

Continues

Example 8.4: *Continued*

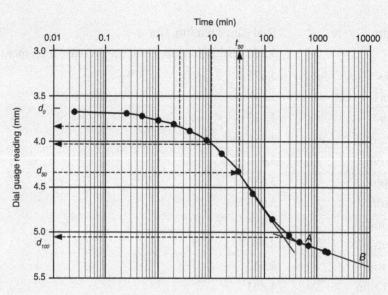

Being doubly drained, $H_{dr} = 15.66/2 = 7.83$ mm:

$$T_{50} = 0.197 = \frac{c_v t_{50}}{H^2_{dr}}$$

$$\therefore c_v = \frac{0.197 \times 7.83^2}{33} = 0.37 \text{ mm}^2/\text{min}$$

Taylor's graphical construction to determine t_{90} is shown in the figure on page 163, from which $\sqrt{t_{90}} = 11.95$ min$^{0.5}$. Hence, $t_{90} = 143$ minutes:

$$d_0 = 3.60 \text{ mm, } d_{90} = 4.86 \text{ mm}$$

$$\therefore d_{50} = 3.60 + (4.86 - 3.60) \times \frac{5}{9} = 4.30 \text{ mm}$$

$\therefore$ Average thickness of the sample during consolidation =

$$20 - 4.30 = 15.70 \text{ mm}$$

$$\therefore H_{dr} = 15.70/2 = 7.85 \text{ mm}$$

$$T_{90} = 0.848 = \frac{c_v t_{90}}{H^2_{dr}} \rightarrow$$

$$c_v = \frac{0.848 \times 7.85^2}{143} = 0.36 \text{ mm}^2/\text{min}$$

close to Casagrande's c_v.

Continues

Example 8.4: *Continued*

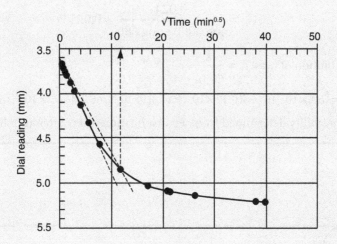

b.

$$\Delta\sigma' = 543.0 - 272.6 = 270.4 \text{ kPa}$$

$$\Delta H = d_{100} - d_0 = 1.42 \text{ mm}$$

$$H_0 = 16.41 \text{ mm}$$

$$\therefore m_v = \frac{\Delta H / H_0}{\Delta\sigma'} = \frac{1.42}{16.41 \times 270.4}$$

$$= 0.32 \times 10^{-3} \text{ kPa}^{-1} = 0.32 \text{ MPa}^{-1}$$

c. Let's consider the two points A and B on the tail of the Casagrande plot, and find the void ratios at these points. From the very beginning of the consolidation test to A:

$$H_0 = 20.0 \text{ mm}, e_0 = 1.71, \Delta H = 5.15 \text{ mm}$$

$$\therefore \Delta e_A = \frac{5.15}{20.0} \times (1 + 1.71) = 0.698$$

From the very beginning of the consolidation test to B:

$$H_0 = 20.0 \text{ mm}, e_0 = 1.71, \Delta H = 5.33 \text{ mm}$$

$$\Delta e_B = \frac{5.33}{20.0} \times (1 + 1.71) = 0.722$$

$$\therefore e_A = 1.710 - 0.698 = 1.012; e_B = 1.710 - 0.722 = 0.988$$

$\therefore$ Change in void ratio between A and B = 0.024. $t_A = 680$ min, $t_B = 6000$ min:

Continues

Example 8.4: *Continued*

$$\therefore C_\alpha = \frac{0.024}{\log 6000 - \log 680} = 0.025$$

d. From the definition of c_v as $c_v = \dfrac{k}{m_v \gamma_w}$:

$$k = (6.2 \times 10^{-9} \, \mathrm{m^2/s})(0.32 \times 10^{-6} \, \mathrm{Pa^{-1}})(9810 \, \mathrm{N/m^3}) = 1.95 \times 10^{-11} \, \mathrm{m/s}$$

Note: The permeability determined from a consolidation test is often unreliable.

Reminder

❖ During consolidation, water is squeezed from the clay over a long time. During this time, the applied load is slowly transferred from the pore water to the soil skeleton (the excess pore water pressure decreases and effective stress increases).

❖ Consolidation is all about the *changes* ($\Delta\sigma$, $\Delta\sigma'$, and Δu) to the initial values σ_{v0}, σ'_{v0}, and u_0 respectively.

❖ The virgin consolidation line is unique for a clay; where the current state lies in $e - \log \sigma_v'$ space with respect to the VCL defines the overconsolidation ratio.

❖ σ'_{v0}, $\Delta\sigma'$, and Δe vary with depth even within a homogeneous clay layer; compute the final consolidation settlement s_c using *mid-depth layer values*.

❖ Drainage and strains have to be one-dimensional in a one-dimensional consolidation.

❖ m_v and c_v are stress-dependent variables.

❖ c_v is larger when the clay is overconsolidated; the larger the c_v, the faster the consolidation process.

WORKED EXAMPLES

1. The void ratio and effective vertical stress data from a consolidation test are summarized:

$\sigma_v{}'$ (kPa)	e
1.4	2.14
6	2.08
13	2.03
26	1.95
38	1.88
58	1.81
86	1.70
130	1.55
194	1.45
110	1.47
26	1.53
52	1.52
104	1.49
208	1.43
416	1.22

The sample was taken from a depth of 2.6 m below the ground level in a soft clay deposit where the water table coincides with the ground level. The initial void ratio was 2.20 and $G_s = 2.70$.

a. Draw the laboratory e versus $\log \sigma_v{}'$ plot and determine the preconsolidation using Casagrande's procedure. Is the clay normally consolidated?
b. Carry out Schmertmann's procedure and determine the in situ virgin consolidation line.
c. Determine the compression index and recompression index.

Solution:

$$e_0 = 2.20 \text{ and } G_s = 2.70 \rightarrow \gamma_{sat} = 15.0 \text{ kN/m}^3$$

At 2.6 m depth, $\sigma'_{v0} = 2.6 \times (15.0 - 9.81) = 13.5$ kPa

The $e - \log \sigma'_v$ plot is shown on page 166, along with Casagrande's construction to determine σ'_p, which is about 43 kPa.

∴ The clay is overconsolidated with an OCR of $\dfrac{43}{13.5} = 3.2$

Schmertmann's graphical procedure for determining in situ VCL is also shown in the following figure. The in situ virgin consolidation line is shown as a thick solid line, where the slope C_c is 0.87. The recompression index is determined from the unload-reload path as 0.10.

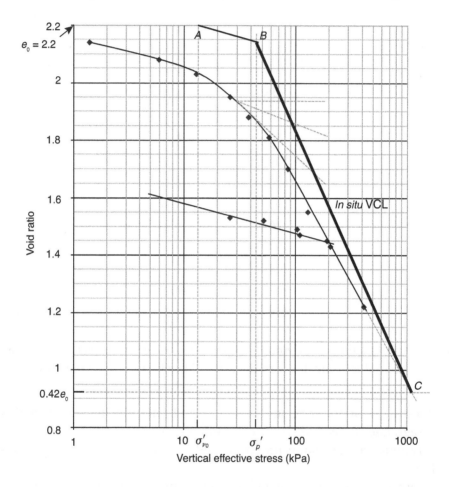

2. The soil profile at a site consists of a 3 m-thick sand layer ($\gamma_m = 16.5$ kN/m³, $\gamma_{sat} = 18.5$ kN/m³) underlain by a 6 m-thick clay layer ($w = 27\%$, $G_s = 2.70$, $m_v = 0.31$ MPa^{-1}, $c_v = 2.6$ m²/year), which is underlain by a gravel layer as shown in the following figure. A 3 m compacted fill with a unit weight of 20 kN/m³ is required to be placed at the ground level.

   a. What would be the final consolidation settlement?
   b. How long will it take for 50 mm of consolidation settlement?
   c. What would be the consolidation settlement in one year?

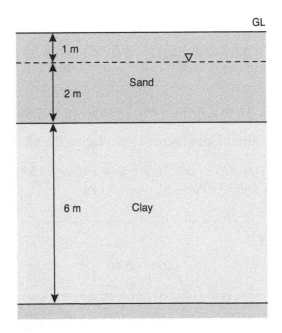

d. What would be the values of σ_v, σ'_v, and u at a 2 m depth within the clay after one year?

e. Plot the variation of pore water pressure and effective stress with depth after one year.

Solution:

a. From Equation 8.3:

$$s_c = m_v \, \Delta\sigma' \, H = 0.31 \times (3 \times 20/1000) \times 6000 = 111.6 \text{ mm}$$

b. $s(t) = 50 \text{ mm} \rightarrow$

$$U_{avg} = \frac{50}{111.6} = 0.448, \, H_{dr} = 3.0 \text{ m}$$

From Figure 8.9b, $T = 0.15$:

$$T = \frac{c_v t}{H^2_{dr}} \rightarrow t = \frac{0.15 \times 3^2}{2.6} \text{ years} = 6.23 \text{ months}$$

c. $t = 1 \text{ year} \rightarrow$

$$T = 1 \text{ year} \rightarrow T = \frac{c_v t}{H^2_{dr}} = \frac{2.6 \times 1.0}{3.0^2} = 0.289$$

From Figure 8.9b:

$$U_{avg} = 0.60 \rightarrow s(1 \text{ year}) = 0.60 \times 111.6 = 67 \text{ mm}$$

d. At the clay layer:

$$w = 27\%, G_s = 2.70 \rightarrow e_0 = 0.729 \text{ and } \gamma_{sat} = 19.5 \text{ kN/m}^3$$

At 2 m depth within the clay layer, before placing the fill:

$$\sigma_{v0} = 1 \times 16.5 + 2 \times 18.5 + 2 \times 19.5 = 92.5 \text{ kN/m}^3$$
$$u_0 = 4 \times 9.81 = 39.2 \text{ kPa} \rightarrow \sigma'_{v0} = 53.3 \text{ kPa}$$

At $t = 1$ year, $T = 0.289$; at depth $z = 2$ m, $Z = z/H_{dr} = 2/3 = 0.67$. From Figure 8.9a:

$$U_z(t) = 0.46$$

$\Delta u_0 = 60$ kPa, which is distributed between $\Delta\sigma'$ and Δu at any time:

$$\therefore \Delta u = 60 \times (1 - 0.46) = 32.4 \text{ kPa}; \Delta\sigma' = 60 \times 0.46 = 27.6 \text{ kPa}$$
$$\sigma'_v = 53.3 + 27.6 = 80.9 \text{ kPa}; u = 39.2 + 32.4 = 71.6 \text{ kPa}; \text{ and}$$
$$\sigma_v = 92.5 + 60 = 152.5 \text{ kPa}$$

e. The values of σ'_v and u computed at depths of 0, 1, 2, 3, 4, 5, and 6 m are plotted on the figure below. Dashed lines are used for pore water pressures and solid lines for vertical effective stresses.

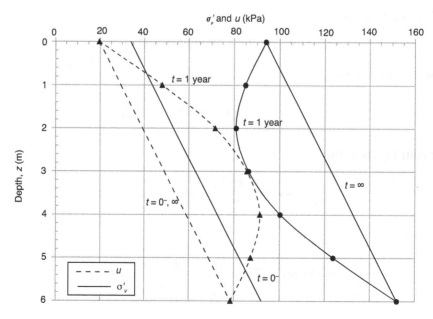

Note that the pore water pressure variation is the same at t = 0⁻ (before loading) and at ∞ (end of consolidation). See how the effective stress variation plot changes during consolidation.

3. A clay layer consolidates after 6 years when its thickness is 5.70 m and the void ratio is 1.08. Assuming $C_\alpha = 0.04$, estimate the secondary compression settlement in the next 15 years.

Solution: Using Equation 8.24:

$$s_s = 0.04 \times \frac{5.70}{1+1.08} \times \log\frac{21}{6} \text{ m} = 59 \text{ mm}$$

4. A 3 m-thick sand layer is underlain by a thick clay layer. The water table lies 1 m below the ground level. Bulk and saturated unit weights of sand are 16 kN/m³ and 20 kN/m³ respectively. Two undisturbed clay samples were taken from depths of 5 m and 11 m below the ground level. The water contents of both samples were 35% and the specific gravity of the soil grains is 2.75. The virgin consolidation line for the clay as determined from previous tests is shown. Calculate the in situ values of the void ratio and the effective vertical stress, and mark the locations of the two samples. Is the clay normally consolidated or overconsolidated at the two depths? Assuming the recompression index is about 1/10 of the compression index, estimate the overconsolidation ratios.

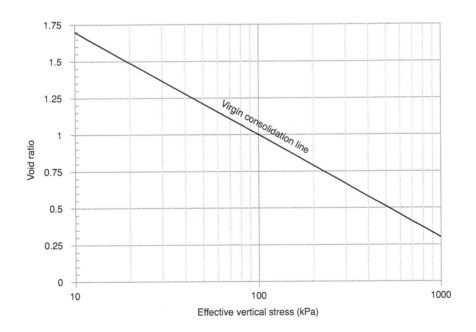

Solution: Natural water content, $w_n = 35\%$, $G_s = 2.75$

$\therefore$ Assuming $S = 100\%$, $e_0 = 0.963$ and $\gamma_{sat} = 18.6$ kN/m³

The slope of VCL is 0.70, which is the compression index C_c. Therefore, $C_r \approx 0.07$

At 5 m depth below GL:

$$\sigma'_{v0} = 1 \times 16 + 2 \times (20 - 9.81) + 2 \times (18.6 - 9.81) = 54.0 \text{ kPa}$$

At 11 m depth below GL:

$$\sigma'_{v0} = 1 \times 16 + 2 \times (20 - 9.81) + 8 \times (18.6 - 9.81) = 106.7 \text{ kPa}$$

The in situ values are shown in the figure below. At an 11 m depth below the ground, the point lies on the VCL. The clay, therefore, is normally consolidated (OCR = 1). At a 5 m depth, the point lies below the VCL, showing that the clay is overconsolidated. To determine the preconsolidation pressure, a line (dashed) is drawn from this point with a slope of 0.07, and it meets the VCL at the preconsolidation pressure, which is about 120 kPa.

$\therefore$ OCR at 5 m depth below the ground level $= \dfrac{120}{54} = 2.2$

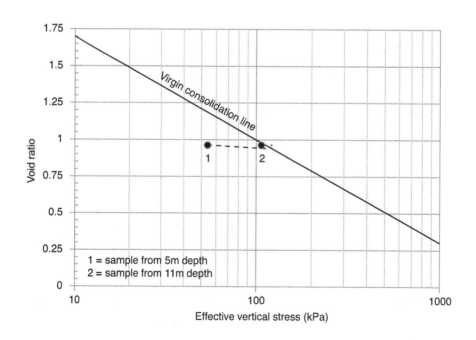

5. A 75 mm diameter clay specimen was consolidated in an oedometer under 200 kPa. At the end of consolidation, the void ratio is 0.863 and the specimen thickness is 18.51 mm. When a stress increment of 200 kPa was added to the current vertical stress of 200 kPa, the specimen consolidated to a thickness of 17.56 mm. Assuming that the clay was normally consolidated under the vertical stress of 200 kPa, find the coefficient of volume compressibility and the compression index of the clay.

Determine if Equation 8.6 relating m_v and C_c holds here.

If the vertical stress is increased from 400 kPa to 800 kPa, what would be the thickness of the specimen at the end of consolidation? Determine this separately, using both m_v and C_c. Why are they different?

Solution:

$$\frac{\Delta e}{1+e_0} = \frac{\Delta H}{H_0} \rightarrow \Delta e = \frac{(18.51-17.56)}{18.51} \times (1+0.863) = 0.0956$$

$$\therefore e_{new} = 0.863 - 0.0956 = 0.767$$

$$C_c = \frac{\Delta e}{\Delta \log \sigma'_v} = \frac{0.0956}{\log \dfrac{400}{200}} = 0.32$$

$$m_v = \frac{\Delta H / H}{\Delta \sigma'} = \frac{0.95}{18.51 \times 200} = 0.257 \times 10^{-3} \text{ kPa}^{-1} = 0.257 \text{ MPa}^{-1}$$

$$m_v = \frac{0.434 \, C_c}{(1+e_0)\sigma'_{average}} \tag{8.6}$$

$$\text{RHS} = \frac{0.434 \times 0.32}{(1+0.863) \times 300} = 0.248 \times 10^{-3} \text{ kPa}^{-1} = 0.248 \text{ MPa}^{-1}$$

Yes, Equation 8.6 is valid.

When σ'_v is increased from 400 kPa to 800 kPa, we will compute the change in thickness using m_v and C_c.

Using m_v:

$$\Delta H = m_v \Delta \sigma' H_0 = 0.257 \times 0.400 \times 17.56 = 1.805 \text{ mm}$$
$$\therefore \text{New thickness} = 17.56 - 1.805 = 15.75 \text{ mm}$$

Using C_c:

$$\Delta e = C_c \log \frac{\sigma'_{v0} + \Delta \sigma'}{\sigma'_{v0}} = 0.32 \log \frac{400 + 400}{400} = 0.096$$

$$\Delta H = \frac{\Delta e}{1 + e_0} H_0 = \frac{0.096}{1 + 0.767} \times 17.56 = 0.954 \text{ mm}$$

$\therefore$ New thickness $= 17.56 - 0.954 = 16.61$ mm

ΔH computed by the two methods (1.805 mm and 0.954 mm) are quite different. The problem is from m_v, which is stress-dependent. The value computed for 200–400 kPa range will not be the same for 400–800 kPa range as we have assumed. Therefore, the C_c method is more reliable unless we have the right values for m_v.

6. Two undisturbed clay samples were taken from the middle of the overconsolidated and normally consolidated clay layers in the soil profile shown. The water table is at the top of the overconsolidated clay layer. Consolidation tests were carried out on the two samples and the results are summarized.

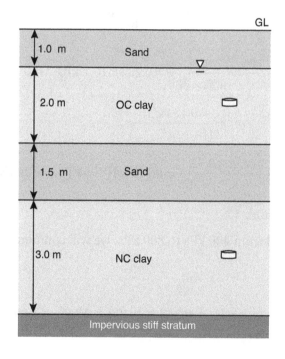

	O.C. Clay	N.C. Clay
Natural water content (%)	20.0	29.0
Preconsolidation pressure (kPa)	50.0	65.0
Compression index	0.55	0.60
Recompression index	0.06	0.07
Coefficient of consolidation (m²/year)	13.0	2.5

Assume that the bulk and saturated unit weights of the sand are 16.0 and 19.0 kN/m³ respectively. Specific gravity of the clay soil grains is 2.70. A 2 m-high compacted fill with a unit weight of 20 kN/m³ is placed at the ground level.

a. What would be the final consolidation settlement?
b. What would be the consolidation settlement after one month?
c. What would be the pore water pressures and effective stresses at the middle of the layers after one month?
d. Plot the variation of consolidation settlement with time and find the time taken for 200 mm of consolidation settlement.

Solution:

a. For OC clay:

$$e = 0.20 \times 2.70 = 0.540 \rightarrow \gamma_{sat} = 20.6 \text{ kN/m}^3$$

For NC clay:

$$e = 0.29 \times 2.70 = 0.783 \rightarrow \gamma_{sat} = 19.2 \text{ kN/m}^3$$

For OC clay:

$$\sigma'_{v0} = 1 \times 16.0 + 1.0(20.6 - 9.81)$$
$$= 26.8 \text{ kPa at the mid-layer}$$

For NC clay:

$$\sigma'_{v0} = 1 \times 16.0 + 2.0(20.6 - 9.81) + 1.5(19.0 - 9.81) + 1.5(19.2 - 9.81)$$
$$= 65.5 \text{ kPa at the mid-layer}$$

The increase in vertical normal stress, $\Delta\sigma' = 2 \times 20 = 40$ kPa, is the same at all depths.

At mid-depth of OC clay:

$$\sigma'_{v0} = 26.8 \text{ kPa}, \Delta\sigma' = 40 \text{ kPa} \rightarrow \sigma'_{v0} + \Delta\sigma' (= 66.8 \text{ kPa}) > \sigma'_p (= 50 \text{ kPa})$$

$$\therefore \Delta e = C_r \log\frac{\sigma'_p}{\sigma'_{v0}} + C_c \log\frac{\sigma'_{v0} + \Delta\sigma'}{\sigma'_p} = 0.06 \log\frac{50}{26.8} + 0.55 \log\frac{66.8}{50} = 0.0855$$

$$\frac{\Delta e}{1+e_0} = \frac{\Delta H}{H_0} \rightarrow \Delta H_{OC} = \frac{0.0855}{1+0.540} \times 2000 = 111.0 \text{ mm}$$

At mid-depth of NC clay:

$$\sigma'_{v0} = 65.5 \text{ kPa}, \Delta\sigma' = 40 \text{ kPa} \rightarrow \sigma'_{v0} + \Delta\sigma' = 105.5 \text{ kPa}$$

$$\therefore \Delta e = C_c \log\frac{\sigma'_{v0} + \Delta\sigma'}{\sigma'_{v0}} = 0.60 \log\frac{105.5}{65.5} = 0.1242$$

$$\frac{\Delta e}{1+e_0} = \frac{\Delta H}{H_0} \rightarrow \Delta H_{NC} = \frac{0.1242}{1+0.783} \times 3000 = 209.0 \text{ mm}$$

$\therefore$ The final consolidation settlement $= 111.0 + 209.0 = 320$ mm

b. After one month, let's find the time factors in both clays:

$$\text{OC: } T = \frac{c_v t}{H^2_{dr}} = \frac{13 \times \left(\dfrac{1}{12}\right)}{1^2} = 1.08 \rightarrow U_{avg} = 94\% \rightarrow \text{settlement} = 0.94 \times 111 = 104.3 \text{ mm}$$

$$\text{NC: } T = \frac{c_v t}{H^2_{dr}} = \frac{2.5 \times \left(\dfrac{1}{12}\right)}{3^2} = 0.023 \rightarrow U_{avg} = 17\% \rightarrow \text{settlement} = 0.17 \times 209 = 35.5 \text{ mm}$$

$\therefore$ Consolidation settlement after one month $= 104.3 + 35.5 = 139.8$ mm

c. In both clays, $\Delta u_0 = 40$ kPa.
At the middle of OC clay layer:

$Z = z/H_{dr} = 1/1 = 1$, $T = 1.08 \rightarrow U_z(t) = 91\%$
$\quad \Delta u = 40 \times 0.09 = 3.6$ kPa and $\Delta\sigma' = 40 \times 0.91 = 36.4$ kPa
$\sigma'_{v0} = 26.8$ kPa, $u_0 = 9.8$ kPa
$\therefore \sigma'_v = \sigma'_{v0} + \Delta\sigma' = 26.8 + 36.4 = 63.2$ kPa, and
$u = u_0 + \Delta u = 9.8 + 3.6 = 13.4$ kPa

At the middle of NC clay layer:

$Z = z/H_{dr} = 1.5/3.0 = 0.5$, $T = 0.023 \rightarrow U_z(t) = 5\%$
$\quad \Delta u = 40 \times 0.95 = 38.0$ kPa and $\Delta\sigma' = 40 \times 0.05 = 2.0$ kPa
$\sigma'_{v0} = 65.5$ kPa, $u_0 = 5 \times 9.81 = 49.1$ kPa
$\therefore \sigma'_v = \sigma'_{v0} + \Delta\sigma' = 65.5 + 2.0 = 67.5$ kPa, and
$u = u_0 + \Delta u = 49.1 + 38.0 = 87.1$ kPa

d. The consolidation settlements of the two layers after different times are summarized in the table on the following page, followed by the plot. It can be seen that a 200 mm settlement occurs after 6 months.

Time	OC layer			NC layer			Total sett
(months)	T	U_{avg} (%)	Sett (mm)	T	U_{avg} (%)	Sett (mm)	(mm)
0	0	0	0	0	0	0	0
1	1.083	94	104.34	0.023	17	35.5	139.9
3	3.25	99.99	110.99	0.069	29	60.6	171.6
6	6.5	100	111	0.139	42	87.8	198.8
12	13	100	111	0.278	59	123.3	234.3
24	26	100	111	0.556	79	165.1	276.1
36	39	100	111	0.833	89	186.0	297.0
60	65	100	111	1.389	97	202.7	313.7
120	130	100	111	2.778	99.5	208.0	319.0

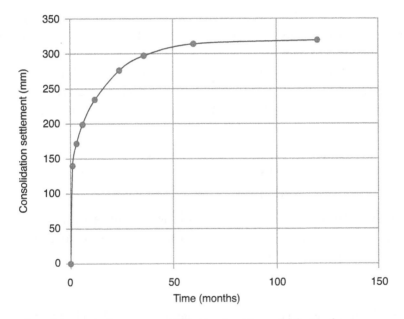

Note that the upper layer consolidates significantly faster for two reasons: (a) it is over consolidated and (b) it is doubly drained.

REVIEW EXERCISES

1. Compression index C_c is often related to the natural water content, liquid limit, and initial void ratio of the clay. List some empirical correlations relating C_c with any of the above.

2. List all assumptions in Terzaghi's one-dimensional consolidation theory and show that:

$$\frac{\partial u}{\partial t} = c_v \frac{\partial^2 u}{\partial z^2}$$

where u is the excess pore water pressure. List the boundary conditions of the above governing differential equation for a doubly drained clay layer of thickness H.

3. Prove Equation 8.6 from the first principles.

4. A 3 m saturated clay layer is covered by 1 m-thick sand and is underlain by sand as well. The water table is 0.5 m below the ground level, and for vertical stress computations, all layers may be assumed to have unit weights of 18 kN/m³. An oedometer test was performed on an undisturbed clay sample obtained from a depth 1.9 m below the ground level. The initial water content of the clay was 35.7% and the specific gravity of the soil grains was 2.65. The sample thicknesses after 24 hours at each load increment in the consolidation test are summarized:

σ_v' (kPa)	—	50	100	200	400	50
H (mm)	19.05	18.44	18.03	17.63	17.21	17.33

a. Plot e and m_v against σ_v' (log).
b. Calculate both the compression index and the recompression index of the clay.
c. Is the clay normally consolidated or overconsolidated at the depth where the sample was taken from? Why?

Answer: 0.14, 0.06; Normally consolidated.

5. The soil profile at a site consists of 3 m of sand ($\gamma_m = 17.5$ kN/m³, $\gamma_{sat} = 18.9$ kN/m³) underlain by 6 m of clay ($w = 27\%$, $G_s = 2.70$, $m_v = 0.32$ MPa⁻¹, $c_v = 4.9$ mm²/min), which is underlain by bedrock. The water table lies 1 m below the ground level. A 3 m-high compacted fill ($\gamma_m = 20$ kN/m³) is placed on the ground in an attempt to raise the ground level.

a. What would be the final consolidation settlement of the clay layer?
b. How long will it take for a 50 mm consolidation settlement to occur?
c. What would be the consolidation settlement in one year?

Answer: 115 mm, 2.23 years, 35 mm

6. A saturated clay sample in an oedometer is under vertical pressure of 120 kPa and is at a normally consolidated state. The void ratio and the sample height at this stage are 1.21 and 18.40 mm respectively. When the vertical stress was increased to 240 kPa at the end of the consolidation, the thickness of the sample was reduced to 16.80 mm. When the vertical pressure was reduced to the original value of 120 kPa, the sample heaved to a thickness of 16.95 mm. Estimate both the compression index and recompression index of the clay.
   What would be the reduction in thickness from now if the vertical pressure was increased by 200 kPa? What is the average coefficient volume compressibility during this pressure increment?

   Answer: 0.64, 0.06, 0.82 mm, 0.24 MPa^{-1}

7. The top 10 m at a site consists of sandy silt ($\gamma_m = 17$ kN/m^3 and $\gamma_{sat} = 19$ kN/m^3). The water table lies at 1 m below the ground level. The sandy silt layer is underlain by a 2 m-thick clay layer ($\gamma_{sat} = 19.5$ kN/m^3, $m_v = 1.2$ MPa^{-1}), which is underlain by sand. If the water table is lowered by 3 m, what would be the consolidation settlement?

   Answer: 56 mm

8. The soil profile at a site consists of a top 4 m layer of dense sand followed by 2 m of clay, which is underlain by a stiff stratum. The water table is at 2 m below the ground level. The following data was obtained from a consolidation test on an undisturbed sample obtained from the middle of the clay layer: water content = 36%, specific gravity of the clay grains = 2.72, compression index = 0.72, recompression index = 0.07, preconsolidation pressure = 85 kPa. The bulk and saturated unit weights of the sand are 17 kN/m^3 and 18.5 kN/m^3 respectively. The ground level was raised by placing 2 m of compacted fill with a unit weight of 20 kN/m^3. Estimate the final consolidation settlement.
   It is proposed to construct a warehouse covering a large area on top of the raised ground, which is expected to impose a pressure of 25 kPa. What would be the additional consolidation settlement?

   Answer: 62 mm, 71 mm

9. A large area of soft clay along the coast is to be reclaimed for a new tourist development. The site investigation shows that the soil profile consists of:
   a. 0–1 m depth: Loose silty sand ($\gamma_{sat} = 18$ kN/m^3)
   b. 1–6 m depth: Soft clay
   c. 6–10 m depth: Very stiff low permeability clay

The current average water level is 1 m above the silty sand (i.e., the area is tidal and hence submerged, except when at low tide). An oedometer test was carried out on an undisturbed clay sample obtained from the middle of the clay layer, and the results are:

Initial (in situ) water content	56%
Specific gravity of the soil grains	2.71
Compression index	0.50
Recompression index	0.06
Preconsolidation pressure	50 kPa

a. Is the clay normally consolidated or overconsolidated?
b. If the site is filled to a 3 m depth with a sandy soil ($\gamma_m = 18.0$ kN/m^3 and $\gamma_{sat} = 20.0$ kN/m^3), estimate the final consolidation settlement of the clay.
c. Once the consolidation due to the above fill is completed, a warehouse will be constructed on top of the fill, imposing a uniform surcharge of 30 kPa over a large area (i.e., one-dimensional consolidation). What would be the additional consolidation settlement due to this warehouse?

Answer: Overconsolidated, 188 mm, 152 mm

10. The soil profile at a site consists of 2 m of sand underlain by 6 m of clay, which is underlain by very stiff clay that can be assumed to be impervious and incompressible. The water table lies 1.5 m below the ground level. The soil properties are as follows:

Sand: $\gamma_{sat} = 18.5$ kN/m^3, $\gamma_m = 17.0$ kN/m^3
Clay: $e_0 = 0.810$, $\gamma_{sat} = 19.0$ kN/m^3, $c_v = 4.5$ m^2/year

When the ground is surcharged with 3 m-high compacted fill with a bulk unit weight of 19 kN/m^3, the settlement was 160 mm in the first year.

a. What would be the settlement in two years?
b. After one year since the fill was placed, what would be the pore water pressure and the effective stress at the middle of the clay layer?
c. If the clay is normally consolidated, estimate the compression index and the coefficient of volume compressibility.

Answer: 230 mm; 73 kPa and 76 kPa; 0.42 and 1.20 MPa^{-1}

11. Two clay layers are separated by a 1 m-thick sand layer as shown. The water table lies 1 m above the ground in this low-lying area. The soil characteristics are summarized in the table on next page.

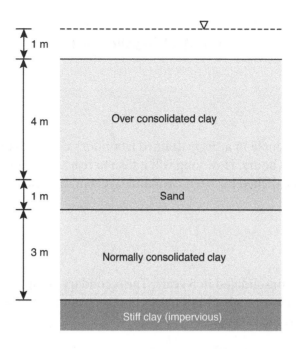

Soil parameter	O.C. Clay	N.C. Clay	Gravelly sand
Saturated unit weight (kN/m³)	20.5	20.0	19.7
Water content (%)	30.0	33.0	29.0
Compression index	0.4	0.35	NA
Recompression index	0.04	0.04	NA
Coeff. of consolidation (mm²/min)	4.5	2.3	NA
Overconsolidation ratio	2	1	NA

A 2.5 m fill ($\gamma_m = 18$ kN/m³, $\gamma_{sat} = 20$ kN/m³) was placed on the ground to raise the ground level.

a. Taking into consideration the settlements in both layers, find the final consolidation settlement.
b. How long will it take for 160 mm?
c. Using a spreadsheet, plot the variation of σ_v' and u with depth for the top 8 m of the soil at:

- $t = 0^-$ (just before the fill was placed)
- $t = 0^+$ (just after the fill was placed)
- $t = 1$ year
- $t = \infty$

Answer: 250 mm, 1 year

Quiz 4. Consolidation

Duration: 20 minutes

1. A 20 mm-thick sample in a *singly* drained laboratory consolidation test reaches 75% consolidation in 5 hours. How long will it take to reach 75% consolidation for a 5 m-thick clay sandwiched between two sand layers in the field?

(4 points)

2. A clay layer has consolidated in 5 years. The secondary compression in the next 7 years is 40 mm. How much additional secondary compression settlement would you expect within the next 10 years?

(4 points)

3. The larger the c_v, the faster is the consolidation. Why?

(2 points)

This book has free material available for download from the
Web Added Value™ resource center at *www.jrosspub.com*

Shear Strength 9

9.1 INTRODUCTION

In engineering applications, when working with steel, concrete, or timber, it is necessary to ensure that they do not fail in tension, compression, or shear. Here, we design them such that their tensile strength is greater than the tensile stresses within the material, that the compressive strength is greater than the compressive stresses within the material, and that the shear strength is greater than the shear stresses within the material. *In soils, failure almost always occurs in shear.*

Soil consists of an assemblage of grains. Failure takes place when the shear stresses exceed the shear strength along the failure surface within the soil mass. Along the failure surface when the shear strength is exceeded, the soil grains slide over each other and failure takes place. There will rarely be a failure of individual soil grains. Shear failure occurs well before the crushing or breaking of individual grains. Figure 9.1 shows the failure of an embankment. *Shear stress* is denoted by τ, and *shear strength* (or shear stress at failure) is denoted by τ_f. The soil wedge shown by the darker zone will be stable and will remain in equilibrium only if $\tau < \tau_f$. When τ becomes equal to τ_f, failure takes place where the soil wedge slides down along the failure surface. Such shear failure can occur within the backfills behind retaining walls or in the soil mass underlying a foundation.

9.2 MOHR CIRCLES

At this stage, let's have a brief overview of *Mohr circles*, which are generally covered in subjects such as Engineering Mechanics or Strength of Materials. A Mohr circle is used for graphically presenting the state of stress *at a point* in a two-dimensional problem. Figure 9.2a shows the state of stress at point A with respect to a Cartesian coordinate system where σ_x and σ_y are the normal stresses acting along the x and y direction on y and x planes respectively and the shear stresses are τ_{xy}. Our sign convention is:

- compressive stresses are positive (hence tensile stresses are negative)
- shear stresses producing counterclockwise couples are positive

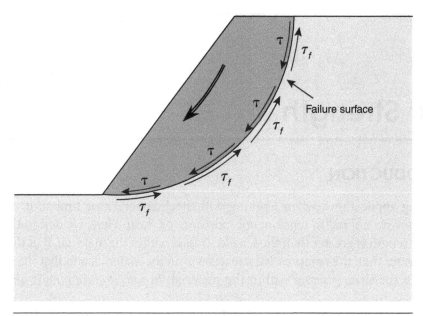

Figure 9.1 Shear failure of an embankment

The million-dollar question is, what would be the normal $\sigma_{x'}$ and shear $\tau_{x'y'}$ stresses on a plane at A inclined at θ to vertical (see Figure 9.2a)? In other words, if the coordinate axes are rotated counterclockwise by an angle θ, what would be the new normal and shear stresses with respect to x' and y' directions? These values would also represent the normal and shear stresses on two different orthogonal planes at A. Remember, we are still referring to the same point A.

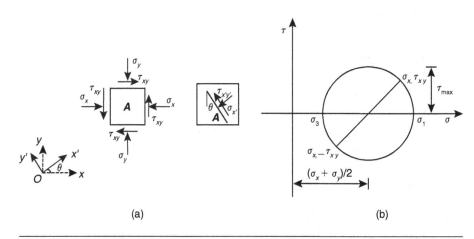

(a)   (b)

Figure 9.2 Stress transformation and Mohr circle for state of stress at point A:
(a) stresses at the point and (b) Mohr circle

It can be shown by equilibrium considerations that they are given by:

$$\sigma_{x'} = \left(\frac{\sigma_x + \sigma_y}{2}\right) + \left(\frac{\sigma_x - \sigma_y}{2}\right)\cos 2\theta - \tau_{xy}\sin 2\theta \qquad (9.1)$$

$$\tau_{x'y'} = \left(\frac{\sigma_x - \sigma_y}{2}\right)\sin 2\theta + \tau_{xy}\cos 2\theta \qquad (9.2)$$

From Equations 9.1 and 9.2, it can be shown that the *major* and *minor principal stresses* at A are:

$$\sigma_{1,3} = \left(\frac{\sigma_x + \sigma_y}{2}\right) \pm R \qquad (9.3)$$

where

$$R = \sqrt{\left(\frac{\sigma_x - \sigma_y}{2}\right)^2 + \tau_{xy}^2} \qquad (9.4)$$

Here, σ_1 and σ_3 are the major (larger) and minor (smaller) *principal stresses*, respectively. They are the maximum and minimum possible values for the normal stress at point A. Remember that principal stress occurs on a plane having no shear stress. The planes on which the principal stresses occur are known as *principal planes*. The two principal planes are perpendicular to each other. Using Equations 9.1 and 9.2, the normal $\sigma_{x'}$ and shear $\tau_{x'y'}$ stresses with respect to the new coordinate axes Ox' and Oy' can be determined for any value of θ. These are simply the stresses acting on a plane through point A, inclined at an angle of θ to vertical.

From Equations 9.1 and 9.2:

$$\left[\sigma_{x'} - \left(\frac{\sigma_x + \sigma_y}{2}\right)\right]^2 = \left(\frac{\sigma_x - \sigma_y}{2}\right)^2 \cos^2 2\theta + \tau_{xy}^2 \sin^2 2\theta - 2\tau_{xy}\left(\frac{\sigma_x - \sigma_y}{2}\right)\sin 2\theta \cos 2\theta \quad (9.5)$$

$$[\tau_{x'y'} - 0]^2 = \left(\frac{\sigma_x - \sigma_y}{2}\right)^2 \sin^2 2\theta + \tau_{xy}^2 \cos^2 2\theta + 2\tau_{xy}\left(\frac{\sigma_x - \sigma_y}{2}\right)\sin 2\theta \cos 2\theta \qquad (9.6)$$

$$\therefore \left[\sigma_{x'} - \left(\frac{\sigma_x + \sigma_y}{2}\right)\right]^2 + [\tau_{x'y'} - 0]^2 = \left(\frac{\sigma_x - \sigma_y}{2}\right)^2 + \tau_{xy}^2 = R^2 \qquad (9.7)$$

The above is an equation of a circle in $\sigma_{x'} = \tau_{xy}'$ space where R, σ_x, σ_y, and τ_{xy} are known constants. The circle has a radius of R and the coordinates of the center are $(\sigma_x + \sigma_y)/2$ and 0. Such a circle drawn on σ-τ space (see Figure 9.2b), is called a *Mohr circle*. It is a convenient, graphical way of determining the normal and shear stresses at any plane passing through point A.

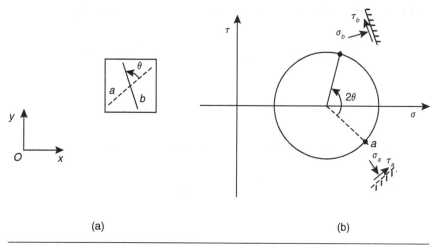

(a)

(b)

Figure 9.3 Rotation of a plane at a point: (a) point (b) Mohr circle

It can be seen from the Mohr circle in Figure 9.2b that the maximum shear stress at A is the same as the radius of the Mohr circle R. Equation 9.3, which gives the principal stress values, is even clearer from looking at the figure.

The state of stress at every point (e.g., point A in Figure 9.2a) within the soil mass can be represented by a unique Mohr circle. Figure 9.3a shows a point for which the state of stress is represented by a Mohr circle shown in Figure 9.3b. The normal σ_a and shear τ_a stress on plane-a are shown by point-a on the Mohr circle. What would be the values of σ_b and τ_b on plane-b inclined at an angle of θ counterclockwise to plane-a? They can be obtained by going counterclockwise by 2θ from point-a on the Mohr circle, as shown in Figure 9.3b. This is a key feature of a Mohr circle.

Example 9.1: Draw a Mohr circle for the state of stress at a point shown in the illustration and find the principal stresses and the maximum shear stress at the point. What would be the inclinations of these planes?

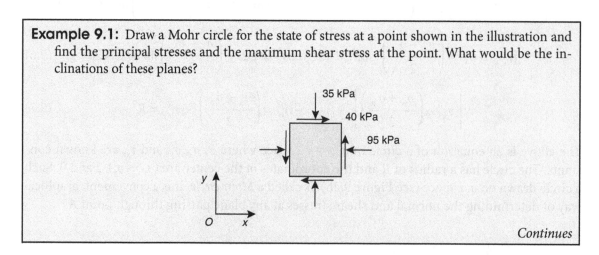

Continues

Example 9.1: *Continued*

What would be the normal and shear stresses on a plane inclined at 30° to vertical, counter-clockwise?

Solution:

$$R = \sqrt{\left(\frac{\sigma_x - \sigma_y}{2}\right)^2 + \tau_{xy}^2} = \sqrt{\left(\frac{95 - 35}{2}\right)^2 + 40^2} = 50 \text{ kPa}$$

$$\text{Center} = (\sigma_x + \sigma_y)/2, 0 = 65, 0$$

A Mohr circle is drawn from the above. The coordinates of the vertical and horizontal planes in σ-τ space are (95, 40) and (35, −40) respectively. They can be marked on the Mohr circle as points P and Q.

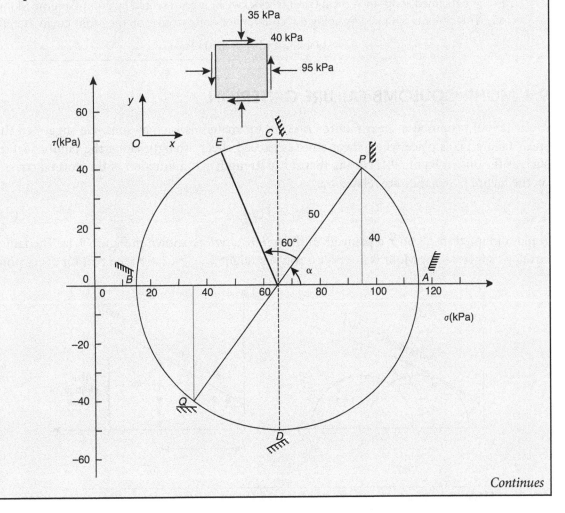

Example 9.1: *Continued*

From the Mohr circle, $\alpha = \tan^{-1}(40/30) = 53.13° \rightarrow \alpha/2 = 26.57°$.

Major and minor principal stresses are: $\sigma_1 = 115$ kPa, $\sigma_3 = 15$ kPa.

They are shown as points A and B on the Mohr circle. The major principal plane is inclined at 26.57° to vertical (clockwise) and the minor principal plane is inclined at 26.57° to horizontal (clockwise). These planes are shown in the figure on page 185.

Maximum shear stress ($\tau_{max} = 50$ kPa) is represented by the two points C and D at the ends of the dashed vertical line. It occurs on two planes: one inclined at 18.43° to vertical (counterclockwise) and the other inclined at 18.43° to horizontal (counterclockwise). These planes are shown in the figure.

The plane inclined at 30° to vertical (counterclockwise) is represented by point E on the Mohr circle. This point is obtained by going 60° counterclockwise from P on the Mohr circle. Here:

$$\sigma_E = 45.4 \text{ kPa and } \tau_E = 46.0 \text{ kPa}$$

9.3 MOHR-COULOMB FAILURE CRITERION

Mohr (1900) proposed a shear failure criterion for materials such as soils. He suggested that shear failure takes place when shear stresses exceed shear strength along the failure surface, such as the one in Figure 9.1. Noting that shear strength τ_f is a function of the normal stress σ_f on the failure plane, they are related by:

$$\tau_f = f(\sigma_f) \tag{9.8}$$

A plot of Equation 9.8 on τ-σ plane gives the *failure envelope* shown in Figure 9.4a. The failure envelope suggested by Mohr is not necessarily a straight line. We have seen that for every point

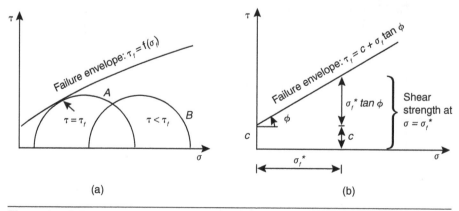

Figure 9.4 Failure criterion: (a) Mohr's (b) Coulomb's

within the soil mass, the state of stress is represented by a unique Mohr circle. Therefore, the soil mass remains stable if all the Mohr circles are contained within the envelope. The two circles in Figure 9.4a represent the states of stress at two different points within the soil: A and B. Circle A touches the failure envelope where $\tau = \tau_f$; hence shear failure takes place at point A. Circle B is well within the envelope ($\tau < \tau_f$); therefore, point B is stable. You may note that we are only showing the upper half of the Mohr circle due to symmetry about the horizontal axis. This will be the case in future discussions as well.

Coulomb (1776) suggested that τ_f is proportional to σ_f, and related them by:

$$\tau_f = c + \sigma \, \tan \phi \qquad (9.9)$$

where c and ϕ are the shear-strength parameters, known as the *cohesion* and *friction angle* respectively. Large parameters equate to more strength. Tan ϕ is similar to the friction coefficient μ that you may have encountered in physics. The friction angle is also known as the *angle of internal friction* or the *angle of shearing resistance*. For now, cohesion can be seen as the stickiness of the soil.

The Mohr-Coulomb failure criterion is the same as Equation 9.9; we replace the slightly curved Mohr's failure envelope (Figure 9.4a) with Coulomb's straight line (Figure 9.4b), which is a reasonable approximation, particularly when the normal stresses are not very high.

It can be seen in Figure 9.4b and Equation 9.9 that the soil derives its shear strength from two separate components: cohesion and friction. The contribution from cohesion is c, which remains the same at all stress levels. The frictional contribution $\sigma_f \tan\phi$, however, increases with the increasing value of σ_f. In granular soils, ϕ is slightly larger for angular grains than it is for rounded grains due to better interlocking between grains. In granular soils, it can vary in the range of 28°–45°; the lower end of the range for loose soils and the upper end for dense soils. Relative density D_r is directly related to the friction angle with a higher D_r, implying a higher ϕ. Understandably, granular soils have no cohesion (i.e., $c = 0$) and consequently, the failure envelope will pass through the origin in τ-σ plane. You can feel the grittiness in a granular soil, but it is never sticky. The stickiness comes only when the soil is cohesive, as is the case with clays. Typical values of cohesion can range from 0 to more than 100 kPa, depending on whether we are talking in terms of total stresses or effective stresses, which we will discuss later.

9.4 A COMMON LOADING SITUATION

Let's consider a soil element (or point) as shown in Figure 9.5a where the point is initially under an isotropic state of stress under a confining pressure of σ_c. In other words, the stresses are equal all around, and the Mohr circle is simply a point at R in Figure 9.5b (Think!). Keeping the confining pressure σ_c, let's apply an additional vertical normal stress $\Delta\sigma$ and increase it from zero. At any stage of loading, the principal stresses are: $\sigma_3 = \sigma_c$ and $\sigma_1 = \sigma_c + \Delta\sigma$, acting on vertical and horizontal planes respectively. A Mohr circle can be drawn at any stage of loading using the

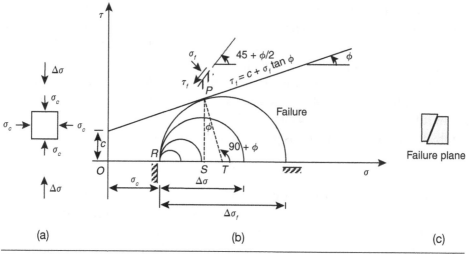

Figure 9.5 A common loading situation: (a) state of stress (b) Mohr circle representation (c) failure plane

above values, where the diameter of the Mohr circle is $\Delta\sigma$ (also the *principal stress difference* at that instant). When $\Delta\sigma$ increases, the Mohr circle becomes larger, and this continues until the Mohr circle touches the failure envelope (at P) and failure takes place. Let's ignore the smaller Mohr circles and take a closer look at the failure circle.

The minor principal stress σ_c remains constant throughout the loading and is represented by a fixed point R. The radius of the Mohr circle at failure is $\Delta\sigma_f/2$. T is the center of this circle, which touches the envelope at P. Therefore, TP is perpendicular to the failure envelope. PS is perpendicular to the σ-axis. Therefore, $\angle TPS = \phi$. Noting that the major and minor principal planes are horizontal and vertical respectively (see Figure 9.5b), it can be deduced that the failure plane, represented by point P on the Mohr circle, is inclined at $45 + \phi/2$ to horizontal or $45 - \phi/2$ to vertical (Figure 9.5c). OS and SP give the values of normal σ_f and shear τ_f stresses on the failure plane. They are:

$$\sigma_f = \sigma_c + \frac{\Delta\sigma_f}{2}(1 - \sin\phi) \tag{9.10}$$

$$\tau_f = \frac{\Delta\sigma_f}{2}\cos\phi \tag{9.11}$$

The maximum shear stress at the point is given by:

$$\tau_{max} = \frac{\Delta\sigma_f}{2} \tag{9.12}$$

Example 9.2: A granular soil specimen is initially under an isotropic stress state where the all-around confining pressure is 50 kPa. The specimen is subjected to additional vertical stress that is gradually increased from zero. The specimen failed when the additional vertical stress was 96 kPa. What is the friction angle of the soil?

Another specimen of the same granular soil at an isotropic confining pressure of 80 kPa is subjected to similar loading. Find the following:

a. The additional vertical stress required to fail the sample
b. Major and minor principal stresses at failure
c. Orientation of the failure plane
d. Normal and shear stresses on the failure plane
e. Maximum shear stress within the sample and orientation of this

Solution: At failure, $\sigma_3 = 50$ kPa and $\sigma_1 = 50 + 96 = 146$ kPa. The Mohr circle (dashed) is shown with center at T and radius $(\Delta\sigma_f/2)$ of 48 kPa. In granular soils, $c = 0$. Therefore, the failure envelope passes through the origin. The envelope is tangent to the Mohr circle at P.

$$\therefore \angle OPT = 90°$$

$$\sin\phi = \frac{PT}{OT} = \frac{48}{98} = 0.490 \rightarrow \phi = 29.3°$$

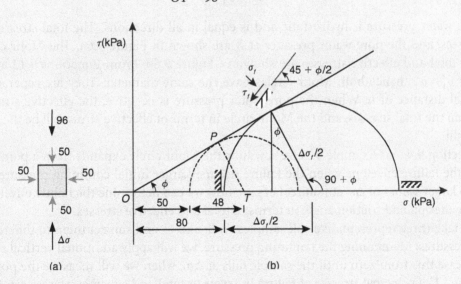

(a)   (b)

For the second specimen, we can use the friction angle calculated here.
Now, $\sigma_{3f} = 80$ kPa and $\sigma_{1f} = 80 + \Delta\sigma_f$, where $\Delta\sigma_f$ is unknown. The subscript f denotes failure.

a. $\sin\phi = 0.490 = \dfrac{\Delta\sigma_f/2}{80 + \Delta\sigma_f/2} \rightarrow \Delta\sigma_f = 153.8$ kPa

Continues

Example 9.2: *Continued*

    b. $\sigma_{3f} = 80$ kPa and $\sigma_{1f} = 80 + 153.8 = 233.8$ kPa

    c. Failure plane is oriented at $45 + 29.3/2 = 59.7°$ to horizontal

    d. $\sigma_f = 80 + 76.9 - 76.9 \sin 29.3 = 119.2$ kPa; $\tau_f = 76.9 \cos 29.3 = 67.0$ kPa

    e. $\tau_{max} = \Delta\sigma_f/2 = 76.9$ kPa is represented by the top of the Mohr circle

Therefore, the inclination of this plane would be 45° to horizontal.

9.5 MOHR CIRCLES AND FAILURE ENVELOPES IN TERMS OF σ AND σ'

Let's consider the state of stress at point X within a saturated soil mass. The normal stresses in saturated soils are carried by the soil grains and pore water, and we could separate the total stress into effective stress and pore water pressure as (see Section 5.2 in Chapter 5):

$$\sigma_1 = \sigma_1' + u \tag{9.13}$$

and

$$\sigma_3 = \sigma_3' + u \tag{9.14}$$

The pore water pressure is hydrostatic and is equal in all directions. The total stresses, effective stresses, and the pore water pressure at X are shown in Figure 9.6a. The Mohr circles in terms of total and effective stresses are shown in Figure 9.6b. From Equations 9.13 and 9.14, $\sigma_1 - \sigma_3 = \sigma_1' - \sigma_3'$, hence both Mohr circles have the same diameter. They are separated by a horizontal distance of u. When the pore water pressure is negative, the effective stresses are larger than the total stresses, and the Mohr circle in terms of effective stress will be the furthest to the right.

In Section 9.4 and Example 9.2, we saw how the Mohr circle expands from a point until it touches the failure envelope when the failure occurs. Larger initial confining pressures correspond to larger values of $\Delta\sigma_f$ at failure. Let's see how we can determine the failure envelope and find the cohesion and friction angle in terms of total and effective stresses.

Let's take three representative soil samples A, B, and C, and subject them to different confining pressures. Maintaining the confining pressure, we will apply additional vertical stress $\Delta\sigma$ and increase this from zero until the sample fails at $\Delta\sigma_f$, when we will measure the pore water pressure u_f. The principal stresses at failure in terms of total and effective stresses can be computed for each sample as follows:

$$\sigma_{3f} = \sigma_c \tag{9.15}$$

$$\sigma_{3f}' = \sigma_c - u_f \tag{9.16}$$

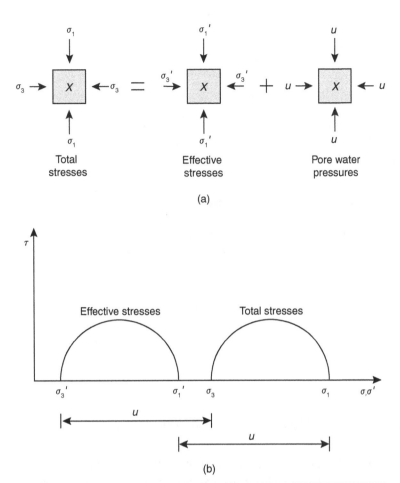

Figure 9.6 Total and effective stresses: (a) state of stress (b) Mohr circle representation

$$\sigma_{1f} = \sigma_c + \Delta\sigma_f \qquad\qquad (9.17)$$

$$\sigma'_{1f} = \sigma_c + \Delta\sigma_f - u_f \qquad\qquad (9.18)$$

From the above values, separate Mohr circles and failure envelopes can be drawn in terms of total and effective stresses, as shown in Figure 9.7. The shear strength parameters can be determined in terms of total (c, ϕ) and effective (c', ϕ') stresses.

9.6 DRAINED AND UNDRAINED LOADING SITUATIONS

Figure 9.8 shows an embankment being built on the ground, which will impose stresses ($\Delta\sigma_1$, $\Delta\sigma_3$) and pore water pressures Δu at every point, in addition to the stresses and pore water

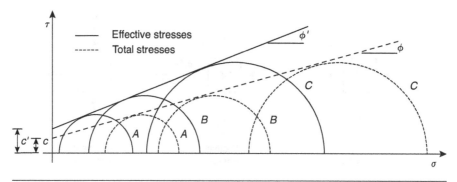

Figure 9.7 Mohr circles and failure envelopes in terms of σ and σ'

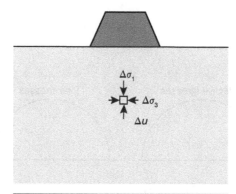

Figure 9.8 Drained and undrained loadings

pressures existing initially. Let's look at two extreme situations: (a) immediately after construction, known as *short-term*, (b) very long time after construction, known as *long-term*. It is necessary to ensure that the soil mass remains stable at all times: short-term, long-term and at any time in between.

If the embankment was built slowly such that there was no buildup of excess pore water pressure and there was adequate time available for drainage, the loading is known as *drained loading*. This situation is far from reality—engineers cannot wait that long. On the other hand, if the entire embankment is placed instantaneously, there will be buildup of pore water pressure with hardly any time allowed for drainage in the short-term. Such loading is known as *undrained loading*. In reality, the loading rate falls somewhere between the two situations, and is neither fully drained nor fully undrained. Most of the time, the short-term loading is assumed to be instantaneous, hence undrained, especially in clays. In granular soils, which have high permeability, even short-term loading is drained.

Irrespective of the loading rate, all the excess pore water pressure would have eventually dissipated over time (i.e., long-term) after the embankment had been placed. This situation can be analyzed as drained loading. For all soils, drained loading can be assumed for long-term analysis.

The total stress or short-term analysis is generally carried out in terms of total stresses using undrained shear strength parameters c_u and ϕ_u. Here, soil is treated as a continuum without separating it into soil skeleton and pore water. It is not necessary to know the pore water pressures. The effective stress or long-term analysis is carried out in terms of effective stresses using the drained shear strength parameters c' and ϕ'. The laboratory test procedures for determining the undrained and drained shear strength parameters are discussed in the following sections.

9.7 TRIAXIAL TEST

A triaxial test apparatus is used to carry out what we have been discussing in Sections 9.4, 9.5, and 9.6. It is used to apply a confining pressure to a cylindrical soil specimen and apply a vertical stress, which is increased until the specimen fails. There are provisions to measure the pore water pressures. A schematic diagram of a triaxial test setup is shown in Figure 9.9.

Triaxial tests are carried out generally on 38–50 mm diameter soil specimens with length-diameter ratios of 2:1. On special occasions, larger diameter samples are used. The specimen is wrapped in an impermeable rubber membrane, and the O-rings at the top and bottom provide a watertight seal, thus allowing drainage from only the top and/or bottom of the sample. The sample is placed on a pedestal (with provisions for drainage and pore water pressure measurement) and enclosed in a cylindrical Perspex cell filled with water. Cell pressure applied to the water within the Perspex cell applies the isotropic all-around confining pressure σ_c to the specimen. The additional vertical stress $\Delta\sigma$ or the principal stress difference $(\sigma_1 - \sigma_3)$, sometimes called *deviator stress*, is applied in the form of a load using a piston.

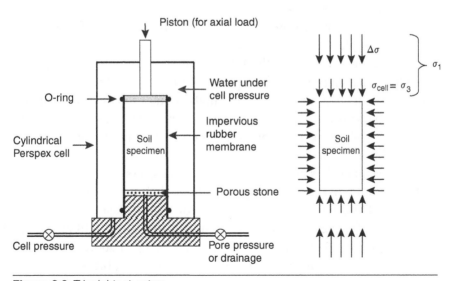

Figure 9.9 Triaxial test setup

The test consists of two stages: (a) application of isotropic confining pressure σ_c and (b) application of the deviator stress $\Delta\sigma$. Depending on whether the drainage is allowed or not during these two stages, we simulate different loading scenarios. While applying the confining pressure, if drainage is allowed, the soil specimen *consolidates*. When the drainage valve is closed, thus not allowing any drainage, the specimen *cannot consolidate* irrespective of the magnitude of the confining pressure. Here, the confining pressure is carried solely by the pore water. While applying the deviator stress, allowing drainage simulates *drained loading*, and not allowing any drainage simulates *undrained loading*. This gives three possible combinations that are commonly used. They are:

a. Consolidated drained (CD) triaxial test (ASTM D4767)
b. Consolidated undrained (CU) triaxial test (ASTM D4767; AS 1289.6.4.2)
c. Unconsolidated undrained (UU) triaxial test (ASTM D2850; AS1289.6.4.1)

You may ask, *Why not include the unconsolidated drained triaxial test too?* It just has no practical relevance.

9.7.1 Consolidated Drained (*CD*) Triaxial Test

In a consolidated drained triaxial test, the drainage is allowed throughout the entire test during the application of both σ_c and $\Delta\sigma$. The specimen is *consolidated* under all-around confining pressure of σ_c and then loaded under *drained* conditions. The loading rate is generally slow enough (e.g., axial strain of 0.1% per hour) to ensure there is no buildup of excess pore water pressure at any stage. If there is no initial pore water pressure such as *backpressure*, total stresses are the same as the effective stresses at all times. Therefore, the envelopes are the same in terms of total and effective stresses, hence $\phi = \phi'$.

Even when sampling below the water table, the saturation level can fall below 100% due to the stress relief of the sample. Compacted clay samples are difficult to saturate by simply soaking in a tank for a few days. To ensure the full saturation of the samples, especially in clays, sometimes we apply a constant pressure through the drainage line into the sample and maintain it throughout the test. This pressure is generally high enough to dissolve any remaining pore air. This is known as backpressure u_0, which is simply an initial constant pore water pressure that remains within the soil. Any excess pore water pressure that is developed during undrained loading will be in addition to this backpressure, which is like a datum. Increasing the cell pressure and backpressure equally has no effect on the effective stress.

It is interesting to note that for normally consolidated clays, $c' \approx 0$. Average values of ϕ' for normally consolidated clays can range from 20° for highly plastic clays to more than 30° for silty or sandy clays. If overconsolidated, ϕ' will be lower and c' will be higher. For compacted clays, ϕ' is typically 25–30° and can be as high as 35°. Laboratory test data suggest that ϕ' decreases with an increasing plasticity index (Kenny 1959; Bjerrum and Simons 1960; Ladd et al. 1977).

9.7.2 Consolidated Undrained (*CU*) Triaxial Test

In a consolidated undrained triaxial test, drainage is allowed only during the isotropic confinement, thus allowing the sample to *consolidate*. As in the CD triaxial test, backpressure can be applied during the consolidation process and turned off during shear. At the end of consolidation, there will be no excess pore water pressure, and the sample is ready for loading. When the additional vertical stress $\Delta\sigma$ is being applied, drainage is not allowed, and thus the sample is being loaded under *undrained* conditions at relatively high strain rates (e.g., axial strain of 1% per minute). During the undrained loading, which typically takes about 10–20 minutes, there will be development of excess pore water pressure, which is measured continuously throughout the loading. The total and effective stresses are different at failure, and separate Mohr circles can be drawn, giving failure envelopes in terms of total and effective stresses. The test gives c, ϕ, c', and ϕ'. The values of c' and ϕ' derived from a CU triaxial test are the same as those obtained from a CD triaxial test. The total stress parameters c and ϕ are of little value.

9.7.3 Unconsolidated Undrained (*UU*) Triaxial Test

An unconsolidated, undrained triaxial test is carried out almost exclusively on cohesive soils. Here, no drainage is allowed at any stage of the test. The isotropic confining pressure is applied with the drainage valve closed, so (provided the sample is saturated) *no consolidation* can take place, however large the confining pressure is. The entire cell pressure is carried by the pore water. The sample is then loaded under *undrained* conditions. During the test, there will be pore water pressure developments, which are not measured. Therefore, the effective stresses remain unknown. Mohr circles are only drawn in terms of total stresses, which enable the failure envelope to be drawn in terms of total stresses, giving c_u and ϕ_u. The subscript u denotes undrained loading. The undrained loading to failure takes about 10–20 minutes. It can be deduced (see Figure 9.10) that the deviator stress at failure $\Delta\sigma_f$ would be the same at any confining pressure. For the three total-stress Mohr circles in Figure 9.10, the effective-stress Mohr circle is the same. Increasing the confining pressure simply increases the pore water pressure by the same value, leaving the effective stresses unchanged. The failure envelope, in terms of total stresses,

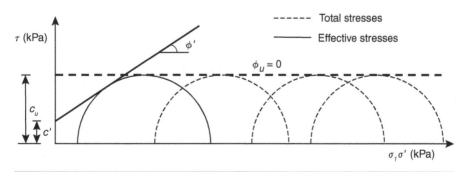

Figure 9.10 Mohr circles for a UU triaxial test

is horizontal for a saturated soil, implying that $\phi_u \approx 0$. c_u is known as *undrained shear strength* of the clay.

During undrained loading, the volume of the soil sample remains constant. Therefore, when the sample is compressed, the length decreases and the cross-sectional area increases. In computing the additional vertical stresses, the corrected area should be used. If A_0 = initial cross-sectional area of the sample, and ε = axial strain at present, the corrected area can be computed as $A_0/(1 - \varepsilon)$.

Being relatively quick and inexpensive, UU triaxial tests are quite popular in geotechnical engineering practice for deriving the undrained shear strength of the clay. However, the test does not provide the shear strength parameters in terms of effective stresses c' and ϕ', which are required for carrying out an effective stress analysis.

Now that we have means of deriving c', ϕ', c_u, and ϕ_u, determining when to use which one may be a bit confusing. Recall our discussion on drained and undrained loading in Section 9.6. In cohesionless soils, always use ϕ' and $c' = 0$, and carry out the analysis in terms of effective stresses. For long-term analysis in clays, assuming drained conditions, use c' and ϕ' to carry out an effective stress analysis. For short-term analysis in saturated clays, assuming undrained conditions, use c_u and $\phi_u = 0$ to carry out total stress analysis.

Example 9.3: The shear strength parameters in terms of effective stresses are: $c' = 15$ kPa and $\phi' = 30°$. In an unconsolidated, undrained UU triaxial test on a sample of this clay, the cell pressure was 250 kPa and the deviator stress at failure was 136 kPa. What would have been the pore water pressure at failure?

Another specimen of the same clay consolidated under a cell pressure of 120 kPa and backpressure of 50 kPa was slowly loaded to failure under drained conditions. The backpressure was maintained during the shearing as well. What would have been the additional vertical stress at failure?

Solution: Let's draw the envelope first with $c' = 15$ kPa and $\phi' = 30°$

For the first sample, at failure:

$$\sigma_{3f} = 250 \text{ kPa}, \Delta\sigma_f = 136 \text{ kPa}$$

$\therefore \sigma'_{3f} = 250 - u_f$ where u_f is the pore water pressure at failure.
The Mohr circle at failure is shown in part (a) of the illustration on page 197.

For $\triangle APT$:

$$\sin 30 = \frac{68}{26 + (250 - u_f) + 68} \rightarrow u_f = 208 \text{ kPa}$$

Continues

Example 9.3: *Continued*

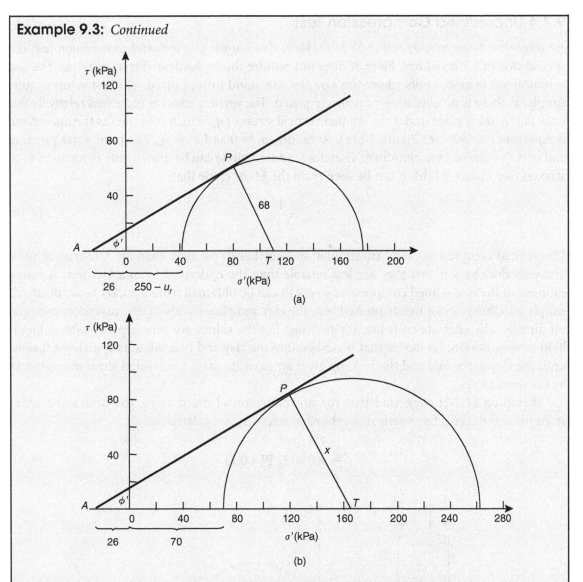

(a)

(b)

The second is a CD test with a constant back pressure of 50 kPa throughout:

$$u_0 = 50 \text{ kPa}, \sigma_{3f} = 120 \text{ kPa}$$

$$\sigma'_{3f} = 70 \text{ kPa}; \Delta\sigma_f = x \text{ (unknown)}$$

The Mohr circle is shown in part (b) of the illustration.

For ΔAPT:

$$\sin 30 = \frac{x}{26 + 70 + x} \rightarrow x = 96 \text{ kPa} \rightarrow \Delta\sigma_f = 192 \text{ kPa}$$

9.7.4 Unconfined Compression Test

An *unconfined compression test* (ASTM D2166), also known as a *uniaxial compression test*, is a special case of a triaxial test. Here, it does not require the sophisticated triaxial setup. The test is mainly for cohesive soils where the samples can stand unsupported. The test setup is quite simple, as there is no confining pressure required. The vertical stress is increased relatively fast until failure takes place under the applied vertical stress of q_u, which is known as the *unconfined compressive strength* (see Figure 9.11a). At failure, $\sigma_3 = 0$ and $\sigma_1 = q_u$. The pore water pressure and effective stresses are unknown. Therefore, a Mohr circle can be drawn only in terms of total stresses (see Figure 9.11b). It can be seen from the Mohr circle that:

$$c_u = \frac{1}{2} q_u \qquad (9.19)$$

Unconfined compression tests are simpler and quicker to perform than are UU triaxial tests. The only drawback is that they are less reliable than the c_u derived from a UU test. A rough estimate of the unconfined compressive strength can be obtained from a *pocket penetrometer*; a simple handheld device that is pushed into the clay sample or walls of an excavation and read off directly. The estimate costs literally nothing, but the values are very approximate. A handheld *torvane* is a similar device that is pushed into the clay and twisted, thus applying a torque, until the clay is sheared and the reading gives an estimate of q_u. Undrained shear strength can be obtained as $\frac{1}{2} q_u$.

Skempton (1957) suggested that for normally consolidated clays, the undrained shear strength and the effective vertical overburden stress σ'_{v0} are related by:

$$\frac{c_u}{\sigma'_{v0}} = 0.0037 \, \text{PI} + 0.11 \qquad (9.20)$$

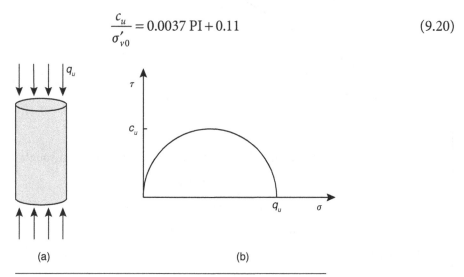

Figure 9.11 Unconfined compression test: (a) loading (b) Mohr circle

For overconsolidated clays (Ladd et al. 1977):

$$\left(\frac{c_u}{\sigma'_{v0}}\right)_{OC} = \left(\frac{c_u}{\sigma'_{v0}}\right)_{NC} (OCR)^{0.8} \tag{9.21}$$

Jamiolokowski et al. (1985) suggested that:

$$\left(\frac{c_u}{\sigma'_{v0}}\right)_{OC} = (0.23 \pm 0.04)(OCR)^{0.8} \tag{9.22}$$

Mesri (1989) suggested that for all clays, $c_u/\sigma'_p = 0.22$ where σ'_p is the preconsolidation pressure (see Section 8.3 in Chapter 8). These empirical correlations are useful in estimating the undrained shear strength of clays. On the basis of c_u or q_u, clays can be classified as shown in Figure 9.12.

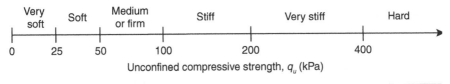

Figure 9.12 Classification of clays based on q_u

Example 9.4: Two 50 mm diameter undisturbed samples A and B are taken from the clay at the depths shown. It is expected that sample A is slightly overconsolidated and B is normally consolidated. For the clay LL = 65 and PL = 27:

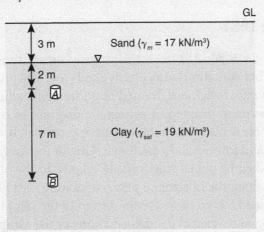

a. Estimate the undrained shear strength of sample B
b. Assuming OCR = 2 for sample A, estimate its undrained shear strength
c. If an unconfined compression test is carried out on sample B, what would be the failure load?

Continues

Example 9.4: *Continued*

Solution:

a. For sample *B*, using Equation 9.20:

$$\frac{c_u}{\sigma'_{v0}} = 0.0037 \times 38 + 0.11 = 0.251$$

At this depth:

$$\sigma'_{v0} = 3 \times 17 + 9 \times (19 - 9.81) = 133.7 \text{ kPa}$$
$$\therefore c_u = 0.251 \times 133.7 = 34 \text{ kPa}$$

b.

$$(c_u/\sigma'_{v0})_{NC} = 0.251$$

From Equation 9.21:

$$(c_u/\sigma'_{v0})_{OC} = 0.251 \times 2^{0.8} = 0.437$$

At the depth of sample *A*:

$$\sigma'_{v0} = 3 \times 17 + 2 \times (19 - 9.81) = 69.4 \text{ kPa}$$

$\therefore$ Undrained shear strength of sample *A*, $c_u = 0.437 \times 69.4 = 30$ kPa

c. Unconfined compressive strength of sample *A*, $q_u = 2 c_u = 60$ kPa

$\therefore$ Failure load $= 60 \times (\pi \times 0.025^2)$ kN $= 118$ N

9.8 DIRECT SHEAR TEST

A direct shear test (ASTM D3080; AS1289.6.2.2) is fairly simple in principle. It is carried out mostly on granular soils, but sometimes on cohesive soils too. The problem with cohesive soils is in controlling the strain rates to achieve drained or undrained loading. In the case of granular soils, loading is always assumed drained. A schematic diagram of the shear box is shown in Figure 9.13a. The soil sample is placed in a square box approximately 60 mm × 60 mm in plan, which is split into upper and lower halves as shown. One of the halves (lower in the figure) is fixed and the other is pushed or pulled horizontally relative to the other half, thus *forcing* the soil sample to shear (fail) along the horizontal plane separating the two halves. Under a specific normal load *N*, the shear load *S* is increased from zero until the sample is fully sheared. During the test, the horizontal δ_{hor} and vertical δ_{ver} deformations of the sample are recorded continuously along with the shear load. Normal stress and shear stress on the horizontal failure plane are calculated as $\sigma = N/A$ and $\tau = S/A$, where *A* is the plan area of the sample, which decreases slightly with the horizontal deformation.

Generally, the shear stress is plotted against the horizontal displacement and the vertical displacement is plotted directly under it against the horizontal displacement (Figure 9.13b). For

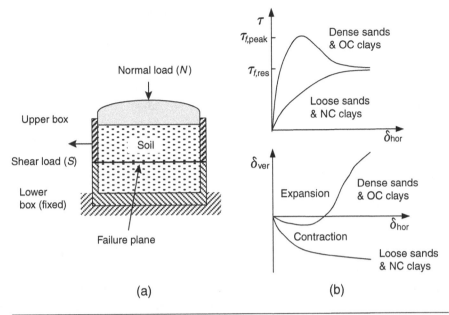

Figure 9.13 Direct shear test: (a) schematic diagram (b) τ-δ_{hor}-δ_{ver} variations

loose sands and normally consolidated clays, the shear stress increases to a maximum value τ_f at large strain. While shearing, the sample contracts; hence the vertical displacement is downward. In dense sands and overconsolidated clays, shear stress increases to a maximum value $\tau_{f,peak}$ and decreases to a lower value $\tau_{f,residual}$ at larger strains. The maximum value of shear stress is known as the *peak shear strength*, and the value at large strain is known as the *residual shear strength*. Here, we can define the failure in terms of *peak* or *residual* values of shear stress. In loose sands and normally consolidated clays, they are the same. The test can be repeated for three or more different values of normal load N, and shear stresses at failure and the corresponding normal stresses can be plotted on τ-σ space where they lie on a straight line, which is the failure envelope. The cohesion and friction angle can be determined from this envelope.

As the loading progresses in dense sands or overconsolidated clays, the sample compresses initially, but only up to the point where it cannot compress any further. Then the grains start sliding over each other, enabling the sample to expand as seen in Figure 9.13b. This is known as *dilation*. Irrespective of the initial relative density, at very large strains, all samples would reach the same void ratio, known as the *critical void ratio*, and the soil would be said to have reached *critical state*. For all practical purposes, residual values can be taken as the critical state values. In dense sands or overconsolidated clays, ϕ'_{peak} is greater than $\phi'_{residual}$; the denser the sand, the larger the difference. At large strains, the cohesive bonds are destroyed and the residual strength is purely frictional. Therefore, $c'_{residual} \approx 0$ in cohesive soils. Typical values of peak and residual friction angle for granular soils are given in Table 9.1. Which friction angles do we use in practice—peak or residual? It depends on the situation. In most geotechnical engineering problems,

Table 9.1 Friction angles of granular soils (after Lambe and Whitman 1979)

Soil type	Friction angle, ϕ (degrees)	
	Residual	Peak
Medium-dense silt	26–30	28–32
Dense silt	26–30	30–34
Medium-dense, uniform fine-to-medium sand	26–30	30–34
Dense, uniform fine-to-medium sand	26–30	32–36
Medium-dense, well-graded sand	30–34	34–40
Dense, well-graded sand	30–34	38–46
Medium-dense sand and gravel	32–36	36–42
Dense sand and gravel	32–36	40–48

strains are small and peak values are appropriate. In problems involving large strains (e.g., landslides), residual values may be more appropriate.

Clays have a fabric that comes from the particle orientations and the bonds between them. Two extreme situations are flocculated and dispersed fabrics (see Section 3.3). Most of the time, it is in between these two. When a clay is *remolded* (i.e., highly disturbed), some of the bonds are broken and the fabric is partly destroyed. This leads to a reduction in strength and stiffness. *Sensitivity* S_t is defined as the ratio of the undisturbed to the remolded shear strength. At very large strains, clay becomes remolded; therefore, the ratio of peak to residual shear strength is approximately equal to the sensitivity. Highly sensitive clays have flocculated fabric. In highly sensitive clays, sensitivity can be as high as 10 or even more, where the clay will lose its strength almost completely when remolded. Some clays will regain their strength after some time since remolding. They are known to be *thixotropic*. This is common in bentonite, which is commonly used as drilling fluid to support the boreholes.

9.9 SKEMPTON'S PORE PRESSURE PARAMETERS

Sir Alec Skempton (1954), a professor at Imperial College–United Kingdom, introduced a simple concept to estimate the change in pore water pressure Δu in a soil element due to the *changes* in major and minor *total* principal stresses ($\Delta\sigma_1$ and $\Delta\sigma_3$) in undrained loading. This is widely used in engineering practice due to its simplicity and for its practical value.

Figure 9.14 shows the major $\Delta\sigma_1$ and minor $\Delta\sigma_3$ total principal stress increments applied on point X, which results in a pore water pressure change of Δu. This can be separated into two scenarios shown on the right: (a) an isotropic loading where $\Delta\sigma_3$ is applied in all directions, leading to a pore water pressure change of Δu_1, and (b) a deviator stress of $\Delta\sigma_1 - \Delta\sigma_3$ applied only in the vertical direction, which changes the pore water pressure by Δu_2. There-

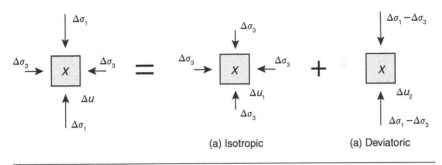

Figure 9.14 Pore water pressure buildup due to principal stress increments

fore, $\Delta u = \Delta u_1 + \Delta u_2$. Skempton (1954) expressed the change in pore water pressure due to $\Delta\sigma_1$ and $\Delta\sigma_3$ as:

$$\Delta u = B[\Delta\sigma_3 + A(\Delta\sigma_1 - \Delta\sigma_3)] \qquad (9.23)$$

where $\Delta u_1 = B\Delta\sigma_3$ and $\Delta u_2 = BA(\Delta\sigma_1 - \Delta\sigma_3)$ where BA is sometimes denoted by $\bar{A}$. The constants A and B are known as Skempton's pore pressure parameters.

B is the ratio of the pore water pressure increase to the increase in confining pressure in undrained loading. In a fully saturated clay, $B \approx 1$. Even with a slightly lower degree of saturation, B can be significantly less than 1. A typical variation of B with the degree of saturation is shown in Figure 9.15a. This B-parameter is often used in triaxial tests to determine if the sample is fully saturated. A value for B greater than 0.95 is often a good indication that the sample is fully saturated. If the soil skeleton is very stiff (e.g., very dense sands or very stiff clays), B can be significantly less than 1 even when fully saturated.

In clays, A is a function of the overconsolidation ratio OCR, stress path, anisotropy, strain rate, etc. It varies during the loading. The value of A at failure is denoted by A_f, the variation of which with OCR is shown in Figure 9.15b. For normally consolidated clays, A_f is generally close to 1, but can be as low as 0.5. For lightly overconsolidated clays, A_f is in the range of 0–0.5. Highly overconsolidated clays dilate under deviator loading where A_f can be negative, implying that negative pore water pressures develop. For very sensitive clays, A_f can be greater than 1.

It should be noted that $\Delta\sigma_1$ and $\Delta\sigma_3$ are not necessarily the changes to σ_1 and σ_3. They are the *algebraically* larger and smaller values, respectively, of the two principal stress increments. Compressive stress increments are positive.

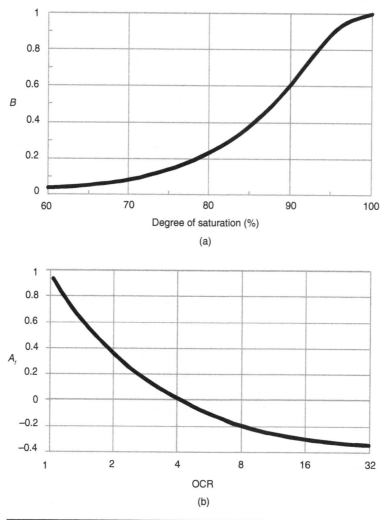

Figure 9.15 Typical values of pore pressure parameters: (a) B (b) A_f (adapted from Bishop and Henkel 1962, Craig 2004)

Example 9.5: A saturated, normally consolidated clay sample is subjected to a consolidated, undrained triaxial compression test under a backpressure of 50 kPa. The cell pressure during consolidation is 200 kPa.

When the sample is fully consolidated, the drainage valve is closed and the additional vertical stress is increased from zero to 110 kPa when the sample failed. During this period of shearing, the pore water pressure increased by 90 kPa. Find the effective friction angle and Skempton's A-parameter at failure.

Continues

Example 9.5: *Continued*

Solution: At the end of consolidation, the 50 kPa backpressure is locked in when the drainage valve is closed. Let's summarize the values at the (a) start of shearing and (b) end of shearing:

(a) Start of shearing

$\sigma_1 = \sigma_3 = 200$ kPa

$u = 50$ kPa

(b) End of shearing (failure)

$\sigma_3 = 200$ kPa, $\sigma_1 = 310$ kPa

$u = 140$ kPa

$\therefore$ During shear [i.e., between (a) and (b)]

$$\Delta\sigma_3 = 200 - 200 = 0$$
$$\Delta\sigma_1 = 310 - 200 = 110 \text{ kPa}$$
$$\Delta u = 140 - 50 = 90 \text{ kPa}$$

Assuming $B = 1$ (Saturated) and substituting these in Equation 9.23:

$$90 = 0 + A_f(110 - 0) \rightarrow A_f = 0.82$$

At failure:

$$\sigma'_{3f} = 200 - 140 = 60 \text{ kPa}; \; \sigma'_{1f} = 310 - 140 = 170 \text{ kPa}$$

Clay is normally consolidated $\rightarrow c' = 0$

Drawing the Mohr circle in terms of effective stresses with the envelope passing through the origin, ϕ' can be calculated as 28.6°.

9.10 $\sigma_1 = \sigma_3$ RELATIONSHIP AT FAILURE

Let's see how the major and minor principal stresses at failure are related. The Mohr circle at failure is shown in Figure 9.16 along with the failure envelope.

Radial line OP is perpendicular to the failure envelope at P:

$$\sin\phi = \frac{\text{OP}}{\text{OA}} = \frac{(\sigma_{1f} - \sigma_{3f})/2}{c\cot\phi + (\sigma_{1f} + \sigma_{3f})/2}$$

$$\frac{\sigma_{1f} - \sigma_{3f}}{2} = \left(\frac{\sigma_{1f} + \sigma_{3f}}{2}\right)\sin\phi + c\cos\phi$$

$$\sigma_{1f} = \sigma_{3f}\left(\frac{1+\sin\phi}{1-\sin\phi}\right) + 2c\sqrt{\left(\frac{1+\sin\phi}{1-\sin\phi}\right)} \qquad (9.24)$$

and

$$\sigma_{3f} = \sigma_{1f}\left(\frac{1-\sin\phi}{1+\sin\phi}\right) - 2c\sqrt{\left(\frac{1-\sin\phi}{1+\sin\phi}\right)} \qquad (9.25)$$

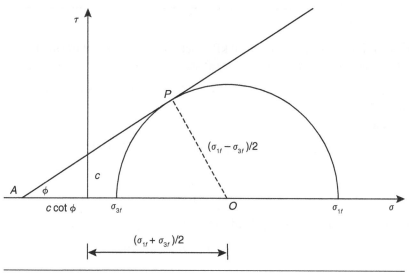

Figure 9.16 Mohr circle at failure

It is useful to note that:

$$\left(\frac{1-\sin\phi}{1+\sin\phi}\right) = \tan^2\left(45 - \frac{\phi}{2}\right)$$

and

$$\left(\frac{1+\sin\phi}{1-\sin\phi}\right) = \tan^2\left(45 + \frac{\phi}{2}\right)$$

The above derivations, including Equations 9.24 and 9.25, are applicable in terms of effective stresses and total stresses.

9.11 STRESS PATHS

Stress paths are very useful for tracking the progress in loading. For example, when a sample is loaded in a triaxial apparatus or in a situation where we want to monitor the state of stress at a point under an embankment, we can always draw a series of Mohr circles representing every change. This can become messy with a cluster of Mohr circles. A stress path is a neat way around it—we only mark the top of the Mohr circle. The entire Mohr circle is represented by a point, known as a *stress point*, as shown in Figure 9.17a.

In most geotechnical engineering applications, the vertical σ_v and horizontal σ_h normal stresses are the principal stresses. For now we will assume $\sigma_v = \sigma_1$ and $\sigma_h = \sigma_3$. The top of the

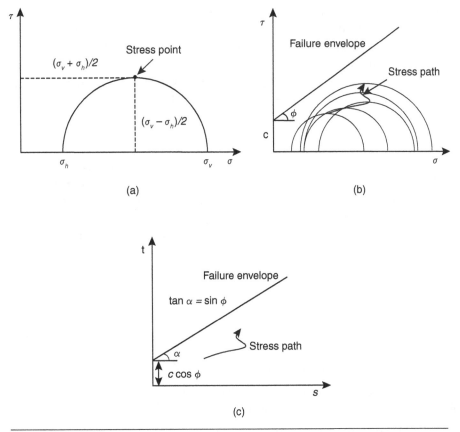

Figure 9.17 Stress path concept: (a) stress point (b) Mohr circles and stress path (c) stress path and failure envelope in *s-t* plane

Mohr circle has coordinates of $\frac{(\sigma_v + \sigma_h)}{2}$ and $\frac{(\sigma_v - \sigma_h)}{2}$ in τ-σ plane. We will call them s and t respectively, defining $s = \frac{(\sigma_v + \sigma_h)}{2}$ and $t = \frac{(\sigma_v - \sigma_h)}{2}$. We will reserve the notations p and q for three-dimensional representations, which are used in critical-state soil mechanics but not discussed here.

A *stress path* is the locus of the stress point as shown in Figures 9.17b and 9.17c. Here, we will just show the top of every Mohr circle and connect them as the loading progresses. Instead of drawing the Mohr circles on σ-τ plane, we will draw stress paths on s-t plane. In σ-τ plane, the failure envelope is $\tau_f = c + \sigma_f \tan\phi$. What would be the failure envelope in s-t plane?

From the Mohr circle at failure, shown in Figure 9.16:

$$\frac{\sigma_{1f} - \sigma_{3f}}{2} = \left(\frac{\sigma_{1f} + \sigma_{3f}}{2} \right) \sin\phi + c \cos\phi$$

i.e., $t_f = s_f \sin\phi + c \cos\phi$

Therefore, the slope of the failure envelope in *s-t* plane is $\sin\phi$ and the intercept on *t*-axis is c $\cos\phi$. When the stress path meets the failure envelope on *s-t* plane, failure takes place.

As in the case of Mohr circles, stress paths can also be drawn in terms of effective stresses where s can be replaced with s', where $s' = \frac{(\sigma'_v + \sigma'_h)}{2}$. There is no t', since t is the same as t'. Remember, there is nothing called τ'—water cannot carry shear stress. Generally, total and effective stress paths are plotted on the same graph where both s and s' are shown on the horizontal axis, preferably using the same scale for all s, s', and t.

Example 9.6: A consolidated, undrained triaxial test on a specimen of normally consolidated saturated clay ($c' = 0$) was carried out under an all-around confining pressure of 500 kPa. Consolidation took place against a backpressure of 100 kPa. During undrained loading, the additional vertical stress $\Delta\sigma_f$ was increased to failure and the test data are summarized.

$\Delta\sigma_f$ (kPa)	0	68	134	182	237	272*
u (kPa)	100	129	177	218	288	333*

*Failure

a. Draw total and effective stress paths.
b. Find the effective friction angle.
c. What is Skempton's *A*-parameter at failure?

Another specimen of the same clay, consolidated under 500 kPa and backpressure of 100 kPa, is subjected to a drained loading to failure.

d. Draw the effective stress path in the above plot.
e. Find the principal stress difference at failure.

Solution: a. The computed values during the undrained loading are summarized in the table.

σ_h (kPa)	$\Delta\sigma$ (kPa)	σ_v (kPa)	u (kPa)	s (kPa)	s' (kPa)	t (kPa)
500	0	500	100	500	400	0
500	68	568	129	534	405	34
500	134	634	177	567	390	67
500	182	682	218	591	373	91
500	237	737	288	618.5	330.5	118.5
500	272	772	333	636	303	136

The stress paths are shown on page 209.

Continues

Example 9.6: *Continued*

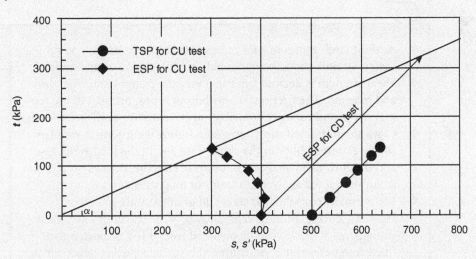

b.

$$\sin\phi' = \tan\alpha = \frac{136}{303} = 0.45 \rightarrow \phi' = 26.7°$$

c.

$$A_f = \frac{\Delta u_f}{\Delta \sigma_f} = \frac{233}{272} = 0.86$$

d. For the drained loading, the stress path (with 45° to s'-axis) intersects the failure envelope at:

$$s' = 726 \text{ kPa and } t = 326 \text{ kPa}$$

$$s' = \frac{\sigma'_v + \sigma'_h}{2} = 726 \ kPa$$

and

$$t = \frac{\sigma'_v - \sigma'_h}{2} = 326 \ kPa$$

Solving these two equations, at failure:

$$\sigma'_{hf} = 400 \text{ kPa and } \sigma'_{vf} = 1052 \text{ kPa}$$
$$u_f = 100 \text{ kPa}, \ \sigma_{hf} = 500 \text{ kPa}$$
$$\sigma_{vf} = 1152 \text{ kPa and } \Delta\sigma_f = 652 \text{ kPa}$$

❖ A Mohr circle represents the state of stress at a point. Due to symmetry, we only show the upper half in geotechnical engineering.

❖ Shear strength is derived from two separate components: *friction* and *cohesion*. The frictional contribution is proportional to the normal stress, and cohesive contribution is a constant at all stress levels.

❖ Clays are undrained short-term and drained long-term. Granular soils are drained both in the short-term and in the long-term. Use c' and ϕ' for drained analysis (in terms of effective stresses) and c_u and ϕ_u for undrained analysis (in terms of total stresses).

❖ For normally consolidated clays and granular soils, $c' = 0$.

❖ For clays, during undrained loading, $\phi_u = 0$. The undrained shear strength c_u $(= 1/2\ q_u)$ can be obtained from a UU triaxial, unconfined compression test, or estimated by using a pocket penetrometer or empirical correlations.

❖ Failure can be defined in terms of *peak* or *residual* values. $\phi'_{peak} > \phi'_{residual}$ and $c'_{res} \approx 0$.

❖ Skempton's pore pressure equation relates the *changes* in σ_1, σ_3, and u, under *undrained* conditions, irrespective of the initial state of stress.

❖ c and ϕ in τ-σ plane are similar to $c\cos\phi$ and $\tan^{-1}(\sin\phi)$ in s-t plane; in τ-σ plane we draw Mohr circles, and in s-t plane we draw stress paths.

❖ When plotting Mohr circles, use the same scales for both (σ and τ) axes; otherwise a circle would look like an ellipse. In stress path plots, the same scale for both axes is recommended.

❖ Empirical correlations are useful for preliminary estimates. They are very approximate.

❖ Clays are classified as soft, medium, etc. based on q_u (see Figure 9.12).

WORKED EXAMPLES

1. A saturated clay sample was consolidated in the triaxial cell under a cell pressure of 150 kPa without any backpressure. The drainage valve was then closed and the deviator stress was gradually increased from zero to 200 kPa when failure occurred. If $c' = 15$ kPa and $\phi' = 20°$, find the pore water pressure and Skempton's A-parameter at failure.

   Solution: This is a CU triaxial test. At failure, $\sigma'_{3f} = \sigma'_c = 150 - u_f$ and $\Delta\sigma_f = 200$ kPa where u_f is the pore water pressure at failure.

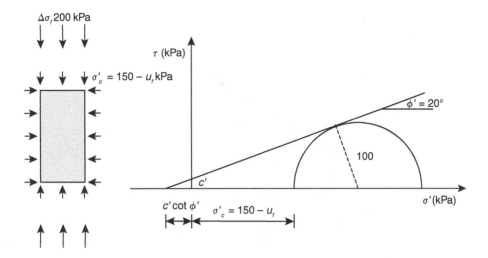

From the Mohr circle:

$$\sin 20 = \frac{100}{15\cot 20 + 150 - u_f + 100} \rightarrow u_f = -1.2 \text{ kPa}$$

During the entire shear:

$$\Delta\sigma_3 = 0, \Delta\sigma_1 = 200 \text{ kPa}, \Delta u = -1.2 \text{ kPa}$$

Substituting these in Equation 9.23, with $B = 1$:

$$A_f = \frac{-1.2}{200} = -0.006$$

2. A conventional, consolidated drained triaxial test was carried out on a normally consolidated clay sample. The consolidation pressure was 150 kPa and the deviator stress at failure was 320 kPa. Find the effective friction angle.

An identical specimen of the same clay was consolidated to 150 kPa and was subjected to a conventional, undrained triaxial test where the deviator stress at failure was 100 kPa. Find the pore water pressure and Skempton's A-parameter at failure.

Solution: The clay is normally consolidated.

$$\therefore c' = 0$$

In the CD test, at failure, $\sigma'_{3f} = 150$ kPa, $\Delta\sigma_f = 320$ kPa:

$$\sin\phi' = \frac{160}{150 + 160} \rightarrow \phi' = 31.1°$$

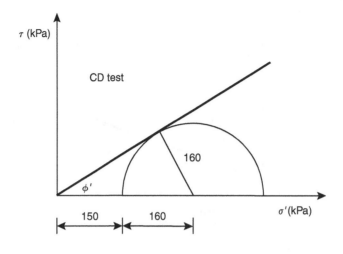

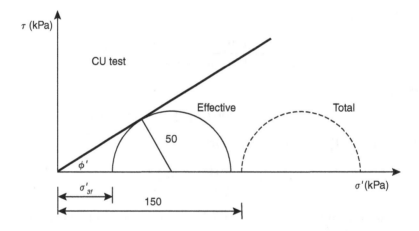

The friction angle ϕ' must be the same in the CU triaxial test, where at failure, $\Delta\sigma_f = 100$ kPa, and $\sigma_{3f} = 150$ kPa. The pore water pressure at failure u_f is unknown.

$$\sin 31.1 = \frac{50}{\sigma'_{3f} + 50} \rightarrow \sigma'_{3f} = 46.9 \text{ kPa} \rightarrow u_f = \sigma_{3f} - \sigma'_{3f} = 103.1 \text{ kPa}$$

During the entire shear in the CU test:

$$\Delta\sigma_3 = 0, \Delta\sigma_1 = 100 \text{ kPa}, \Delta u = 103.1 \text{ kPa}$$

∴ Substituting these in Equation 9.23, $A_f = 1.03$

3. A series of consolidated, undrained triaxial tests were carried out on three identical saturated clay specimens. The results are:

Specimen No.	Cell pressure (kPa)	At failure (kPa)	
		Deviator stress	Pore water pressure
1	100	170	40
2	200	260	95
3	300	360	135

Determine c, ϕ, c', and ϕ' using (a) Mohr circles, and (b) stress points at failure.

Solution: The values of σ_3, σ_1, u, σ'_3, σ'_1, s, s', and t at failure are summarized:

No.	σ_3 (kPa)	$\Delta\sigma_f$ (kPa)	u_f (kPa)	σ_1 (kPa)	σ'_3 (kPa)	σ'_1 (kPa)	s (kPa)	s' (kPa)	t (kPa)
1	100	170	40	270	60	230	185	145	85
2	200	260	95	460	105	365	330	235	130
3	300	360	135	660	165	525	480	345	180

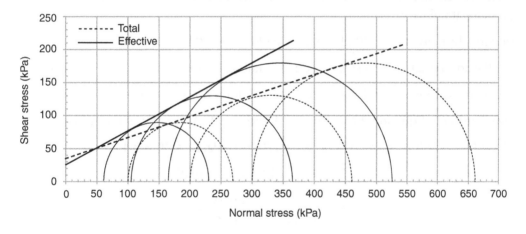

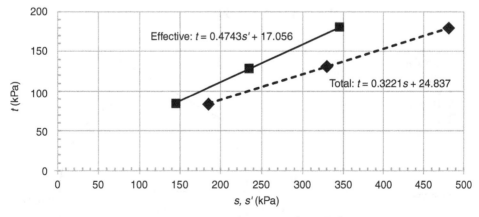

    a. Mohr circles: From the tangents to the Mohr circles, $c = 34$ kPa, $\phi = 18°$; and $c' = 25$ kPa and $\phi' = 28°$

    b. Stress points: From the stress points envelope shown below, $c = 26$ kPa, $\phi = 19°$; and $c' = 19$ kPa, $\phi' = 28°$

4. The current state of stress at a saturated clay element in the ground is:

Total vertical stress, $\sigma_{v0} = 120$ kPa

Total horizontal stress, $\sigma_{h0} = 70$ kPa

Pore water pressure, $u_0 = 30$ kPa

The friction angle and cohesion in terms of effective stresses are 30° and 10 kPa respectively. Skempton's A and B parameters are both 1. Due to some loading at the ground level, the total vertical stress is rapidly increased under undrained conditions, while the total horizontal stress remains the same. *Using Mohr circles*, find the maximum additional vertical stress that the soil element can take before failure is reached and the pore water pressure at failure is reached.

Solution: Initially, $\sigma_{v0} = 120$ kPa, $\sigma_{h0} = 70$ kPa, $u_0 = 30$ kPa.

Let the additional vertical stress applied at failure be x kPa.

Therefore, during the undrained loading to failure, $\Delta\sigma_3 = 0$, $\Delta\sigma_1 = x$.

Substituting in Equation 9.23 with $A = 1$ and $B = 1$, Δu can be estimated as x.

Therefore, at failure: $\sigma_{3f} = 70$ kPa, $\sigma_{1f} = 120 + x$ kPa, $u_f = 30 + x$ kPa

$$\sigma'_{3f} = 40 - x \text{ kPa}, \sigma'_{1f} = 90 \text{ kPa}$$

$$\sin 30 = \frac{25 + 0.5x}{17.3 + (40 - x) + (25 + 0.5x)}$$

$$x = 21.5 \text{ kPa and } u_f = 30 + x = 51.5 \text{ kPa}$$

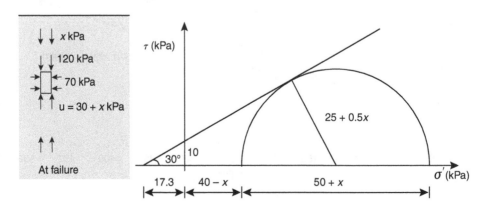

5. Repeat Problem 4 *using stress paths.*

Solution:

Initially, σ_{v0} = 120 kPa, σ_{h0} = 70 kPa and u_0 = 30 kPa.

$$\therefore s_0 = \frac{\sigma_{v0}+\sigma_{h0}}{2} = 95 \text{ kPa}; t = \frac{\sigma_{v0}-\sigma_{h0}}{2} = 25 \text{ kPa}; s_0' = \frac{\sigma_{v0}'+\sigma_{h0}'}{2} = 65 \text{ kPa}$$

Let's apply a small vertical stress increment y and calculate the changes in s, t, u, and s'. We can draw the total and effective stress paths from these:

$$\Delta\sigma_v = y, \Delta\sigma_h = 0$$

Substituting these in Equation 9.23, $\Delta u = y$:

$$\Delta s = \frac{\Delta\sigma_v + \Delta\sigma_h}{2} = 0.5y$$

$$\Delta t = \frac{\Delta\sigma_v - \Delta\sigma_h}{2} = 0.5y$$

$$\Delta s' = \Delta s - \Delta u = 0.5y - y = -0.5y$$

i.e., when the vertical stress is increased by y, s and t both *increase* by 0.5y while s' *decreases* by 0.5y. The changes will be in the same proportion when the loading continues.

Now that we have the initial values and the changes, we can show them in s-t and s'-t planes:

$$c' \cos\phi' = 10 \cos 30 = 8.7 \text{ kPa}; \alpha = \tan^{-1}(\sin 30) = 26.6°$$

The initial state, in terms of effective and total stresses, is represented by the points E (65, 25) and T (95, 25) respectively.

The effective and total stress paths, starting from E and T, are drawn as straight lines, inclined at 45° to horizontal, as shown in the figure on page 216. When the effective stress path meets the effective failure envelope at F, failure takes place.

At failure (see figure on page 216):

$$\tan 26.6 = 0.5 = \frac{25+z}{17.3+65-z} \rightarrow z = 10.8 \text{ kPa}$$

i.e., t has increased by 10.8 kPa to failure. Therefore, the additional vertical stress that was placed was 2 × 10.8 = 21.6 kPa. The pore water pressure at failure is 30 + 2z = 51.6 kPa.

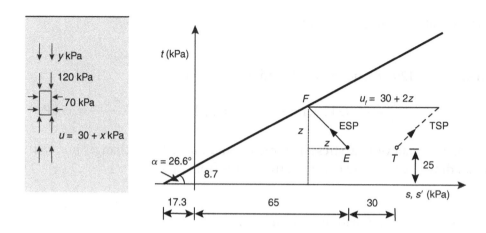

6. Skempton's A and B parameters of a saturated clay deposit are 0.8 and 0.97 respectively. Due to the construction of an embankment on this clay, the total horizontal and vertical stresses at a point increased by 40 kPa and 60 kPa respectively. What would be the increase in pore water pressure? The above clay has $c' = 0$ and $\phi' = 27°$. A triaxial sample is consolidated under a cell pressure of 300 kPa and backpressure of 100 kPa. Once the consolidation was completed, the sample was sheared undrained by applying a vertical load. What would be the principal stress difference and pore water pressure at failure?

Solution:

$$\Delta\sigma_h = 40 \text{ kPa}; \Delta\sigma_v = 60 \text{ kPa} \rightarrow \Delta\sigma_3 = 40 \text{ kPa and } \Delta\sigma_1 = 60 \text{ kPa}$$

Substituting in Equation 9.23:

$$\Delta u = 0.97[40 + 0.8(60 - 40)] = 54.3 \text{ kPa}$$

At failure, let $\Delta\sigma_f = x$.

With $\Delta\sigma_3 = 0$ and $\Delta\sigma_1 = x$, from Equation 9.23:

$$\Delta u_f = 0.97(0 + 0.8x) = 0.776\, x$$
$$\therefore u_f = 100 + 0.776\, x, \sigma_{3f} = 300 \text{ kPa; and } \sigma_{1f} = 300 + x$$

The Mohr circle at failure is shown in the figure on page 217:

$$\sin 27 = \frac{0.5x}{200 - 0.776x + 0.5x} \rightarrow x = 145.3 \text{ kPa}$$

Deviator stress at failure = 145.3 kPa

Pore water pressure at failure = $100 + 0.776\, x = 212.8$ kPa

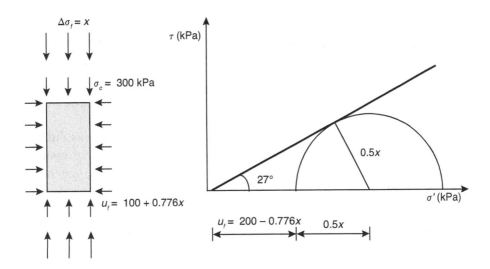

REVIEW EXERCISES

1. A consolidated, drained triaxial test was conducted on a normally consolidated clay under a confining pressure of 276 kPa. The deviator stress at failure was 276 kPa.
   a. Find the friction angle.
   b. What is the inclination of the failure plane to horizontal?
   c. Determine the normal and shear stresses acting on the failure plane.
   d. Determine the normal stress on the plane of maximum shear stress.
   e. Explain why the failure took place along a plane as determined in (b) and not along the plane where the shear stress is the maximum.

   Answer: 19.5°; 54.7°; 368 kPa, 130 kPa; 414 kPa

2. A series of consolidated, undrained triaxial tests were carried out on specimens of a saturated clay under no backpressure. The test data *at failure* are summarized:

Confining pressure (kPa)	Deviator stress (kPa)	Pore water pressure (kPa)
150	192	80
300	341	154
450	504	222

   a. Draw the Mohr circles and find the cohesion and friction angles in terms of effective stresses.
   b. Compute Skempton's A-parameter at failure for all three specimens.

c. Is the soil normally consolidated or overconsolidated? Why?

d. Another specimen of the same clay that was consolidated under a cell pressure of 250 kPa was subjected to a consolidated, drained triaxial test. What would be the deviator stress at failure?

Answer: 32 kPa, 27.9°; 0.42, 0.45, 0.44; c' ≠ 0 and A_f ≈ 0.45 → overconsolidated; 546 kPa

3. A consolidated, drained triaxial test was carried out on a normally consolidated clay. The specimen was consolidated under a cell pressure of 100 kPa and backpressure of 30 kPa. The axial deviator stress was slowly increased to failure so that there was no excess pore water pressure development while shearing. The specimen failed under a deviator stress of 130 kPa. The backpressure of 30 kPa was maintained throughout the test. Find the effective friction angle and the normal and shear stresses on the failure plane.

Answer: 28.8°; 104 kPa, 57 kPa

4. Consolidated, undrained triaxial tests were carried out on three samples with no backpressure. The test results at failure are summarized:

Cell pressure (kPa)	300	400	600
Principal stress difference at failure (kPa)	186	240	360
Pore water pressure at failure (kPa)	159	222	338

Using (a) Mohr circles and (b) stress points, determine the shear strength parameters in terms of total and effective stresses.

Answer: 5 kPa, 13°; 7 kPa, 23°

5. A series of unconsolidated, undrained triaxial tests were carried out on three samples of clay. The confining pressures and the additional vertical stresses that are required to fail the samples are summarized below. Draw the Mohr circles in terms of total stresses, and determine c_u and ϕ_u.

Confining pressure (kPa)	100	300	600
Additional vertical stress at failure (kPa)	252	271	290

Answer: 120 kPa, 2.4°

6. The failure envelope obtained in an unconsolidated, undrained triaxial test is shown on page 219, along with the Mohr circle from an unconfined compression test. Show that:

$$q_u = 2\left(\frac{\cos\phi_u}{1-\sin\phi_u}\right)c_u$$

From the above, deduce that when $\phi_u = 0$, $q_u = 2\,c_u$.

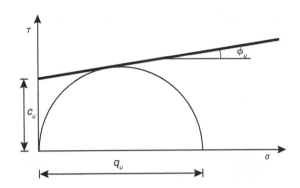

7. A consolidated, undrained triaxial test is being carried out on a normally consolidated clay where $c' = 0$ and $\phi' = 26°$. The triaxial specimen was consolidated under a cell pressure of 300 kPa and backpressure of 80 kPa. Skempton's A-parameter at failure is estimated to be 0.80. The drainage valve has since been closed and the vertical deviator stress increased to failure. What would be the deviator stress and pore water pressure at failure?

Answer: 153 kPa, 202 kPa

8. A normally consolidated soft clay specimen is consolidated in the triaxial cell under an all-around pressure of 200 kPa with no backpressure. The drainage valve is then closed and the cell pressure increased by 300 kPa, and the pore water pressure increased to 300 kPa. Then, the vertical deviator stress was increased from 0 to 110 kPa when the sample failed, and the pore water pressure was 420 kPa. Find the effective friction angle and Skempton's pore pressure parameters B and A_f. A second specimen of the same clay is consolidated under an all-around pressure of 70 kPa. Under undrained conditions, the vertical stress is increased to failure. Find the vertical deviator stress and pore water pressure at failure.

A third specimen of the same clay was isotropically consolidated under 70 kPa and was subjected to a vertical deviator stress that was increased to failure under drained conditions. What would be the deviator stress at failure?

Answer: 24.0°, 1.0, 1.09; 39 kPa, 42 kPa; 96 kPa

9. A 50 mm diameter normally consolidated clay sample with $\phi' = 27°$ was subjected to an unconfined compression test where it failed under the axial load of 157 N. Find the undrained shear strength and the pore water pressure within the sample.

Answer: 40 kPa, −48.1 kPa

10. A direct shear test is carried out on a sandy soil, and the normal loads and the peak and residual shear loads at failure are summarized below. Assuming that the cross-section area

of the direct shear sample remains the same in all tests, determine the peak and residual effective-friction angles:

Normal load (N)	100	200	350
Peak failure shear load (N)	75	153	262
Residual failure shear load (N)	60	118	212

Answer: 31°, 37°

11. A clay sample was consolidated in a triaxial cell under a backpressure of 50 kPa and cell pressure of 150 kPa. The drainage valve was then closed and the cell pressure was increased to 200 kPa when the pore pressure increased to 98 kPa. What is Skempton's B-parameter? The above sample was then subjected to a vertical deviator stress, which was increased from zero under undrained conditions. The sample failed when the pore water pressure was 160 kPa and the deviator stress was 70 kPa. What is Skempton's A-parameter at failure? Assuming the clay is normally consolidated, find the friction angle in terms of effective stresses.

Answer: 0.96; 0.92, 27.8°

12. A consolidated, undrained triaxial test was carried out on a 73.0 mm diameter and 146.6 mm-long decomposed granodiorite sample at an initial water content of 26%. The sample was obtained from Palmerston Highway, North Queensland, Australia, to back-analyze a slope failure, and was initially consolidated under a cell pressure of 200 kPa and backpressure of 150 kPa. The drainage valve was closed and the cell pressure was increased to 254 kPa when the pore water pressure increased to 182 kPa. Find Skempton's B-parameter. The nature of the soil sample is such that it was not possible to achieve a higher B value. The drainage valve was opened and the sample was consolidated further under the cell pressure of 254 kPa and backpressure of 150 kPa. At the end of consolidation, the drainage valve was closed, locking in the backpressure in preparation for the undrained loading. The axial strain ε, additional vertical stress applied to the sample under undrained conditions $\Delta\sigma$, and the pore water pressure u measured during the test are summarized in the table on page 221:
   a. Plot the total and effective stress paths
   b. Plot $\Delta\sigma$ and pore water pressure against the axial strain on the same plot
   c. Find the peak and residual shear stresses at failure, and the corresponding values of A_f

Answer: 0.59; 133 kPa, 117 kPa; 0.06, −0.09

ε (%)	Δσ (kPa)	u (kPa)
0	2	155
0.4	10	155
0.5	24	157
1	22	161
1.5	51	165
2	70	175
2.5	100	185
3	121	191
3.5	145	195
4	166	197
4.5	186	197
5	200	197
5.5	215	195
6	227	193
6.5	236	190
7	243	187
7.5	249	184
8.5	256	179
9.5	264	170
10.5	266	165
12.5	264	158
14.5	259	152
16.5	253	149
18.5	248	145
20.5	242	143
22.5	233	142

13. A direct shear test was carried out on a sand sample under normal stress of 450 kPa. The shear stress at failure was 310 kPa. Assuming that the failure plane was horizontal, draw a Mohr circle and find the principal stresses and the orientations of the major and minor principal planes.

    Answer: 34.6°; 1040 kPa, 287 kPa; inclined at 117.7° and 27.7° respectively to horizontal

14. The following test data were obtained from three consolidated, undrained triaxial tests on a saturated clay with no backpressure:

Confining cell pressure, σ_c (kPa)	100	200	300
Deviator stress at failure, $\Delta\sigma_f$ (kPa)	146	191	239
Pore water pressure at failure, u_f (kPa)	56	133	176

a. Plot the stress points at failure and determine the shear strength parameters c' and ϕ'.
b. Compute Skempton's A-parameters at failure for all three samples. Why are they different?
c. Is the clay normally consolidated or overconsolidated?
d. Three further samples of the same clay, A, B, and C, are consolidated under a confining pressure of 150 kPa with no backpressure. Sample A is sheared slowly under drained conditions with the drainage valve open to ensure there is no pore water pressure building up. Sample B was sheared quickly under undrained conditions with the drainage valve closed. In the case of Sample C, the drainage valve was closed and the confining pressure was increased to 250 kPa. Then the deviator stress was quickly applied to failure under undrained conditions. Find the deviator stress at failure for all three samples. Assume an appropriate value of A_f for samples B and C.

Answer: 42 kPa, 19°; 0.38, 0.70 and 0.74; OC; 263 kPa, 173 kPa, 173 kPa

15. The state of stress at a point within a saturated clay is given as: $\sigma_{v0} = 140$ kPa, $\sigma_{h0} = 100$ kPa, $u_0 = 40$ kPa. Skempton's A and B parameters for this clay are 0.5 and 1 respectively. Shear strength parameters are: $c' = 0$ and $\phi' = 26°$.
a. Calculate the initial values s_0, s'_0, and t_0 and show the total and effective stress points, along with the failure envelope on s-s'-t plane (see Worked Example 5).
b. When the following stress changes take place at this point under *undrained conditions*, calculate the changes in s, s', and t.
   i.   Both σ_v and σ_h increased by 10 kPa.
   ii. σ_v increased by 10 kPa and σ_h remained the same.
   iii. σ_h decreased by 10 kPa and σ_v remained the same.
c. From the above values from (b), plot the stress points for the three situations. Assuming the loading continues with further increments, draw the stress paths in terms of total and effective stresses.
d. Determine the maximum shear stress and the corresponding pore water pressure in the soil element at failure for scenarios (ii) and (iii).
e. Discuss the stress paths for loading scenario (i).

Answer: (a) 120 kPa, 80 kPa, 20 kPa; (b) 10 kPa, 0, 0; 5 kPa, 0, 5 kPa; −5 kPa, 0, 5 kPa (d) 35 kPa, 25 kPa; 35 kPa, 55 kPa

16. The state of stress at a point within a saturated clay is given as: $\sigma_{v0} = 140$ kPa, $\sigma_{h0} = 100$ kPa, $u_0 = 40$ kPa. Skempton's A and B parameters for this clay are 0.5 and 1 respectively. Shear strength parameters are: $c' = 0$ and $\phi' = 26°$.

a. Calculate the initial values s_0, s'_0, and t_0. Show the total and effective stress points along with the failure envelope on s-s'-t plane (see Worked Example 5).

b. When the following stress changes take place at this point under *drained conditions*, calculate the changes in s, s', and t.

   i.   Both σ_v and σ_h increased by 10 kPa.
   ii. σ_v increased by 10 kPa and σ_h remained the same.
   iii. σ_h decreased by 10 kPa and σ_v remained the same.

c. In which of the above scenarios will there be no failure?

d. In scenario (iii), what would be the vertical and horizontal stresses at failure?

Answer: (a) 120 kPa, 80 kPa, 20 kPa; (b) 10 kPa, 0, 10 kPa; 5 kPa, 5 kPa, 5 kPa; −5 kPa, 5 kPa, −5 kPa; (c) scenario (i); (d) 140 kPa, 79 kPa

Quiz 5. Shear Strength

Duration: 20 minutes

1. In a direct shear test on a sandy soil, the shear load at failure was 135 N when the normal load was 190 N. What is the friction angle of the sand?

(1 point)

2. In a consolidated, drained triaxial test on a sandy soil, the principal stress difference at failure was twice the confining pressure. What is the effective friction angle?

(2 points)

3. An unconsolidated, undrained triaxial test was carried out on three clay samples from a homogeneous, saturated clay at confining pressures of 100 kPa, 200 kPa, and 300 kPa. In all three cases, an additional vertical stress of 110 kPa was required to fail the samples, suggesting that $\phi_u = 0$ and $c_u = 55$ kPa. If it is known that the clay has $c' = 15$ kPa and $\phi' = 25°$, what would be the pore water pressures at failure for the above three samples?

(7 points)

Web
Added
Value™

This book has free material available for download from the
Web Added Value™ resource center at *www.jrosspub.com*

Lateral Earth Pressures 10

10.1 INTRODUCTION

Pressure at a point within a liquid is the same in all directions (e.g., pore water pressure). Due to friction between the grains, this is not the case in soils where normal stress varies with direction. The lateral earth pressure can be quite different from the vertical normal stress that we have been calculating in the previous chapters.

Very often in geotechnical engineering, we encounter problems that require the computation of the lateral loadings on structures such as retaining walls, braced excavations, sheet piles, basement walls, etc. Now that we know how to compute the vertical stresses at a point within the soil mass—including the vertical stress increases caused by various loadings—it is time to look at the *horizontal* loadings. Figure 10.1 shows examples of a few typical geotechnical applications where it is required to know the horizontal loading. Figure 10.1a shows a concrete cantilevered retaining wall that prevents the soil on the right from entering the highway; to assess the retaining wall's stability, it is necessary to know the horizontal loadings on both sides. Figure 10.1b shows a cantilevered sheet pile that supports the walls of the excavation. Sheet piles are sheets of concrete, timber, or steel that interlock and are driven into the ground to form a continuous wall. To ensure the excavation's stability, it is required to know the horizontal earth pressures on both sides of the sheet pile. When excavating narrow trenches for the purposes of laying pipelines etc., the excavation walls are supported with timber or steel sheets and horizontal struts as shown in Figure 10.1c. A good understanding of the horizontal earth pressures is necessary for computing the loadings on the struts and for designing the bracing system.

The focus of this chapter is to determine the horizontal normal stresses and their variations with depth under special circumstances. The total and effective horizontal stresses are denoted by σ_h and σ'_h respectively. The three special circumstances are *at-rest state*, *active state*, and *passive state*. The state of at-rest is very stable, whereas the active and passive states occur when the soil fails. We generally force most of our geotechnical problems into one of these three situations, which are typically easier to solve. There are no simple analytical solutions to the problem when it lies outside these three states.

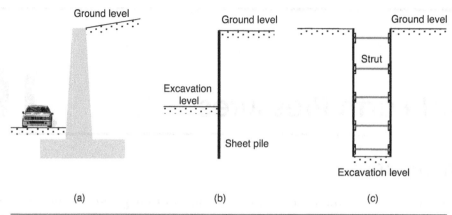

Figure 10.1 Geotechnical applications: (a) cantilevered retaining wall (b) cantilevered sheet pile (c) braced excavation

10.2 AT-REST STATE

Figure 10.2a shows a homogeneous soil mass where A, B, and C are three points that show the vertical and horizontal *effective* stresses. It is interesting to note that the ratio of σ'_h to σ'_v is the same at all three points. This ratio, known as the coefficient of earth pressure at rest K_0, is a unique constant for the homogeneous soil mass. When the soil is *at-rest*, there are no horizontal strains or deformations, the main criterion defining an at-rest situation. An *at-rest state* is also known as a K_0-*state* or K_0-*condition*. The Mohr circles representing the states of stresses at the three points are shown in Figure 10.2b where the circles lie well below the failure envelope. In saturated soils, in the presence of pore water pressure where the total and effective stresses are different, σ_h/σ_v is not a constant. Figure 10.2c shows a soil profile that consists of three different soils with their specific values of K_0. One-dimensional consolidation in an oedometer takes place under K_0 condition—any strain is only vertical.

K_0 is a very useful parameter in geotechnical engineering computations. It can be measured in a special triaxial apparatus where σ'_h and σ'_v are increased such that there is no lateral strain on the sample during consolidation. Such consolidation, different than the isotropic consolidation discussed in Chapter 9, is known as K_0-*consolidation*. K_0-consolidation is more realistic than the isotropic consolidation in representing the in situ state of stress. In the field, K_0 can be measured by a pressuremeter, dilatometer, or K_0 stepped blade test, which will be discussed in Chapter 11. Nevertheless, these tests are often costly for the client, and are not always justified. Generally, K_0 is estimated using empirical correlations, which are discussed below; these estimates literally cost nothing. If we assume that soil is a perfectly elastic isotropic continuum, it can be shown that:

$$K_0 = \frac{\nu}{1 - \nu} \tag{10.1}$$

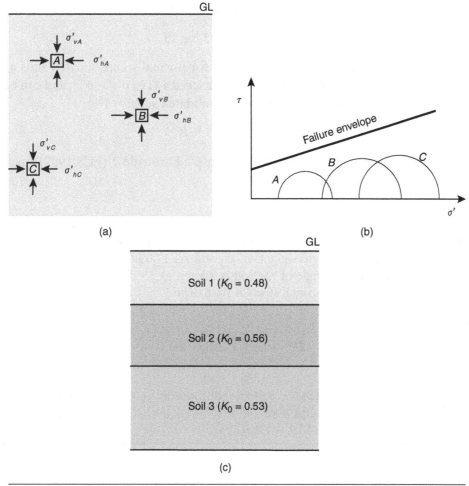

Figure 10.2 At-rest state: (a) stresses at different points (b) Mohr circles (c) K_0 for different soils

where ν is the Poisson's ratio of the soil. There are few empirical correlations for estimating K_0. The most popular of these is the one proposed by Jaky (1948) for normally consolidated clays and sands, shown as:

$$K_0 = 1 - \sin \phi' \tag{10.2}$$

where ϕ' is the effective friction angle. For normally consolidated clays, Massarsch (1979) showed that:

$$K_0 = 0.44 + 0.0042 \, PI \tag{10.3}$$

For normally consolidated clays, Alpan (1967) suggested that:

$$K_0 = 0.19 + 0.233 \log PI \tag{10.4}$$

The above equations show that typical values of K_0 for normally consolidated soils are in the range of 0.4 to 0.6. For overconsolidated soils, it can exceed 1 (i.e., $\sigma'_h > \sigma'_v$), and can be as high as 3 for heavily overconsolidated clays. For overconsolidated soils:

$$(K_0)_{OC} = (K_0)_{NC} \, OCR^m \tag{10.5}$$

Mayne and Kulhawy (1982) suggested that $m = \sin \phi'$. Eurocode 7 (ECS 1997) suggests that $m = 0.5$ if the OCR is not very large.

Example 10.1: In a normally consolidated sandy clay deposit, the water table lies at a depth of 4 m. The bulk and saturated unit weights of the soil are 17.0 kN/m³ and 18.5 kN/m³ respectively. The effective friction angle of the soil is known as 25° from a consolidated, drained triaxial test. Find the total horizontal stress at 10 m depth.

Solution:

$$\phi' = 25° \rightarrow K_0 = 1 - \sin 25 = 0.58$$

At 10 m depth:

$$\sigma'_v = 4 \times 17.0 + 6 \times (18.5 - 9.81) = 120.1 \text{ kPa}; \, u = 6 \times 9.81 = 58.9 \text{ kPa}$$
$$\therefore \sigma'_h = K_0 \, \sigma'_v = 0.58 \times 120.1 = 69.7 \text{ kPa}$$
$$\sigma_h = \sigma'_h + u = 69.7 + 58.9 = 128.6 \text{ kPa}$$

Example 10.2: A rigid basement wall retains 6 m of backfill as shown below. The K_0 values of the sand and clay are 0.45 and 0.56 respectively. Assuming the entire soil mass is in K_0-state, draw the lateral pressure distribution with depth and determine the magnitude and location of the resultant thrust on the wall.

Solution: Let's compute the values of σ'_h, u, and σ_h at $z = 0$, 2 m, 3 m, and 6 m depth where z is measured from the ground level.

At $z = 0$, $\sigma_v = 0$, $u = 0$, $\sigma_h = 0$, and $\sigma'_h = 0$

At $z = 2$ m:

$$\sigma'_v = 2 \times 16.5 = 33.0 \text{ kPa}$$
$$\sigma'_h = K_0 \, \sigma'_v = 0.45 \times 33.0 = 14.9 \text{ kPa}$$
$$u = 0 \rightarrow \sigma_h = \sigma'_h + u = 14.9 \text{ kPa}$$

Continues

Example 10.2: *Continued*

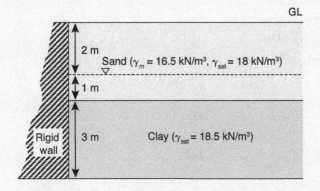

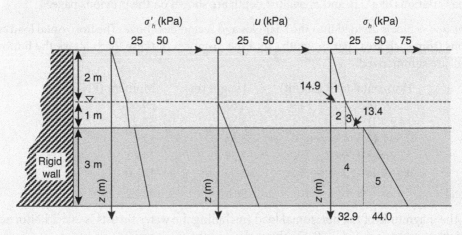

At $z = 3$ m (in sand):

$$\sigma'_v = 2 \times 16.5 + 1 \times (18 - 9.81) = 41.2 \text{ kPa}$$
$$\sigma'_h = K_0 \, \sigma'_v = 0.45 \times 41.2 = 18.5 \text{ kPa, and}$$
$$u = 1 \times 9.81 = 9.8 \text{ kPa}$$
$$\therefore \sigma_h = \sigma'_h + u = 28.3 \text{ kPa}$$

At $z = 3$ m (in clay):

$$\sigma'_v = 41.2 \text{ kPa}$$
$$\sigma'_h = K_0 \, \sigma'_v = 0.56 \times 41.2 = 23.1 \text{ kPa, and}$$
$$u = 9.8 \text{ kPa}$$
$$\sigma_h = \sigma'_h + u = 32.9 \text{ kPa}$$

At $z = 6$ m (in clay):

$$\sigma'_v = 2 \times 16.5 + 1 \times (18 - 9.81) + 3 \times (18.5 - 9.81) = 67.3 \text{ kPa}$$
$$\sigma'_h = K_0 \, \sigma'_v = 0.56 \times 67.3 = 37.7 \text{ kPa, and}$$

Continues

Example 10.2: *Continued*

$$u = 39.2 \text{ kPa} \rightarrow$$
$$\sigma_h = \sigma'_h + u = 76.9 \text{ kPa}$$

These values are summarized:

	σ'_h (kPa)	u (kPa)	σ_h (kPa)
$z = 0$	0	0	0
$z = 2$ m	14.9	0	14.9
$z = 3$ m (sand)	18.5	9.8	28.3
$z = 3$ m (clay)	23.1	9.8	32.9
$z = 6$ m	37.7	39.2	76.9

The variations of σ'_h, u, and σ_h against depth are shown on the previous page.

The $\sigma_h = z$ plot is divided into the triangles and rectangles above. The horizontal load contributions from each area (per m width), and the distances of these loads above the bottom of the wall are summarized:

Zone	Horizontal load (kN/m)	Height (m)	Moment (kN-m/m)
1	$0.5 \times 14.9 \times 2 = 14.9$	4.67	69.6
2	$14.9 \times 1 = 14.9$	3.50	52.2
3	$0.5 \times 13.4 \times 1 = 6.7$	3.33	22.3
4	$32.9 \times 3 = 98.7$	1.50	148.1
5	$0.5 \times 44 \times 3 = 66.0$	2.00	132.0
Total		201.2	424.2

$\therefore$ The magnitude of the horizontal load (including the water thrust) is 201.2 kN/m acting at a height of 2.11 m ($= 424.2/201.2$) above the bottom of the wall.

10.3 RANKINE'S EARTH PRESSURE THEORY

The theories of Rankine (1857) and Coulomb (1776) are two earth pressure theories that we will study in this chapter. These theories are often referred to as the classical earth pressure theories. Rankine's theory is simpler and therefore more popular for computing earth pressures behind retaining walls, basement walls, sheet piles, and braced excavations. This theory assumes that the wall is smooth and vertical with no adhesion or friction along the soil-wall interface. Consequently, there is no shear stress along the wall when the soil slides along the wall at failure. In the absence of shear stresses along the wall, σ'_v and σ'_h are principal stresses (provided the wall is vertical) as shown in Figure 10.3a.

The smooth, vertical wall shown in Figure 10.3a supports an excavation. As the excavation proceeds, the wall slowly deflects toward the left, moving *away* from the soil on the right, and *toward* the soil on the left, below the excavation level. The wall movement leads to a reduction in σ'_h within the soil mass on the right, and an increase in σ'_h within the soil mass on the left. σ'_v remains the same during the wall movement. When the horizontal movement of the wall becomes large, failure takes place within the soil mass on both sides of the wall due to different mechanisms. We will discuss them separately.

10.3.1 Active State

Figure 10.3b shows a smooth, vertical wall retaining a *granular* backfill of γ unit weight. There are no lateral strains, and hence the soil is initially in at-rest state with $\sigma'_{v0} = \gamma z$ and $\sigma'_{h0} = K_0\gamma z$, represented by the dashed Mohr circle as shown. When the wall moves away from the soil, σ'_v remains the same ($= \gamma z$) but σ'_h decreases, and the Mohr circle becomes larger until it touches the failure envelope where failure takes place. We consider this the instant that the soil reaches *active state*. The effective horizontal stress in this new active state is known as the active earth pressure σ'_{ha}. From the Mohr circle, $AP = \frac{\sigma'_v - \sigma'_{ha}}{2}$ and $AO = \frac{\sigma'_v + \sigma'_{ha}}{2}$. Therefore:

$$\sin\phi' = \frac{AP}{AO} = \frac{\sigma'_v - \sigma'_{ha}}{\sigma'_v + \sigma'_{ha}}$$

$$\therefore \sigma'_{ha} = K_A\sigma'_v \tag{10.6}$$

where $K_A = \left(\frac{1-\sin\phi'}{1+\sin\phi'}\right) = \tan^2(45 - \phi'/2)$, known as *Rankine's coefficient of active earth pressure*. In the case of cohesive soils, because of the cohesion intercept on the τ-axis, Equation 10.6 becomes:

$$\sigma'_{ha} = K_A\sigma'_v - 2c'\sqrt{K_A} \tag{10.7}$$

The horizontal and vertical planes on the Mohr circle are shown along with the values of σ'_v and σ'_{ha} in Figure 10.3b. The failure plane is represented by point P on the Mohr circle. It can be deduced that the failure plane is inclined at $45 + \phi'/2$ degrees to horizontal.

10.3.2 Passive State

As in the previous case, for the situation shown in Figure 10.3c, the soil is initially under no lateral strains, and hence is in at-rest state with $\sigma'_{v0} = \gamma z$ and $\sigma'_{h0} = K_0\gamma z$, represented by the dashed Mohr circle. When the wall moves toward the soil (i.e., due to active earth pressure on the right side), σ'_v remains the same ($= \gamma z$), but σ'_h increases, and the Mohr circle becomes a point the instant they become equal. From this point forward, σ'_h exceeds σ'_v, and the Mohr circle continues to expand until the failure envelope is touched; the soil is now considered in a *passive state*. The effective horizontal stress in the passive state is known as the passive earth pressure σ'_{hp}. From the Mohr circle at the passive state (Figure 10.3c):

$$AP = \frac{\sigma'_{hp} - \sigma'_v}{2} \quad \text{and} \quad AO = \frac{\sigma'_{hp} + \sigma'_v}{2}$$

Figure 10.3 (a) lateral movement of a
smooth wall (b) when the wall moves
away from the soil (c) when
the wall moves *toward* the soil

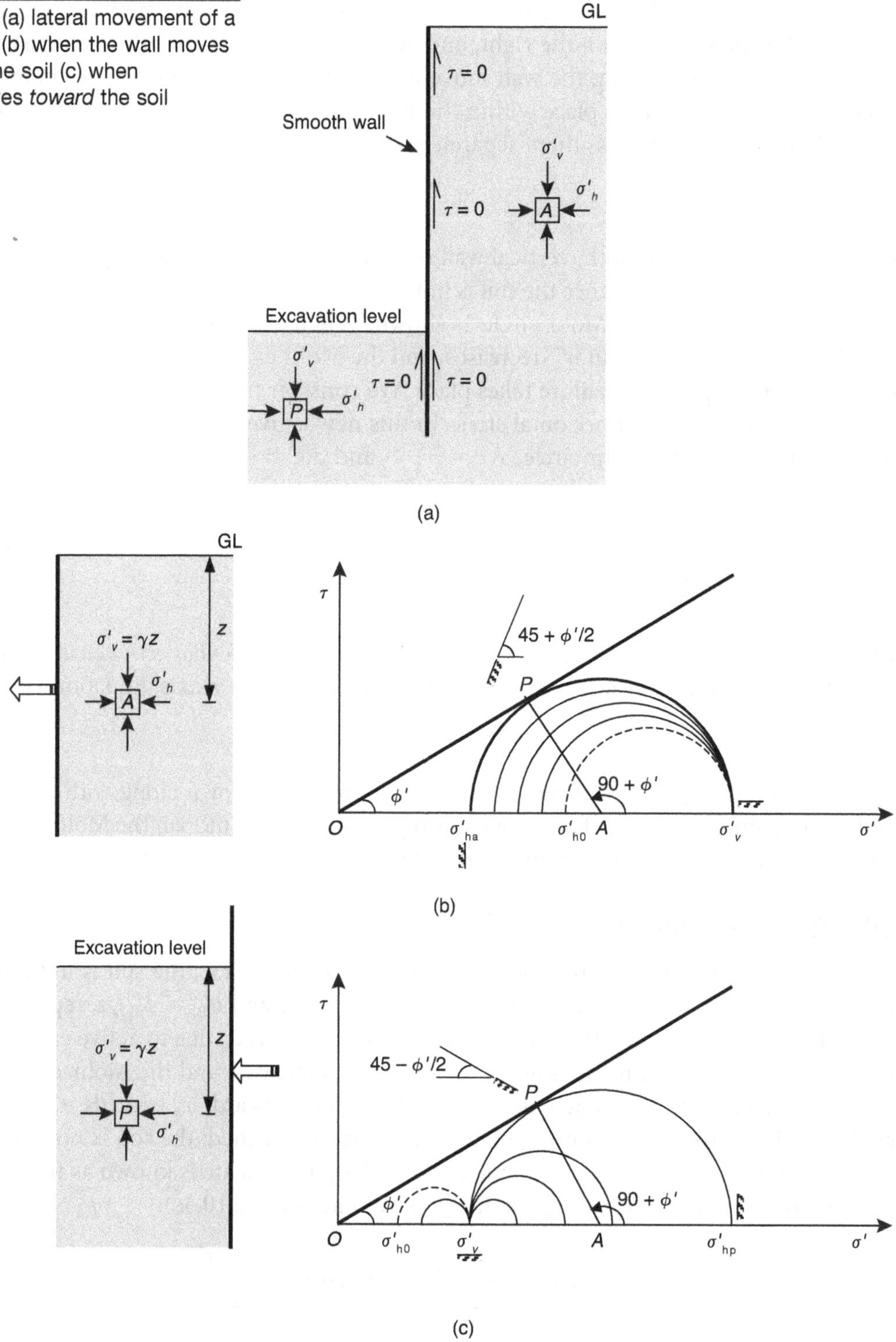

(a)

(b)

(c)

Therefore:

$$\sin\phi' = \frac{AP}{AO} = \frac{\sigma'_{hp} - \sigma'_v}{\sigma'_{hp} + \sigma'_v}$$

$$\therefore \sigma'_{hp} = K_P \sigma'_v \tag{10.8}$$

where $K_P = \left(\frac{1+\sin\phi'}{1-\sin\phi'}\right) = \tan^2(45+\phi'/2)$, known as *Rankine's coefficient of passive earth pressure*. In cohesive soils, Equation 10.8 becomes:

$$\sigma'_{hp} = K_P \sigma'_v + 2c'\sqrt{K_P} \tag{10.9}$$

The horizontal and vertical planes are shown along with the values of σ'_v and σ'_{hp} on the Mohr circle in Figure 10.3c. The failure plane is represented by point P on the Mohr circle. It can be deduced that the failure plane is inclined at $45 - \phi'/2$ degrees to horizontal.

The passive state occurs when the soil is laterally compressed to failure. The active state occurs when the soil is allowed to laterally expand to failure from the initial at-rest state. The active state occurs at every point within the soil mass to the right of the wall, and the passive state occurs at every point within the soil mass to the left of the wall, with the failure planes oriented at $45 + \phi'/2$ and $45 - \phi'/2$ degrees respectively to horizontal as shown in Figure 10.4a.

When the wall moves away from the soil, σ'_h decreases from the initial value of $\sigma'_{h0} (= K_0\sigma'_v)$ to $\sigma'_{ha} (= K_A\sigma'_v)$ at the active state, as shown in Figure 10.4b. When the wall moves toward the soil, σ'_h increases from the initial value of $\sigma'_{h0} (= K_0\sigma'_v)$ to $\sigma'_{hp} (= K_P\sigma'_v)$ at the passive state, as shown in Figure 10.4c. The active and passive earth pressures are the lower- and upper-bound values for the earth pressure at a point within the soil mass. This applies to any loading situation. The lateral movement required to fully mobilize the active (Δ active) or passive (Δ passive) state depends on the soil condition. These values are typically 0.1–2.0% of the wall height. The values are significantly less for the active state than the values for the passive state. In other words, the active state must be fully mobilized before the passive state. The weaker the soil, the larger the horizontal movement required to mobilize active and passive states. The lateral displacement can take place due to translational movement of the wall or rotation around the top or bottom of the wall. The passive earth pressure coefficient is an order of magnitude greater than the active earth pressure coefficient. For example, when $\phi' = 30°$, $K_A = 0.333$, $K_0 = 0.5$, and $K_P = 3$.

10.3.3 Lateral Pressure Distributions in Active and Passive States

The lateral earth pressure distributions on both sides of a smooth wall are shown in Figure 10.5a for a granular soil and Figure 10.5b for a cohesive soil. The heights of the retained soil are H on the right and h on the left. The entire soil masses on the right and left are assumed active and passive respectively. The unit weight of the soil is γ. In granular soil, $\sigma'_{ha} = K_A\sigma'_v = K_A\gamma z$, where z is the depth below the ground level. Therefore, the lateral pressure distribution is linear on both sides of the wall as shown in Figure 10.5a, with values of $K_A\gamma H$ and $K_P\gamma h$ at the bottom.

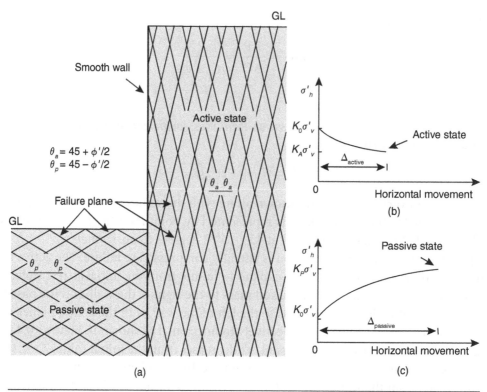

Figure 10.4 (a) failure planes (b) σ'_h variation while wall moves away from the soil (c) σ'_h variation while wall moves toward the soil

The resultant active P_A and passive P_P thrusts on the wall are the areas of the pressure diagrams, given by:

$$P_A = \frac{1}{2}K_A\gamma H^2 \tag{10.10}$$

and

$$P_P = \frac{1}{2}K_P\gamma h^2 \tag{10.11}$$

which act at heights of $H/3$ and $h/3$ respectively from the bottom of the wall.

σ'_h in cohesive soils is given by Equations 10.7 and 10.9 in active and passive states respectively. The variations of σ'_h with depth are shown in Figure 10.5b. For cohesive soils in the active state, the soil is in tension up to a depth of z_0. At the ground level ($z = 0$), the values of σ'_h in the active and passive states are $-2c'\sqrt{K_A}$ and $2c'\sqrt{K_P}$ respectively. In granular soils, they were zero. In the viewpoint of a designer, active thrust is a load and passive thrust is a resistance.

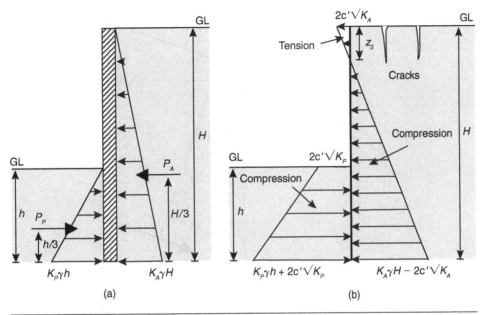

Figure 10.5 Lateral earth pressure distributions: (a) in granular soils (b) in cohesive soils

Theoretically, the tensile stresses near the ground on the right work in favor of the designer, thus reducing the resultant thrust and improving the stability. In reality, tensile cracks are likely to develop up to a depth of z_0, with little contact between the wall and the soil in this zone. Therefore, it is unwise to rely on these tensile stresses. It is a good practice to neglect the tensile zone and to *conservatively* estimate the resultant active thrust as $0.5K_A\gamma(H - z_0)^2$. The depth z_0 can be calculated as $2c'/(\gamma\sqrt{K_A})$. For clays in undrained situations, c_u and $\phi_u = 0$ should be used in Equations 10.7 and 10.9, with $K_A = K_P = 1$. The depth z_0 becomes $2c_u/\gamma$.

10.3.4 Inclined Granular Backfills

Until now, we were looking at smooth, vertical walls retaining granular and cohesive backfills where the ground level was horizontal. Let's have a brief look at smooth, vertical walls retaining granular backfills where the ground is inclined at β to horizontal as shown in Figure 10.6.

The pressure on the wall at depth z from the top, acting parallel to the slope (i.e., inclined at β to horizontal), is $K_A\gamma z$ in the active state (to the right of the wall in Figure 10.6) and $K_P\gamma z$ in the passive state. However, the coefficients K_A and K_P are now different. From Mohr circles, they are given by:

$$K_A = \cos\beta\frac{\cos\beta - \sqrt{\cos^2\beta - \cos^2\phi'}}{\cos\beta + \sqrt{\cos^2\beta - \cos^2\phi'}} \tag{10.12}$$

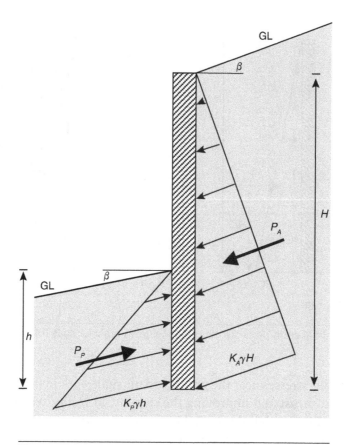

Figure 10.6 Inclined granular backfills

$$K_P = \cos \beta \frac{\cos \beta + \sqrt{\cos^2 \beta - \cos^2 \phi'}}{\cos \beta - \sqrt{\cos^2 \beta - \cos^2 \phi'}} \qquad (10.13)$$

The resultant active and passive thrusts are given by $0.5\, K_A \gamma H^2$ and $0.5\, K_P \gamma h^2$. When $\beta = 0$, Equations 10.12 and 10.13 are the same as the Rankine's coefficients of earth pressure with horizontal backfills. When $c' \neq 0$ (i.e., cohesive soils), the above equations cannot be applied. For a specific friction angle, K_A increases with β, and K_P decreases with β.

10.3.5 Effect of Uniform Surcharge

When the lateral earth pressure distributions are computed on the active and passive sides, sometimes it may be required to assess the effects of having some surcharge at the ground level. A close look at Equations 10.7 and 10.9 shows that the surcharge q at the ground level, spread over a large lateral extent, would increase σ'_v at any depth by q, and hence increase σ'_h at any depth by $K q$, where K can be K_A, K_0, or K_P, depending on the situation.

Example 10.3: A 6 m-high smooth, vertical wall retains 4 m of sandy backfill underlain by 2 m of clayey gravel. The entire soil mass is in the active state. $\phi'_{sand} = 34°$; $\phi'_{clayey\ gravel} = 31°$; and $c'_{clayey\ gravel} = 5$ kPa. If a uniform surcharge of 25 kPa is placed at the ground level on top of the retained soil mass, what would be the magnitude, direction, and location of the *additional* horizontal thrust due to this surcharge?

Solution: $K_{A,\ sand} = \tan^2\left(45 - \dfrac{34}{2}\right) = 0.283$; $K_{A,\ clayey\ gravel} = \tan^2\left(45 - \dfrac{31}{2}\right) = 0.320$

The distribution of additional σ'_h, caused by the surcharge, is shown:

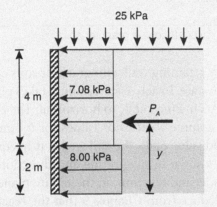

The resultant thrust, $P_A = 4 \times 7.08 + 2 \times 8.00 = 44.32$ kN per m width, acting at a height of y, given by:

$$y = \frac{(4 \times 7.08 \times 4) + (2 \times 8.00 \times 1)}{44.32} = 2.917 \text{ m}$$

10.4 COULOMB'S EARTH PRESSURE THEORY

Coulomb's (1776) limit equilibrium theory was proposed about 80 years before Rankine's, and is a little more complex. The assumptions are closer to reality, however. For example, Coulomb's theory does not assume a smooth wall and allows for friction and adhesion along the wall. It does not require that the wall be vertical. It assumes that the wall moves laterally to allow failure to take place along a plane passing through the toe of the wall (see Figure 10.7). Here, the soil wedge trapped between the retaining wall and the failure plane slides downward along the failure plane in the active state and upward along the failure plane in the passive state. A graphical procedure (discussed on page 238) is required for computing the active and passive earth pressures when the ground surface is irregular.

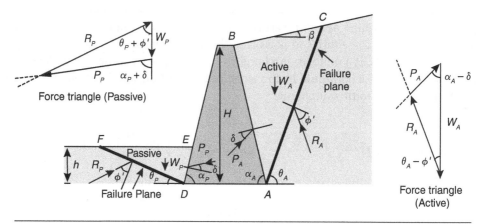

Figure 10.7 Coulomb's failure theory in granular soils

Figure 10.7 shows a gravity retaining wall with granular soils on both sides—right in the active state and left in the passive state. In active state, failure takes place when the soil wedge ABC slides along the failure plane AC inclined at θ_A to horizontal. The exact inclination of the failure plane is not known. We will assume a series of values for θ_A, and will carry out a trial-and-error process. For any assumed value of θ_A, the soil wedge is in equilibrium under three forces: self-weight of the wedge W_A, known in magnitude and direction; active thrust P_A, known in direction but not the magnitude; and reaction on the failure plane R_A, known in direction, but not the magnitude. We can deduce from Chapter 9 that the reaction R_A would be inclined at an angle of ϕ' to the normal to the failure plane. This is true on a soil-soil interface such as AC. When a soil mass slides along another material surface such as AB, this angle would be less, and is known as the angle of wall friction, denoted by δ. This angle of wall friction depends on the friction angle of the soil and the surface characteristics of the material. It can be determined from a direct shear test. For a soil-concrete interface, δ can be taken as 0.5–0.8 ϕ', with $\frac{2}{3} \phi'$ being a popular choice. δ/ϕ' is generally higher for concrete than it is for steel. The lower end of the range applies when soil is in contact with timber, steel, and precast concrete, and the upper end applies to cast-in-place concrete where the interface is relatively rough. Theoretically, $0 \le \delta \le \phi'$, with $\delta = 0$ for very smooth walls and $\delta = \phi'$ for very rough walls.

The active thrust P_A for the assumed value of θ_A can be determined by drawing a force triangle as shown in Figure 10.7. This can be repeated for several values of θ_A, against which the computed values of P_A can be plotted. The highest value of P_A is taken as the resultant active thrust on the wall.

The graphical procedure discussed above is quite similar for the passive side as well. When the computed values of P_P are plotted against the assumed values of θ_P, the lowest value of P_P is taken as the resultant passive thrust on the wall. Remember, active thrust is a load and passive thrust is a resistance. Therefore, taking the maximum value for P_A and the minimum value for P_P makes sense.

When the ground surface is inclined at β to horizontal on the active side, the resultant active thrust P_A can be shown to be $0.5 \, K_A \gamma H^2$, where K_A is given by:

$$K_A = \left(\frac{\sin(\alpha_A + \phi')/\sin \alpha_A}{\sqrt{\sin(\alpha_A - \delta)} + \sqrt{\dfrac{\sin(\phi' + \delta)\sin(\phi' - \beta)}{\sin(\alpha_A + \beta)}}} \right)^2 \tag{10.14}$$

For $\alpha_A = 90°$, $\delta = 0$, $\beta = 0$, K_A reduces to what is given by Rankine's theory for vertical walls with horizontal backfills. Coulomb's theory does not give the location of the active thrust P_A. We can assume it is acting at a height of $H/3$ from the bottom of the wall, inclined at δ to the normal to the wall-soil interface as shown in the figure.

The passive thrust P_P can be written as $0.5 \, K_P \gamma h^2$, where h is the height of point E from the bottom, and K_P is given by:

$$K_P = \left(\frac{\sin(\alpha_P - \phi')/\sin \alpha_P}{\sqrt{\sin(\alpha_P + \delta)} - \sqrt{\dfrac{\sin(\phi' + \delta)\sin(\phi' + \beta)}{\sin(\alpha_P + \beta)}}} \right)^2 \tag{10.15}$$

β is the inclination of the ground level on the passive side. For $\alpha_P = 90°$, $\delta = 0$, $\beta = 0$, K_A reduces to what is given by Rankine's theory for vertical walls with horizontal backfills.

Allowing friction along the soil-wall interface leads to a reduction in P_A and an increase in P_P from what is expected when the wall is smooth. In reality, the failure planes (or more appropriately, surfaces) are curved near the bottom of the wall, which leads to a slight underestimation of the active thrust. The error is more significant on the passive side, especially when $\delta > \phi'/3$, grossly overestimating the passive thrust. More realistic estimates of P_P can be obtained by neglecting the wall friction (i.e., $\delta = 0$) or by using Rankine's theory.

In granular soils, the soil wedges in both active and passive states are in equilibrium under three forces. In cohesive soils, it is necessary to include the *cohesive resistance* along the failure plane within the soil (*AC* or *DF*) and the *adhesive resistance* along the wall-soil interface (*AB* or *DE*). For both forces, the magnitudes and directions are known, and hence the force polygon can be drawn. The cohesive resistance is the product of the length of the failure plane (*AC* or *DF*) and cohesion. The adhesive resistance is the product of the length of the wall-soil contact plane (*AB* and *DE*) and adhesion. We defined the angle of wall friction δ as a fraction of ϕ'. A similar definition is applicable for adhesion. It can be defined as a fraction of cohesion, typically 0.5–0.7, where the fraction depends on the contact surface and whether the soil is in the active or passive state.

Reminder

❖ K_0 is defined in terms of *effective* stresses; σ_h/σ_v is not a constant.

❖ $K_0 = 1 - \sin \phi'$ in normally consolidated clays and sands; it increases with the OCR.

❖ Rankine's theory assumes that the wall is vertical and smooth. Coulomb's theory allows the wall to be inclined and friction and/or adhesion along the soil-wall interface.

❖ Rankine: For a smooth, vertical wall against a horizontal backfill, $\sigma'_{ha} = K_A\sigma'_v - 2c'\sqrt{K_A}$ and $\sigma'_{hp} = K_p\sigma'_v + 2c'\sqrt{K_p}$; failure planes are inclined at $45 + \phi'/2$ to horizontal in the active state and $45 - \phi'/2$ to horizontal in the passive state. $K_A = \left(\frac{1-\sin\phi'}{1+\sin\phi'}\right) = \tan^2(45 - \phi'/2)$ and $K_p = \left(\frac{1+\sin\phi'}{1-\sin\phi'}\right) = \tan^2(45 - \phi'/2)$. Use Equations 10.12 and 10.13 for K_A and K_p of inclined granular backfills.

❖ Coulomb's theory overestimates passive resistance significantly when $\delta > \phi'/3$. Rankine's theory is better for passive resistance, or you can assume $\delta = 0$.

WORKED EXAMPLES

1. The soil profile shown in the figure on page 241 consists of a 6 m-thick sand layer underlain by saturated clay where the water table lies 2 m below the ground level. The entire soil mass is retained by a concrete retaining wall and is in the active state. Find the total horizontal earth pressures at A, B, and C.

Solution:

For sand, $K_A = \tan^2\left(45 - \dfrac{34}{2}\right) = 0.283$

For clay, $K_A = \tan^2\left(45 - \dfrac{25}{2}\right) = 0.406$

At A:

$$\sigma'_v = 1 \times 17 = 17 \text{ kPa}$$
$$\sigma'_h = K_A \sigma'_v = 0.283 \times 17 = 4.81 \text{ kPa, and } u = 0$$
$$\therefore \sigma_h = \sigma'_h + u = 4.81 = 4.8 \text{ kPa}$$

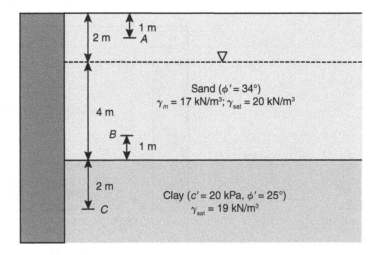

At *B*:

$$\sigma'_v = 2 \times 17 + 3 \times (20 - 9.81) = 64.6 \text{ kPa}$$
$$\sigma'_h = K_A \, \sigma'_v = 0.283 \times 64.6 = 18.3 \text{ kPa}$$
$$u = 3 \times 9.81 = 29.4 \text{ kPa}$$
$$\therefore \sigma_h = \sigma'_h + u = 18.3 + 29.4 = 47.7 \text{ kPa}$$

At *C*:

$$\sigma'_v = 2 \times 17 + 4 \times (20 - 9.81) + 2 \times (19 - 9.81) = 93.1 \text{ kPa}$$
$$\sigma'_h = K_A \, \sigma'_v - 2c'\sqrt{K_A} = 0.406 \times 93.1 - 2 \times 20 \times \sqrt{0.406} = 12.3 \text{ kPa}$$
$$u = 6 \times 9.81 = 58.9 \text{ kPa} \rightarrow \sigma_h = 12.3 + 58.9 = 71.2 \text{ kPa}$$

2. A smooth retaining wall with 2 m of embedment in the clayey sand retains a 6 m-high sandy backfill as shown in part (a) of the figure on page 242. Assuming that the entire soil mass on the right side of the wall is in the active state and the soil on the left is in the passive state, compute the active and passive thrusts on the wall.

Solution:

$$K_{A,\text{sand}} = \tan^2\left(45 - \frac{33}{2}\right) = 0.295$$

$$K_{A,\text{clayey sand}} = \tan^2\left(45 - \frac{25}{2}\right) = 0.406$$

$$K_{P,\text{clayey sand}} = \tan^2\left(45 + \frac{25}{2}\right) = 2.46$$

Let's calculate σ'_h values on the right (active) side.

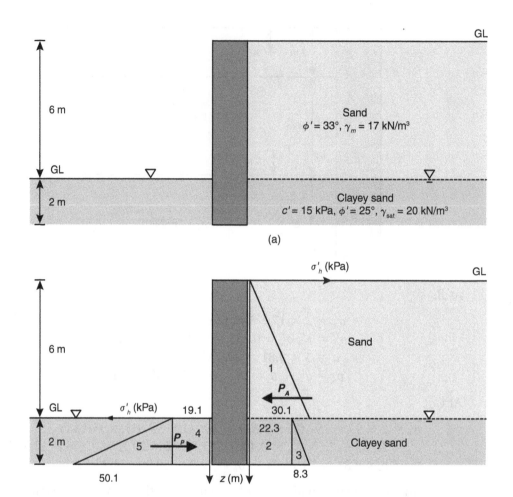

(a)

(b)

Top of sand: $\sigma'_h = 0$

Just above the water table: $\sigma'_h = 0.295 \times 6 \times 17.0 = 30.1$ kPa

Just below the water table:

$$\sigma'_h = K_A \sigma'_v - 2c'\sqrt{K_A}$$
$$= 0.406 \times 6 \times 17 - 2 \times 15 \times \sqrt{0.406} = 22.3 \text{ kPa}$$

At 2 m into the clayey sand:

$$\sigma'_h = K_A \sigma'_v - 2c'\sqrt{K_A}$$
$$= 0.406 \times [6 \times 17 + 2 \times (20 - 9.81)] - 2 \times 15 \times \sqrt{0.406}$$
$$= 30.6 \text{ kPa}$$

Now, let's calculate σ'_h values on the left (passive) side.

Top of clayey sand:

$$\sigma'_h = K_p\,\sigma'_v + 2c'\sqrt{K_p} = 2 \times 15 \times \sqrt{0.406} = 19.1 \text{ kPa}$$

At 2 m into clayey sand:

$$\sigma'_h = K_p\,\sigma'_v + 2c'\sqrt{K_p}$$
$$= 2.46 \times 2 \times (20 - 9.81) + 2 \times 15 \times \sqrt{0.406} = 69.2 \text{ kPa}$$

These values of σ'_h are plotted with depth as shown in part (b) in the figure on page 242.

Zone	Horizontal load (kN/m)	Height (m)	Moment (kN-m/m)
1	$0.5 \times 30.1 \times 6 = 90.3$	4.0	361.2
2	$22.3 \times 2 = 44.6$	1.0	44.6
3	$0.5 \times 8.3 \times 2 = 8.3$	0.667	5.5
4	$19.1 \times 2 = 38.2$	1.0	38.2
5	$0.5 \times 50.1 \times 2 = 50.1$	0.667	33.4

$P_A = 90.3 + 44.6 + 8.3 = 143.2 \text{ kN}$

$P_P = 38.2 + 50.1 = 88.3 \text{ kN}$

P_A acts at a height of $\dfrac{(361.2 + 44.6 + 5.5)}{143.2} = 2.87$ m above the bottom of the wall.

P_P acts at a height of $\dfrac{(38.2 + 33.4)}{88.3} = 0.81$ m above the bottom of the wall.

In addition to P_A and P_P, there is also the water thrust on the wall due to the pore water pressure, which is the same on both sides.

3. A vertical wall retains a granular backfill where the inclination of the ground level to horizontal is expected to be within 20°. Carry out a quantitative assessment of the possible earth pressures, assuming the backfill is in the active state, using Rankine's and Coulomb's lateral earth pressure theories.

Solution: In both Coulomb's and Rankine's earth pressure theories, the magnitude of the resultant active thrust P_A is given by $0.5\,K_A\gamma H^2$. It acts at $H/3$ from the bottom of the wall with inclination of β to horizontal according to Rankine's theory and δ to horizontal according to Coulomb's theory. Let's investigate the K_A values.

The problem below shows the plot of K_A versus ϕ' for different values of β based on Rankine's theory (Equation 10.12) and Coulomb's theory (Equation 10.14). In Equation 10.14, substituting $\alpha_A = 90°$:

$$K_A = \frac{\cos \phi'}{\sqrt{\cos \delta} + \sqrt{\dfrac{\sin(\phi' + \delta)\sin(\phi' - \beta)}{\cos \beta}}}$$

The above expression was used to develop the plot for Coulomb's K_A.

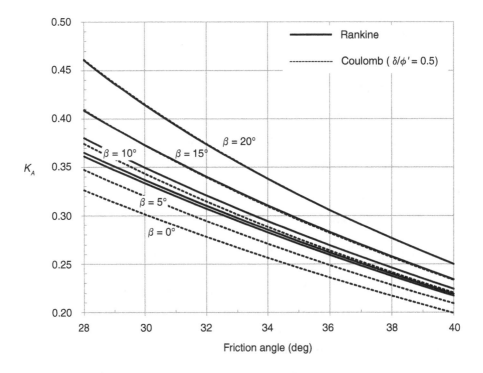

In the case of Coulomb's theory, as expected, the greater the wall friction, the lower the lateral earth pressure. Nevertheless, there is very little difference between $\delta = 0.5\ \phi'$ and $\delta = 0.8\ \phi'$, the difference being less than 2% in K_A. For $\beta = 10 - 20°$, Rankine's and Coulomb's theories give very similar values. For small values of β, Rankine's theory gives larger earth pressures, and hence is more conservative than Coulomb's theory. K_A increases with β.

4. A vertical wall retains a granular backfill where the ground level is horizontal. It is proposed to use Coulomb's earth pressure theory for computing the lateral earth pressure, assuming the backfill is in the active state. Assess the effect of δ/ϕ' on K_A.

Solution: For $\beta = 0$ and $\alpha_A = 90°$, K_A given by Equation 10.14 becomes:

$$K_A = \frac{\cos\phi'}{\sqrt{\cos\delta} + \sqrt{\sin(\phi'+\delta)\sin\phi'}}$$

The above expression for K_A was used to develop the illustration on this page for δ/ϕ' values of 0, 0.25, 0.5, 0.75, and 1.0. For $\delta/\phi' = 0$ (smooth wall), the K_A values are the same as those from Rankine's theory. It is expected that the larger the wall friction δ/ϕ', the lower the K_A. At high friction angles, there is some inconsistency when δ/ϕ' is greater than 0.25. There is about a 10% reduction in K_A when δ/ϕ' increases from 0 to 0.5, and there is little change from 0.5 to 1.0.

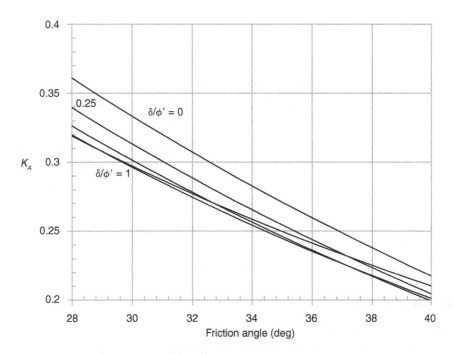

5. A smooth, vertical wall retains an inclined granular backfill. Discuss the difference between the K_A values obtained from using Rankine's (Equation 10.12) and Coulomb's (Equation 10.14) theories.

   Solution: Substituting $\delta = 0$ in Coulomb's equation does not give Rankine's K_A; they are slightly different. They are the same only when $\beta = 0$. Coulomb's K_A from Equation 10.14 becomes:

$$K_A = \frac{\cos\phi'}{1 + \sqrt{\dfrac{\sin\phi'\sin(\phi'-\beta)}{\cos\beta}}}$$

The K_A values generated for $\beta = 0$, 5°, 10°, 15°, and 20° are shown. Both Rankine's and Coulomb's theories suggest that the larger the β, the larger the K_A, which can be seen intuitively. When the wall friction is neglected, Coulomb's K_A values are slightly larger than Rankine's K_A values at all friction angles; they are the same only for $\beta = 0$.

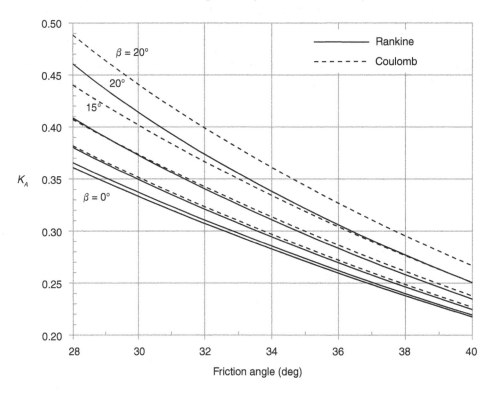

REVIEW EXERCISES

1. State whether the following are true or false.
   a. In the active state, the lateral thrust on a smooth, vertical wall retaining a horizontal backfill is greater in loose sands than it is in dense sands.
   b. In the passive state, the lateral thrust on a smooth vertical wall retaining a horizontal backfill is greater in loose sands than it is in dense sands.
   c. A smooth, vertical wall retains a granular soil, which is at-rest (K_0 state). The lateral thrust is greater if the soil is overconsolidated than if it is normally consolidated.
   d. A smooth wall retains an inclined granular backfill. The larger the inclination of the backfill, the larger the lateral thrust.
   e. Generally, Coulomb's K_A is greater than Rankine's.

   Answer: True, False, True, True, False.

2. A 5 m-high smooth, vertical wall retains a granular backfill with unit weight of 18 kN/m³ and a friction angle of 35°. Find the magnitude and location of the resultant active thrust. If a 10 kPa uniform surcharge acts at the top of the backfill, find the magnitude and location of the active thrust.

Answer: 61.0 kN/m @ 1.67 m above the bottom; 74.5 kN @ 1.82 m

3. An 8 m-high smooth, vertical wall retains a backfill where the ground level is horizontal. The top 3 m of the backfill consists of clay where $c' = 10$ kPa, $\gamma = 19$ kN/m³, and $\phi' = 23°$. The bottom 5 m is sand where $\gamma = 19$ kN/m³ and $\phi' = 33°$. Assuming the backfill is in the active state, estimate the depth up to which tension cracks would be present.

   Neglecting the tensile zone, estimate the magnitude and location of the active thrust that you would recommend.

Answer: 1.59 m; 162.3 kN/m @ 2.29 m above the bottom

4. A 10 m-high gravity retaining wall retains a granular backfill where the ground is inclined to the horizontal at 15°. The friction angle and bulk unit weight of the backfill are 34° and 18 kN/m³. The wall is inclined at 80° to horizontal. Using Coulomb's theory and assuming a wall friction angle of 20°, estimate the magnitude of the active thrust on the wall.

Answer: 371 kN/m

5. A smooth, vertical wall retains a 7 m-high granular backfill with the ground level being horizontal. The water table lies at a depth of 3 m from the top. The friction angle of the backfill is 32°. The bulk and saturated unit weight of the soil are 16.5 kN/m³ and 18.0 kN/m³ respectively. Assuming the soil is in the active state, determine the magnitude and location of the horizontal thrust on the wall.

Answer: 225 kN @ 2.30 m above the bottom

6. A 6 m-high vertical wall retains a granular backfill where the ground level is inclined at 10° to the horizontal. The bulk unit weight of the fill is 18.0 kN/m³, and the friction angle is 33°. Assuming the backfill is in the active state, determine the magnitude of the resultant thrust on the wall assuming the following:
   a. Rankine: Smooth wall
   b. Coulomb: Smooth wall
   c. Coulomb: $\delta/\phi' = 0.5$
   d. Coulomb: $\delta/\phi' = 0.67$

Answer: 99.6 kN/m, 106.2 kN, 97.6 kN, 97.0 kN

7. A 3 m-high vertical wall is pushed against a granular soil where the ground level is horizontal. The bulk unit weight and friction angle of the soil are 18.0 kN/m^3 and 34° respectively. If the soil is in the passive state, determine the horizontal thrust assuming the following:
   a. Rankine: Smooth wall
   b. Coulomb: Smooth wall
   c. Coulomb: $\delta/\phi' = 0.5$
   d. Coulomb: $\delta/\phi' = 0.67$

   Answer: 286.5 kN/m, 286.5 kN/m, 548.1 kN/m, 726.3 kN/m

8. A smooth gravity wall retains a 12 m-high backfill as shown in the figure below. The top 8 m is sand, which is underlain by some clay. The soil properties are as follows:

   Sand: $\gamma_m = 18.9$ kN/m^3, $\gamma_{sat} = 19.8$ kN/m^3; $\phi' = 32°$
   Clay: $\gamma_{sat} = 20.1$ kN/m^3; $\phi' = 18°$, $c' = 20$ kPa

   Assuming that the entire soil is in the active state, find the location and magnitude of the *total thrust* on the wall.

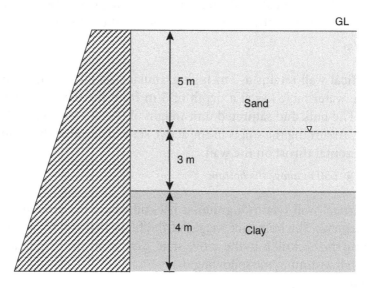

Answer: 603.6 kN/m at 3.46 m above the bottom of the wall

9. The gravity wall shown in the figure on the next page retains medium-dense sandy soil with a friction angle of 35° and a saturated unit weight of 20.0 kN/m^3. The specific gravity of the sand is 2.65 and permeability is 4.5 × 10^{-3} cm/s.
   a. Compute the flow rate beneath the wall in m^3/day per m width
   b. Find the safety factor with respect to piping

c. Compute the pore water pressure and effective vertical stress at A, B, C, D, and E
d. Estimate the total thrust on the right side of the wall, assuming that the entire soil is in the active state

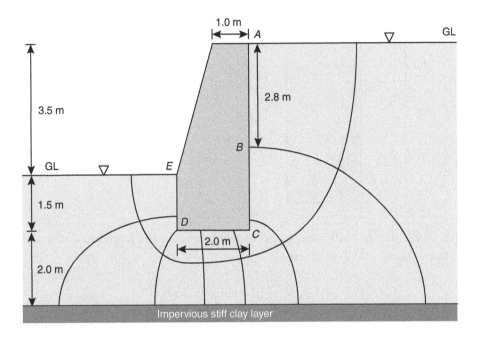

Answer: 3.9 m³/day per m; 2.3; 0 kPa, 23 kPa, 37 kPa, 25 kPa, 0 kPa; 0 kPa, 33 kPa, 63 kPa, 5 kPa, 0 kPa; 139 kN per m at 1.8 m above the bottom of the wall.

10. A rigid basement wall shown in the figure on the following page retains a granular backfill. A strip footing of width b at the ground level applies a uniform pressure of q to the underlying soil. For $q = 50$ kPa, $a = 1.5$ m, $b = 2.0$ m, and $h = 7.0$ m. Assuming the soil to be elastic ($E = 10$ MPa, $\nu = 0.25$), use *SIGMA/W* to assess the horizontal loadings on the basement wall due to the strip load.

    Assuming that the wall does not yield, the literature reports that the horizontal stress at a point A is given by:

    $$\sigma'_h = \frac{q}{\pi}(\beta - \sin\beta\cos 2\alpha)$$

    Determine if your estimates from *SIGMA/W* match the predictions from the above equation.

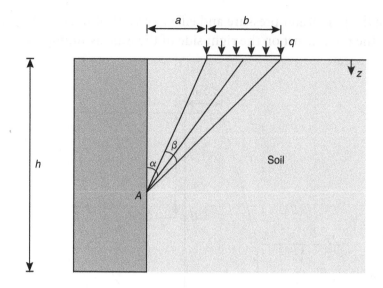

Note that in reality, the wall is expected to yield, making the horizontal stress significantly greater, the value of which is given by:

$$\sigma'_h = \frac{2q}{\pi}(\beta - \sin\beta\cos 2\alpha)$$

Site Investigation 11

11.1 INTRODUCTION

When constructing either a dam or a building at a site, it is essential to know what is beneath the surface. To ensure that the constructed facility is stable and meeting expectations during its design life, we must know the *subsoil profile* and soil characteristics before we can carry out a proper engineering analysis. Unlike most engineering materials such as concrete and steel, soils have a high degree of variability associated with their properties. The soil conditions can vary dramatically within just a few meters, making them difficult to deal with. Another difference is that we typically have the luxury of specifying the grades of steel or concrete that we have determined will meet our requirements. When it comes to soils, however, we are expected to assess and understand the soil conditions and work around them. It is not as simple as calling for a better quality soil to suit your purpose.

Site investigation (also known as *subsurface exploration* or *site characterization*) is a process that can involve many tasks including *desk study, site reconnaissance, drilling, sampling, geophysical surveys, laboratory tests*, and *in situ* (or *field*) *tests*. These tasks attempt to define the subsoil profile and determine the geotechnical characteristics of the different soils that are encountered. Depending on the nature of the project and the available budget, the site investigation can account for 0–1.5% of the total project cost. A good site investigation exercise should gather as much information as possible about the site for a minimal cost.

The desk study is the first stage of the site investigation program. This requires accessing all available information such as aerial photos and geological, topographical, and soil-survey maps. All this is accessible through federal, state, and local governmental agencies. Soil information can also be obtained from the soil data of nearby sites. Today, with Google Earth and online topographical maps from local agencies available through the Internet, substantial information including the contour levels, aerial images, vegetation, and ground water information can easily be obtained. Site reconnaissance involves a site visit with a camera to collect firsthand information on site access, exposed overburden, rock outcrops, nearby rivers or streams, vegetation, previous land use, problems with nearby structures, etc. These two stages can cost literally nothing, but play an important role in planning the detailed site investigation program.

Boreholes and *trial pits* are an integral part of any site investigation program. Boreholes are typically about 50–75 mm diameter holes—usually vertical—advanced into the ground to

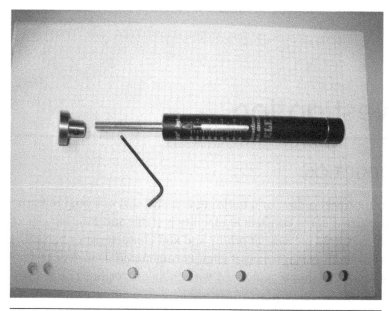

Figure 11.1 Pocket penetrometer with soft clay attachment

depths as high as 50 m or more for the purpose of obtaining samples and identifying the underlying soils. The samples are then transported to the laboratory for a series of tests such as water content and Atterberg limits determination, consolidation, triaxial test, etc. Trial pits, also known as *test pits*, are made at a few locations using an excavator or backhoe. They are relatively inexpensive, but are limited in depth. Beyond 4 m, due to shoring and bracing requirements to support the walls of the pit, the cost of trial pits can increase significantly. The advantage of a trial pit is that it enables visual inspection of the soil. Undisturbed block samples can be cut from the wall or floor of a trial pit. In clays, it is common practice to push a *pocket penetrometer* into the walls of the pit to read unconfined compressive strength. These are approximate, but they are obtained at no additional cost. Figure 11.1 shows a pocket penetrometer with an attachment for soft clays.

A typical layout of boreholes and trial pits at a proposed site is shown in Figure 11.2a. Boreholes are not always advanced to the refusal or bedrock as shown in Figure 11.2b. For smaller structures and lighter loadings, boreholes can be terminated well before reaching the bedrock. Generally, undisturbed samples are collected from clay layers only; it is very difficult to get undisturbed samples from granular soils. Figure 11.2c shows some clay cores recovered from boreholes, which are placed in a *core box* shown on the left, with the depth clearly identified and sealed to prevent moisture loss. Figure 11.2d shows undisturbed clay samples in sampling tubes that are waxed at the ends and sealed in plastic wraps to prevent moisture loss during transportation to the soil-testing laboratory.

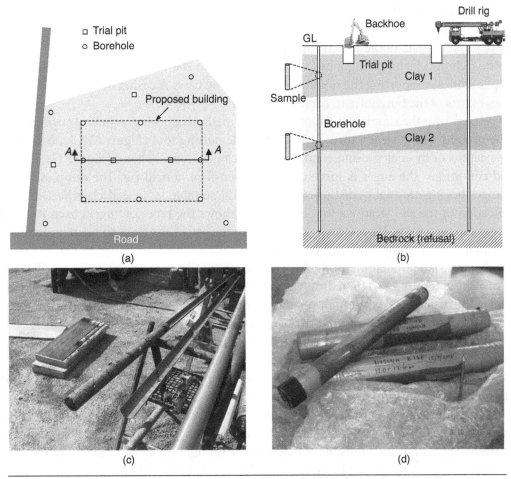

Figure 11.2 Site investigation: (a) site plan and layout of boreholes and trial pits (b) sectional elevation *A-A* (c) soil sample in tube liner and sample tray (d) sealed samples in the lab

11.2 DRILLING AND SAMPLING

Drilling, in situ testing, and sampling go hand-in-hand. Trial pits or trenches are inexpensive and are generally adequate for shallow depths and some preliminary investigations. They enable visual inspection of the stratification of the soils near the ground level. When it comes to detailed soil exploration, it is necessary to drill some boreholes to desired depths and collect samples at various depths.

11.2.1 Drilling

To prevent the borehole walls from *caving* in, especially below the water table, and to prevent the bottom of the borehole from *heaving* due to stress relief, it is common practice to fill the hole with a drilling fluid such as a 6% bentonite-water mix, at least up to the water-table level. The drilling fluid is *thixotropic*, showing very low strength when remolded and relatively high strength while at rest. While the drilling progresses, the agitation within the borehole keeps bentonite in liquid form, giving a hydrostatic pressure to the walls; when the drilling stops, it quickly solidifies, supporting the borehole walls and the base. A casing or liner can be used for the upper parts of the boreholes to prevent caving.

Auger drilling is the simplest and most common method of boring. A helical auger is screwed into the ground with a steady thrust to advance the cutting tool (Figures 11.3a, b and c). For shallow depths or in weak ground conditions, this can be done by hand (Figure 11.3d). In firm ground conditions, the auger is mechanically driven from a drill rig. The samples recovered from auger boring are highly disturbed, but are still suitable for visual classification and for identification of the soil stratification. Figure 11.3 shows the types of augers used in the field. The auger can be removed from the hole along with the soil at any stage to push sampling tubes into the hole for collecting undisturbed samples.

Wash boring is a popular method of drilling in most soils, except in gravels. A drill bit in the shape of a chisel is raised and then dropped into the borehole to cut and loosen the soil. Water is sent down the drill rod to exit at high velocity through the holes in the drill bit, washing the soil trimmings and bringing them to the surface through the annular space between the rod and the borehole wall. The water is recirculated, which allows the soil particles from the cuttings to settle in a sump. Any change in stratification can be detected from the color of the wash water.

Percussion drilling is probably the only method that is applicable in gravelly sites or wherever there is significant presence of boulders and cobbles. Here, a cutting tool in the form of a shell (or baler), clay cutter, or chisel, attached to the end of a drill rod, is repeatedly raised by 1–2 m and then dropped. The shell, clay cutter, and chisel are used in sands, clays, and rocks respectively. The trimmings and soil particles can be brought up by recirculated water.

Rotary drilling is mainly employed in rocks. Here, a drilling tool in the form of a cutting bit or coring bit is attached to the drill rod. It is rotated under pressure to advance into the soil or rock. A drilling fluid is pumped down the drill rod to cool and lubricate the cutting tool and to carry the cuttings to the surface.

Due to budget constraints, it is often necessary to limit the number of boreholes and the depth to which they are extended. Every additional borehole is an added expense for the client. They are generally spaced at intervals of 15 m (for heavy loads) to 50 m (for very light loads). Along highways, boreholes can be located at 150–500 m intervals. The boreholes should be advanced to depths where the average vertical stress increase due to the proposed structure is about 10% of the pressure applied at the surface, or where the additional vertical stress increase is about 5% of the current effective overburden stress. ASCE (1972) suggests using the smaller of the two depths.

11.2.2 Sampling

In granular soils, it is very difficult to obtain undisturbed samples from the field. There are special techniques (e.g., freezing the ground, using resins) for sampling in granular soils, but

Figure 11.3 Types of augers: (a) 500 mm diameter short-flight auger (b and c) mechanical continuous flight auger (d) mechanical handheld auger (Courtesy: Dr. K. Pirapakaran, Coffey Geotechnics)

they are very expensive. Therefore, the common practice is to rely on in situ (or field) tests to determine their geotechnical characteristics. If necessary, *reconstituted samples* can be used in the laboratory. These are laboratory samples prepared to a specific packing density to match the in situ conditions. Therefore, the sampling exercise discussed herein is relevant to sampling in *cohesive soils*.

Atterberg limits, water content, specific gravity, etc., commonly known as index properties, can be determined from remolded samples and trimmings. Nevertheless, consolidation tests and triaxial tests require good quality, undisturbed clay samples, which can come from *tube samples* or *block samples*. Tube samples (see Figures 11.2c and 11.2d) are obtained by pushing thin-walled metal tubes, about 75–100 mm in diameter and 600–900 mm in length, into the soil at desired depths. Under exceptional circumstances, very large diameter boreholes and samples are taken, but only to limited depths. These can be very expensive. Block samples are obtained from the wall or floor of an excavation or trial pit. Samples can be cut from these blocks for consolidation or triaxial tests.

Especially in cohesive soils, any disturbance during sampling can destroy the fabric, which can result in an underestimation of the strength and stiffness. Therefore, it is highly desirable to minimize the soil sample disturbance during sampling and later during the handling and transportation. The disturbance to the soil sample comes in two forms. First, when the sample is brought to the ground from a certain depth, there is a significant *stress relief*. When the sampling tube is manipulated into the borehole, there can be a *mechanical disturbance* in the sample, especially in the annular region near the wall of the sampler. While the stress relief cannot be avoided, the mechanical disturbance can be minimized. The degree of disturbance becomes greater as the wall thickness increases. An area ratio A_R is introduced to quantify the degree of mechanical disturbance as (Hvorslev 1949):

$$A_R \, (\%) = \frac{D_0^2 - D_i^2}{D_i^2} \times 100 \tag{11.1}$$

where D_i and D_o are the inner and outer diameters of the sampler. For a sample to be considered undisturbed, it is suggested that A_R be less than 10%. The most common thin-walled samplers used in practice are the 50–100 mm diameter thin-walled *Shelby tubes*™, which are seamless steel tubes often made of gauge 16 ($\frac{1}{16}$ in or 1.6 mm thick) stainless steel or galvanized steel. There are specialized samplers such as a *piston sampler* that can be used for obtaining high-quality undisturbed samples. Here, a piston at the top of a thin-walled sampler helps to retain the sample through suction while the tube is removed from the ground.

The cutting edge of a thin-walled sampler is so thin that it may not penetrate into some stiffer materials. Here, it may be necessary to use samplers with thicker walls, and thus a larger A_R, such as the split-spoon sampler from a standard penetration test discussed later. More details of samplers and sampling procedures using thin-walled samplers are discussed in ASTM D1587.

> **Example 11.1:** A thin-walled Shelby tube™ has an external diameter of 76.2 mm and a wall thickness of 1.63 mm. What is the area ratio?
>
> *Solution:*
>
> $$A_R = \frac{76.20^2 - 72.94^2}{72.94^2} \times 100 = 9.1\%$$

Figure 11.4 shows the undrained shear strength and Young's modulus data obtained in the site investigation exercise for the proposed 1000 m-high Nakheel Tower in Dubai, where the ground conditions consist of weak rocks. The undrained shear strength and Young's modulus were measured by pressuremeter tests carried out within three boreholes and undisturbed samples recovered from the site. A triple tube PQ3 coring method was used for collecting good quality cores. In spite of all the precautions and testing most of the samples on the same day, mostly due to stress relief, the undrained shear strength and Young's modulus measured from the samples in the laboratory were significantly less than the in situ values measured by the pressuremeter at all depths. In this case, the stress relief effects were quite significant in the carbonate-cemented siltstones.

11.2.3 Locating the Water Table

The location of the water table plays a key role in computing the effective stresses. Locating the water table is one of the objectives of the site investigation exercise. This can be done by observing the water table within the borehole 24 hours after drilling, when any fluctuations have stabilized. Alternatively, the water table elevation can also be measured from nearby wells. Water samples can be taken to the laboratory for a chemical analysis to detect undesirable substances (e.g., sulphates) that might be harmful to concrete.

11.3 IN SITU TESTS

In situ tests consist of inserting a device into the ground and measuring its resistance to penetration or deformation, which is then translated into strength and stiffness parameters. The most common form of in situ tests are the penetration tests (e.g., standard penetration test, cone penetration test) where an open-ended sampler or a solid cone is driven or pushed into the ground, and the resistance to penetration is measured. This resistance is translated into strength and stiffness of the soil. About 80–90% of the in situ testing exercises worldwide consist of penetration tests such as standard penetration tests or cone penetration tests. From the penetration resistance at any depth, the shear strength parameters (e.g., ϕ', c_u) and soil stiffness E can be determined.

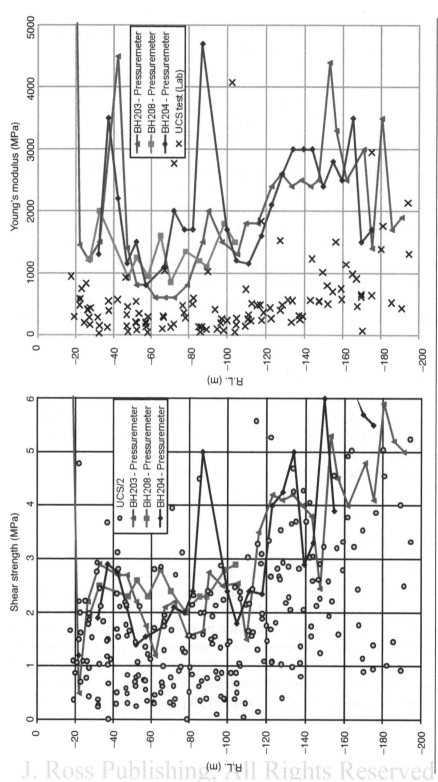

Figure 11.4 Effects of stress relief on shear strength and stiffness (Data courtesy of Dr. Chris Haberfield, Golder Associates, Australia)

In situ or field tests are carried out at the site within or outside the boreholes. Advantages of in situ tests are that they are rapid and provide a continuous record with depth in a relatively short time. There is no sampling, and therefore no sampling disturbance. The soil is tested in its in situ state, representing a larger volume. However, it is not possible to determine the soil parameters from them directly. They are determined indirectly by some empirical or semi-empirical methods. An advantage with the laboratory tests is that we have complete control of the drainage and boundary conditions and have a rational means (e.g., Mohr circles) of analyzing and interpreting the test results. In situ tests are not there to replace laboratory tests. When the in situ test data are used in conjunction with the laboratory test data, they complement each other; one should never be at the expense of the other. Let's have a look at some common in situ tests for soils.

11.3.1 Standard Penetration Test

The standard penetration test (ASTM D1586; AS 1289.6.3.1) is one of the oldest and most commonly used in situ tests in geotechnical engineering. Nicknamed the SPT, it was originally developed in the United States in 1927 for sands. A 35 mm internal diameter and 50 mm external diameter *split-barrel sampler* with a sharp cutting edge is attached to a drill rod and placed at the bottom of the borehole. The sampler is driven into the ground by a 63.5 kg hammer that is repeatedly dropped from a height of 760 mm as shown in Figure 11.5a. The number of blows required to achieve three subsequent 150 mm penetrations is recorded. The number of blows required to penetrate the final 300 mm is known as the *blow count, penetration number*, or *N-value*, and is denoted by N. The blow count for the first 150 mm is ignored due to the end effects and the disturbance at the bottom of the borehole. The split-barrel sampler, also known as the *split-spoon sampler*, is about 450–750 mm long and can be split longitudinally into two halves to recover the samples. With very thick walls and high A_R values, these samples are highly remolded and can be used only for classification purposes. Tests are carried out at 1–1.5 m intervals in a borehole, and the blow count is plotted with depth at each borehole, where the points are connected by *straight lines*.

The schematic arrangement of an SPT setup in Figure 11.5a shows an old-fashioned rotating cathead mechanism for raising and dropping the donut hammer. Today, there is an automatic tripping mechanism as shown in Figure 11.5b. Figure 11.5c shows a dynamic cone penetration test, which is very similar to the SPT, where the split-spoon sampler is replaced by a solid cone and is driven into the ground by a falling hammer. This is effective in gravels where the split-barrel sampler may sustain damage while driving. The test does not give samples.

In very fine or silty sands below the water table, the buildup of excess pore water pressures during driving reduces the effective stresses, causing an overestimation of blow counts. Here, the measured blow count must be reduced using the following equation (Terzaghi and Peck 1948):

$$N = 15 + 0.5(N_{measured} - 15) \tag{11.2}$$

Figure 11.5 Penetration tests: (a) schematic diagram of SPT (b) photograph of SPT (Courtesy of Mr. Mark Arnold) (c) photograph of a dynamic cone penetration test rig

Due to the variability associated with the choice of the SPT equipment and the test procedure worldwide, various correction factors are applied to the measured blow count N. The two most important correction factors are the overburden pressure correction C_N and the hammer efficiency correction E_h. The blow count, corrected for overburden pressure and hammer efficiency, $(N_1)_{60}$, is expressed as:

$$(N_1)_{60} = C_N E_h N \tag{11.3}$$

C_N is the ratio of the measured blow count to what the blow count *would be* at the overburden pressure of ton/sq. ft (approximately 1 kg/cm^2). Several expressions have been proposed for C_N, the most popular one being (Liao and Whitman 1986):

$$C_N = 9.78 \sqrt{\frac{1}{\sigma'_{vo}(kPa)}} \tag{11.4}$$

The actual energy delivered by the hammer to the split-spoon sampler can be significantly less than the theoretical value, which is the product of the hammer weight and the drop. Kovacs and Salomone (1982) reported that the actual efficiency of the system is between 30 and 80%. Most SPT correlations are based on a hammer efficiency of 60%, and therefore, the current practice is to accept this 60% efficiency as the standard (Terzaghi et al. 1996). Assuming that hammer efficiency is inversely proportional to the measured blow count, E_h is defined as:

$$E_h = \frac{\text{Hammer efficiency}}{60} \tag{11.5}$$

N_{60} (= $E_h N$) is the blow count corrected for hammer efficiency, but not corrected for overburden. Two other correction factors are borehole diameter correction C_b and drill rod length correction C_d, which are given in Tables 11.1 and 11.2. These are discussed in detail by Skempton (1986). Blow counts should be multiplied by these factors. When using samplers with liners, the blow count is overestimated; a further multiplication factor of 0.8 is recommended in dense sands and clays, and 0.9 in loose sands (Bowles 1988). *These correction factors must be used when defining* N_{60} *and* $(N_1)_{60}$.

The only parameter measured in the standard penetration test is the blow count and its variation with depth at every test location. In granular soils, the blow count can be translated into effective friction angle ϕ', relative density D_r, or Young's modulus E. There are several empirical correlations relating either N_{60} or $(N_1)_{60}$ to ϕ', D_r, and E.

A very popular correlation used in geotechnical engineering practice is the graphical one proposed by Peck et al. (1974) relating N_{60} and ϕ', which can be approximated as (Wolff 1989):

$$\phi' = 27.1 + 0.3 N_{60} - 0.00054 N_{60}^2 \tag{11.6}$$

The more recent correlations between N_{60} and ϕ' also account for the overburden pressure by incorporating σ'_{vo} in the equation or by simply using $(N_1)_{60}$. Schmertmann's (1975) graphical relation, $N_{60} - \phi' - \sigma'_{vo}$, can be expressed as (Kulhawy and Mayne 1990):

$$\phi' = \tan^{-1}\left[\frac{N_{60}}{12.2 + 20.3\left(\dfrac{\sigma'_{v0}}{p_a}\right)} \right]^{0.34} \tag{11.7}$$

where p_a is the atmospheric pressure (= 101.3 kPa). Hatanaka and Uchida (1996) suggested that for sands:

$$\phi' = \sqrt{20 (N_1)_{60}} + 20 \tag{11.8}$$

Table 11.1 Borehole diameter correction factor C_b (Skempton 1986)

Borehole diameter (mm)	Correction factor C_b
60–115	1.00
150	1.05
200	1.15

Table 11.2 Drill rod length correction factor C_d (Skempton 1986)

Rod length (m)	Correction factor C_d
0–4	0.75
4–6	0.85
6–10	0.95
> 10	1.00

Friction angles estimated from Equation 11.6 are quite conservative (i.e., lower) compared to those derived from Equations 11.7 and 11.8. The differences can be quite large for dense sands.

Skempton (1986) suggested that for sands with a $D_r > 35\%$:

$$\frac{(N_1)_{60}}{D_r^2} \approx 60 \tag{11.9}$$

where $(N_1)_{60}$ should be multiplied by 0.92 for coarse sands and 1.08 for fine sands. Kulhawy and Mayne (1990) suggested that:

$$\frac{(N_1)_{60}}{D_r^2} \approx (60 + 25 \log D_{50})\left(1.2 + 0.05 \log \frac{t}{100}\right) OCR^{0.18} \tag{11.10}$$

where D_{50} is the median grain size in mm, and t is the age of the soil since deposition. This gives slightly higher values for $(N_1)_{60}/D_r^2$ than 60 proposed by Skempton.

Young's modulus is an essential parameter for computing deformations, including settlements of foundations. Leonards (1986) suggested that for normally consolidated sands, E (kg/cm^2) $\approx 8 N_{60}$. Kulhawy and Mayne (1990) suggested that:

$$\frac{E}{p_a} = \alpha N_{60} \tag{11.11}$$

where $\alpha = 5$ for fine sands; 10 for clean, normally consolidated sands; and 15 for clean, over-consolidated sands.

In spite of its simplicity, rugged equipment, and its large historical database, the SPT has numerous sources of uncertainties and errors, making it less reproducible. Lately, static cone penetration tests, using piezocones, are becoming increasingly popular for better rationale, improved reproducibility, and the ability to provide continuous measurements. SPTs are not very reliable in cohesive soils due to the pore-pressure development during driving that may temporarily affect the effective stresses. For this reason, any correlations in clays should be used with caution. A rough estimate of the undrained shear strength can be obtained from (Hara et al. 1971; Kulhawy and Mayne 1990):

$$\frac{c_u}{p_a} = 0.29 N_{60}^{0.72} \tag{11.12}$$

Example 11.2: A standard penetration test was conducted at 6 m depth and the blow counts measured for 150 mm penetration are 11, 13, and 12. The SPT rig used an automatic hammer that was released through a trip mechanism, with a hammer efficiency of 72%. Find N_{60} and $(N_1)_{60}$ at this depth. Assume an average unit weight of 18 kN/m^3 for the soil, and assume that the water table is well below this depth.

Continues

Example 11.2: *Continued*

Solution: The measured $N = 13 + 12 = 25$.

$$N_{60} = 25 \times \frac{72}{60} = 30$$

$$C_N = 9.78 \times \sqrt{\frac{1}{6 \times 18}} = 0.94$$

$$\therefore (N_1)_{60} = 0.94 \times 30 = 28.2$$

In Chapter 3 (see Figure 3.3) we saw how granular soils are classified as loose, dense, etc. Figure 11.6 shows the approximate borderline values of N_{60}, $(N_1)_{60}$, ϕ', and $(N_1)_{60}/D_r^2$ for granular soils.

11.3.2 Static Cone Penetration Test

The static cone penetration test, also known as the Dutch cone penetration test, was originally developed in the Netherlands in 1920 and can be used in most soils (ASTM D3441; AS1289.6.5.1). The split-spoon sampler is replaced by a probe that consists of a solid cone with a 60° apex angle and base area of 10 cm^2, attached to a drill rod with a friction sleeve having a surface area of 150 cm^2. The probe is advanced into the soil, often jacked in by a truck, at the rate of 20 mm/s (see Figure 11.7).

Today, the cones consist of one or more porous stones at various locations (Figure 11.7a) for the measurement of pore water pressures, and are hence known as *piezocones*. Here, the three measurements that are taken continuously as the cone is pushed into the soil are cone resistance q_c, sleeve friction f_s, and pore water pressure u. Granular soils have high q_c and low f_s, while clays

	*Very loose	Loose	Medium dense	Dense	Very dense	
#D_r (%)	0	15	35	65	85	100
*N_{60}		4	10	30	50	
##$(N_1)_{60}$		3	8	25	42	
**ϕ'(deg)		28	30	36	41	
##$(N_1)_{60}/D_r^2$			65	59	58	

*Terzaghi & Peck (1948); #Gibb & Holtz (1957); ##Skempton (1986); **Peck et al. (1974)

Figure 11.6 Borderline values of D_r, N, and ϕ' for granular soils

Figure 11.7 Static cone penetration test: (a) schematic diagram of piezocone
(b) truck-mounted piezocone rig (Courtesy of Mr. Bruce Stewart, Douglas Partners)

have high f_s and low q_c. Figure 11.7b shows a soil-testing truck equipped with a static cone that is carrying out a cone penetration test at a mine site. A close-up view of the cone and the interior of the truck are shown in the insets. The friction ratio f_R at any depth, defined as:

$$f_R(\%) = \frac{f_s}{q_c} \times 100 \tag{11.13}$$

is a useful parameter in identifying the soil. Values for f_R are in the range of 0–10%, with the granular soils at the lower end and cohesive soils at the upper end of the range. Using the pair of values for q_c and f_R, the soil type can be identified from Figure 11.8. There are a few modified versions of this plot available in the literature. A sample datasheet from a piezocone test is shown in Figure 11.9, along with the soil profile, interpreted from the data in Figure 11.8.

The undrained shear strength c_u of clays can be estimated from (Schmertmann 1975):

$$c_u = \frac{q_c - \sigma_{v0}}{N_k} \tag{11.14}$$

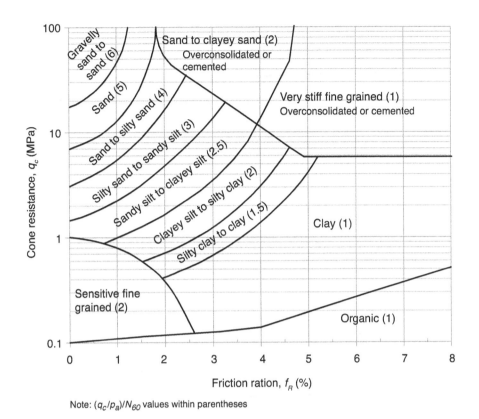

Note: $(q_c/p_a)/N_{60}$ values within parentheses

Figure 11.8 A chart for classifying soil based on static cone penetration test data (adapted from Robertson et al. 1986)

Example 11.3: For the piezocone data shown in Figure 11.9, determine the soil located at 2 m depth below the ground level.

Solution: At 2 m depth, $q_c = 1.1$ MPa and $f_s = 0.04$ MPa:

$$\therefore f_R = \frac{0.04}{1.1} \times 100 = 3.6\%$$

From Figure 11.8, the soil is possibly silty clay to clay.

---

where σ_{v0} is the *total* overburden pressure at the test depth and N_k is the cone factor that varies in the range of 14–25, which can be obtained through calibration. The lower end of the range applies to normally consolidated clays and the upper end to overconsolidated clays. The cone factor depends on the penetrometer and the type of clay, and increases slightly with the plasticity index. Based on the test data from Aas et al. (1986), N_k can be estimated by (Bowles 1988):

$$N_k = 13 + 0.11\ PI \pm 2 \tag{11.15}$$

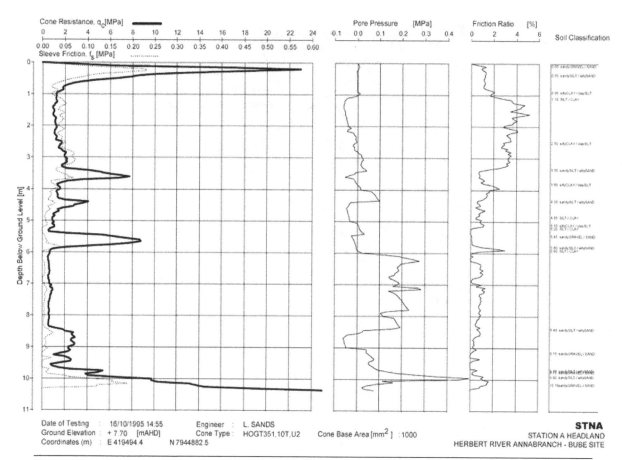

Figure 11.9 Piezocone data and soil classification (Courtesy of Mr. Leonard Sands, Venezuela)

where PI is the plasticity index of the soil. Mayne and Kemper (1988) suggested an N_k of 15 for electric cones and 20 for mechanical cones. The classification of clays based on the undrained shear strength and their corresponding consistency terms are given in Table 11.3. Also given in the table are the approximate borderline values of $(N_1)_{60}$ and q_c/p_a and the field identification guide.

Kulhawy and Mayne (1990) showed that the $q_c - \sigma'_{v0} - \phi'$ relationship in sands, proposed graphically by Robertson and Campanella (1983), can be approximated by:

$$\phi' = \tan^{-1}\left[0.1 + 0.38 \log\left(\frac{q_c}{\sigma'_{v0}}\right)\right] \tag{11.16}$$

In 1970, Schmertmann proposed that $E = 2\, q_c$ in sands, and later (Schmertmann et al. 1978) modified this to $E = 2.5\, q_c$ for axisymmetric loading and $E = 3.5\, q_c$ for plane strain loading.

Geotechnical engineers do not always have the luxury of having both the SPT and CPT data. When only one is available, it is useful to have some means of converting from one to the

Table 11.3 Consistency terms for clays with $(N_1)_{60}$ and q_c values

[#]Consistency	[#]c_u (kPa)	[#]$(N_1)_{60}$	[*]q_c/p_a	[##, **]Field identification guide
Very soft	< 12	0–2	< 5	Exudes between fingers when squeezed in hand; easily penetrated with a fist to a depth of several centimeters
Soft	12–25	2–4	5–15	Can be molded with light finger pressure; easily penetrated with the thumb to a depth of several centimeters
Firm	25–50	4–8		Can be molded with strong finger pressure; can be penetrated with a thumb using moderate effort to a depth of several centimeters
Stiff	50–100	8–15	15–30	Cannot be molded by fingers; can be indented with a thumb, but penetrated only with great effort
Very stiff	100–200	15–30	30–60	Readily indented with a thumbnail
Hard	> 200	> 30	>60	Can be indented with a thumbnail, but with difficulty

[#]Terzaghi & Peck (1948); [*]McCarthy (2007); [##]Australian Standards (1993); [**]Canadian Geotechnical Society (1992)

other. Ratios of q_c/N_{60} for different soils, as given by Sanglerat (1972) and Schmertmann (1970, 1978), are shown in Table 11.4. Robertson et al. (1983) presented the variation of q_c/N_{60} with the median grain size D_{50}, and the upper and lower bounds are shown in Figure 11.10. The soil data were limited to D_{50} less than 1 mm. Also shown in the figure are the upper and lower bounds proposed by Burland and Burbidge (1985), and the average values suggested by *The Canadian Foundation Engineering Manual* (Canadian Geotechnical Society 1992), Kulhawy and Mayne (1990) and Anagnostpoulos et al. (2003). All the curves in Figure 11.8 take the following form:

$$\frac{\left(\dfrac{q_c}{p_a}\right)}{N_{60}} = c \, D_{50}^a \tag{11.17}$$

The values of a and c are shown in Figure 11.10. Kulhawy and Mayne (1990) approximated the dependence of q_c/N_{60} ratio on D_{50} (mm) as:

$$\frac{\left(\dfrac{q_c}{p_a}\right)}{N_{60}} = 5.44 \, D_{50}^{0.26} \tag{11.18}$$

Based on an extensive database of 337 points with test data for D_{50} as high as 8 mm, Anagnostopoulos et al. (2003) noted that for Greek soils:

$$\frac{\left(\dfrac{q_c}{p_a}\right)}{N_{60}} = 7.64 \, D_{50}^{0.26} \tag{11.19}$$

Table 11.4 Ratios of q_c/N (after Sanglerat 1972; Schmertmann 1970, 1978)

Soil	q_c (kg/cm^2)/N_{60}
Silts, sandy silts, slightly cohesive silt-sand mix	2[a] (2–4)[b]
Clean, fine to medium sands and slightly silty sands	3–4[a] (3–5)[b]
Coarse sands and sands with little gravel	5–6[a] (4–5)[b]
Sandy gravel and gravel	8–10[a] (6–8)[b]

[a]Values proposed by Sanglerat (1972) and reported in Peck et al. (1974)
[b]Values suggested by Schmertmann (1970, 1978) reported by Holtz (1991) in parentheses.

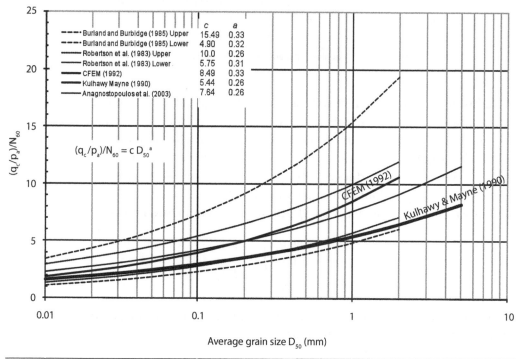

Figure 11.10 The relation between q_c and N_{60}

Kulhawy and Mayne (1990) also suggested that q_c/N_{60} can be related to the fine content in a granular soil as:

$$\frac{\left(\dfrac{q_c}{p_a}\right)}{N_{60}} \approx 4.25 - \frac{\%\ \text{fines}}{41.3} \tag{11.20}$$

In clays, the cone can be paused at any depth for carrying out a pore-pressure dissipation test to determine the consolidation and permeability characteristics. Including a geophone in the piezocone enables the measurement of shear wave velocities from which the dynamic shear modulus can be determined. Such piezocones are known as *seismic cones*.

11.3.3 Vane Shear Test

The vane shear test (ASTM D2573; AS1289.6.2.1) is used for determining undrained shear strength in clays that are particularly soft and hence vulnerable to sample disturbance. The vane consists of two rectangular metal blades that are perpendicular to each other as shown in Figure 11.11 a, b and c. The vane is pushed into the borehole to the required depth where the test is carried out (Figure 11.11a). It is rotated at the rate of 0.1° per second by applying a torque at the surface through a torque meter that measures the torque (Figure 11.11c). This rotation will initiate a shearing of the clay along a cylindrical surface surrounding the vanes. The undrained shear strength of the undisturbed clay can be determined from the applied torque T using the following equation:

$$c_u = \frac{2T}{\pi d^2 (h + d/3)}$$

(11.21)

where h and d are the height and breadth of the rectangular blades (i.e., height and diameter of the cylindrical surface sheared), which are typically of a 2:1 ratio with d in the range of

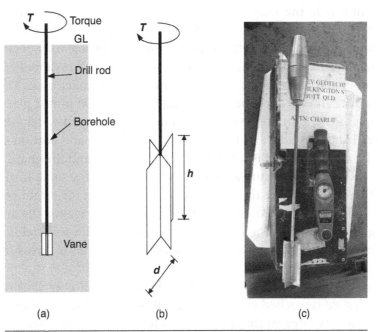

(a)   (b)   (c)

Figure 11.11 Vane shear test: (a) in a bore hole (b) vane (c) vane and torque meter (Courtesy of Dr. K. Pirapakaran, Coffey Geotechnics)

38–100 mm for the field vanes (Figure 11.11c). Miniature vanes are used in laboratories to determine the undrained shear strength of clay samples still in sampling tubes. The test can be continued by rotating the vane rapidly after shearing the clay to determine the remolded shear strength. The test can be carried out at depths as high as 50 m.

A back analysis of several failed embankments, foundations, and excavations in clays has shown that the vane shear test overestimates the undrained shear strength. A reduction factor λ has been proposed to correct the shear strength measured by vane shear test, and the correct shear strength is given by:

$$c_{u \text{ (corrected)}} = \lambda \, c_{u \text{ (vane)}} \tag{11.22}$$

where Bjerrum (1972) has proposed that:

$$\lambda = 1.7 - 0.54 \log (PI) \tag{11.23}$$

Morris and Williams (1994) suggested that for $PI > 5$:

$$\lambda = 1.18 \exp(-0.08 \, PI) + 0.57 \tag{11.24}$$

11.3.4 Pressuremeter Test

The pressuremeter test (ASTM D4719) was originally developed in France and is more popular in Europe than it is in the United States. It has several advantages over the penetration tests due to its well-defined boundary conditions and rational interpretation based on the cylindrical-cavity expansion theory. It removes a lot of empiricism associated with most of the in situ testing devices, and is hence seen as a panacea in soil testing. Pressuremeter tests can be carried out in all types of soils, including fractured or intact rocks and mines. Here, a 32–74 mm diameter cylindrical probe with a length of 400–800 mm is placed in a borehole and expanded against the borehole walls with compressed air and water. The *probe* consists of a measuring cell at the middle and two guard cells at the top and bottom ends as shown in Figure 11.12a. The measuring cell is inflated by water pressure and the guard cells are inflated by gas (typically CO_2 or N_2) pressure such that the pressure is the same in all three cells. The guard cells are there to eliminate the end effects and ensure plane strain conditions for the measuring cell. The volume V of the measuring cell is plotted against the applied pressure p as shown in Figure 11.12b, and the test is terminated when the soil yields and the volume increase is excessive, or when the volume increase is negligible. A pressuremeter probe is shown in Figure 11.12c.

The initial contact between the probe and the borehole wall is established at pressure p_i. The true in situ K_0 state is reached at p_0, and the soil starts yielding at p_f. The limit pressure p_l is achievable only at very large strains and is estimated by some extrapolation. The soil is in a pseudo-elastic state for $p_i < p < p_f$. The soil stiffness, expressed in the form of pressuremeter modulus E_p, is computed as:

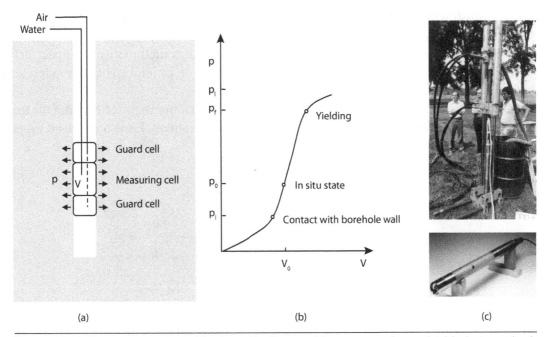

Figure 11.12 Pressuremeter test: (a) schematic diagram (b) pressure-volume plot (c) photograph of test setup and a pressuremeter

$$E_p = 2(1+v)V_0\left(\frac{dp}{dV}\right)_{p=p_0} \tag{11.25}$$

and the in situ shear modulus G is given by:

$$G = \frac{E_p}{2(1+v)} = V_0\left(\frac{dp}{dV}\right)_{p=p_0} \tag{11.26}$$

where V_0 is the volume corresponding to the in situ state where $p = p_0$. From the in situ horizontal stress p_0, the coefficient of earth pressure at rest K_0 can be determined. In sands, the pressuremeter test gives the effective friction angle ϕ'. In clays, the test gives the undrained shear strength c_u and the horizontal coefficient of consolidation c_h. The self-boring pressuremeter has a cutting tool at the bottom and does not require a prebored hole for inserting the probe, thus minimizing the disturbance due to stress relief.

11.3.5 Dilatometer Test

The flat-blade dilatometer (ASTM 6635) was developed in Italy in 1975 by Dr. Silvano Marchetti. It consists of a 240 mm-long, 95 mm-wide, and 15 mm-thick stainless steel blade with a flat, thin, expandable 60 mm diameter and 0.20–0.25 mm-thick circular steel membrane that is mounted flush with one face (Figure 11.13a). The bottom 50 mm of the blade is tapered to provide a sharp cutting edge when penetrating the soil. The blade is advanced into the soil, generally using a cone penetration test rig, at a rate of 20 mm/s, but sometimes using impact-driven hammers similar to those used in a standard penetration test. A general layout of the test setup is shown in Figure 11.13b.

At any depth, three pressure readings are taken: (a) the pressure required to bring the membrane flush with the soil surface, generally after 0.05 mm movement, known as lift-off pressure

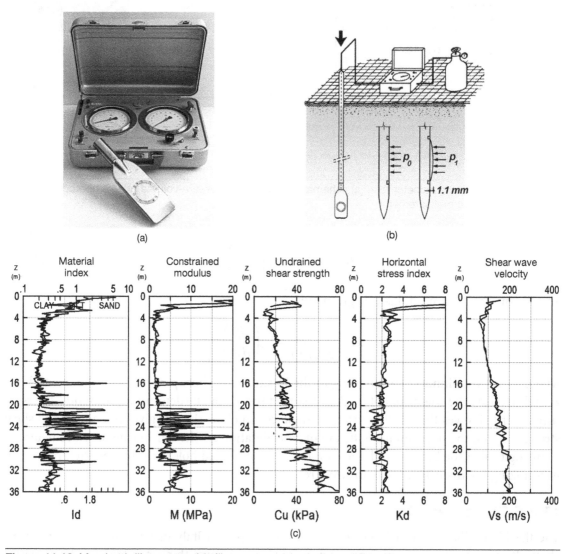

Figure 11.13 Marchetti dilatometer: (a) dilatometer and control unit (b) setup (c) sample data (Courtesy of Professor Marchetti, Italy)

or A pressure, (b) the pressure required to push the membrane laterally by 1.1 mm against the soil, known as B pressure, and (c) the pressure when the membrane is deflated, known as closing pressure or C pressure, which is a measure of the pore water pressure in the soil. The test is conducted at 200 mm depth intervals.

The interpretation of a dilatometer test is rather empirical. A material index I_D, horizontal stress index K_D, and a dilatometer modulus E_D are computed empirically. The material index, which is low for soft clays, medium for silts, and high for sands, is used to identify the soil. A horizontal stress index is used to determine horizontal stress, and hence K_0, the OCR, and the undrained shear strength c_u in clays and the effective friction angle ϕ' in sands. The dilatometer modulus is used to determine the constrained modulus, and hence the modulus of elasticity. A typical datasheet with interpretations from a dilatometer test location is shown in Figure 11.13c.

11.3.6 Borehole Shear Test

The borehole shear test was developed in the United States by Dr. Richard Handy at Iowa State University in the 1960s. Here, a direct shear test is carried out on the borehole walls to measure the drained shear strength of the in situ soil. The shear head, shown in the right of Figure 11.14a, consists of two serrated stainless steel shear plates with a total area of 10 sq. inch (6450 mm^2). The shear head is advanced into a 75 mm diameter borehole to the desired depth, and the shear plates are pushed against the borehole wall, applying a normal stress. After allowing the soil to consolidate under the applied normal stress (5 minutes in sands and 10–20 minutes in clays), the shear head is pulled upward to measure the shear strength of the soil in contact with the shear plates. From three or more test points, the Mohr-Coulomb envelope can be drawn and c' and ϕ' can be determined. Figure 11.14b shows a borehole shear test in progress, with the shear head inside the borehole attached to the control unit on the ground.

11.3.7 K_0 Stepped-Blade Test

In the 1970s the K_0 stepped-blade test for measuring lateral in situ stress and hence K_0 was also developed by Dr. Richard Handy at Iowa State University. The long blade consists of four steps, 100 mm apart, ranging from 3 mm thin to 7.5 mm thick, from its bottom to its top (Figure 11.15). Even the thickest step is thinner than the dilatometer; therefore the soil disturbance is relatively less. Each step carries a pneumatic pressure cell flush with the flat surface that comes in contact with the soil when pushed into it.

The test is conducted in a borehole where the first blade is pushed into the soil at the bottom of the hole and the pressure in the bottom step P_1 is measured. The second blade is pushed into the soil and the pressures in the bottom two steps (P_1 and P_2) are measured. This is repeated until all the steps are in the soil, giving 14 ($=1+2+3+4+4$) pressure measurements. The fifth step has the same thickness as the fourth, but with no pressure cell (see the photograph). As

(a)

(b)

Figure 11.14 Borehole shear test: (a) shear head (b) test in progress
(Courtesy of Professor David White, Iowa State University)

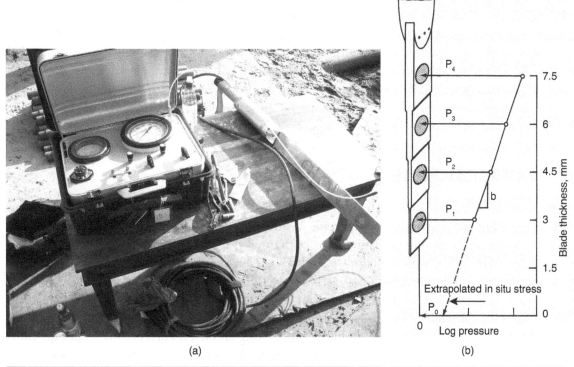

(a)   (b)

Figure 11.15 K_0 stepped-blade test (Courtesy of Professor David White, Iowa State University)

shown in Figure 11.15, the logarithm of pressure is plotted against the blade thickness. The pressure corresponding to zero blade thickness P_0 is extrapolated from the figure and is taken as the *total* in situ horizontal pressure, from which K_0 can be computed once the pore water pressure is known from the groundwater table depth. The pressure should increase with blade thickness. Any data that do not show an increase in pressure with an increase in step thickness must be discarded, and only the remaining data should be used in estimating the in situ horizontal pressure.

11.3.8 Plate Load Test

The plate load test (ASTM D1194 and ASTM D1196) is generally carried out to simulate the loadings on a prototype foundation or pavement. It involves loading a 300–500 mm square or circular plate in the site at a location and elevation where the proposed loads will be applied. The settlement is plotted against the applied pressure from which the *modulus of subgrade reaction* is obtained. The modulus of subgrade reaction is the pressure required to produce a unit settlement. The load is applied through a hydraulic jack against a horizontal reaction beam that is anchored into the ground or loaded by jacking against a *kentledge* carrying heavy weights. A kentledge is a stack of heavy weights used to keep the horizontal reaction beam from moving up while jacking in a plate load test or a pile load test (see Chapter 13). The main problem with the plate load test is the influence depth, which is only about 1–2 times the width of the loaded area. Therefore, the plate load test assesses the load-deformation characteristics at very shallow depths, whereas the actual depth of influence in the prototype structure would be significantly more. In other words, the plate load test can miss some problem soils that are present within the influence zone of the prototype foundation.

11.4 LABORATORY TESTS

Appropriate laboratory tests on disturbed and undisturbed samples collected from a site are an integral part of a site investigation exercise. While the index properties are relatively inexpensive to determine, consolidated drained or undrained triaxial tests and consolidation tests are quite expensive. When working within a limited budget, one should be prudent when selecting the number of samples for laboratory tests and deciding on the types of tests.

Index properties can be determined from the disturbed samples, including trimmings and those samples collected from the split-barrel sampler of a standard penetration test. They are useful for classification purposes, and also when using empirical correlations (e.g., Equations 8.5, 9.20) to estimate the compressibility and strength characteristics of clays, which can be useful in the absence of any other data, especially in the preliminary studies.

High-quality undisturbed samples are necessary for triaxial and consolidation tests. They come from Shelby tubes™ or special samplers such as piston samplers where the disturbance is minimal. The details of laboratory tests are discussed elsewhere. The major laboratory tests, their purposes, and the parameters derived are summarized in Table 11.5.

11.5 SITE INVESTIGATION REPORT

The in situ data pertaining to every borehole or trial pit are summarized in the form of a *bore log*, which shows the soil profile, the different layers, standard penetration test blow counts, water table depth, etc. A typical bore log of a 29 m-deep borehole is shown in Figure 11.16. Some laboratory test data such as water content, unit weight, shear strength, etc. can also be included in the bore log.

All the bore logs are collated and presented in the form of a site investigation report, which should contain the site plan with locations of all boreholes and trial pits, all laboratory test data, and any recommendations.

Table 11.5 Laboratory tests

Purpose	Laboratory test	Parameters derived
	Water content	w
Phase relation calculations	Specific gravity	G_s
	Density	ρ_m, ρ_{sat}, or ρ_d
	Grain size distribution	D_{10}, D_{30}, D_{50}, D_{60}, ...
	Sieve (coarse)	C_u, C_c
	Hydrometer (fines)	% of gravels, sands, and fines
Soil classification	Atterberg limits	
	Liquid limit	LL
	Plastic limit	PL $\rightarrow$ PI
	Linear shrinkage	LS $\rightarrow$ PI
	Compaction	$\rho_{d,max}$ and w_{opt}
Earthwork control	Field density	w and ρ_m
	Maximum/minimum density	e_{max} and $e_{min} \rightarrow D_r$
	Direct shear	c' and ϕ'; c_u and ϕ_u
	Triaxial	
	Consolidated drained	c' and ϕ'
Strength/stability analysis	Consolidated undrained	c' and ϕ'
	Unconsolidated undrained	c_u and ϕ_u
	Unconfined compression	$q_u \rightarrow c_u$
Settlement calculations	Consolidation	m_v, C_c, C_r, σ_p'; c_v; C_α
	Permeability	k
Seepage analysis	Constant head (coarse)	
	Falling head (fines)	

coffey geotechnics

Engineering Log - Borehole

Borehole No.	**XXX**
Sheet	1 of 4
Project No:	

Client:	**PORT OF BRISBANE CORPORATION**	Date started:	**13.9.2006**
Principal:		Date completed:	**13.9.2006**
Project:	**FUTURE RECLAMATION**	Logged by:	**JG**
Borehole Location:	**TERMINAL 11**	Checked by:	

drill model and mounting:	Hydrapower Scout Track	Easting: 4935	slope: -90°	R.L. Surface: 6.7
hole diameter:	100 mm	Northing: 3695	bearing:	datum:

drilling information | **material substance**

Column headers: method | penetration 1 2 3 | support | water | notes, samples, tests, etc | RL | depth metres | graphic log | classification symbol | material — soil type: plasticity or particle characteristics, colour, secondary and minor components | moisture condition | consistency/density index | pocket penetrometer kPa 100 200 300 400 | structure and additional observations

Material substance log entries:

- SP — FILL: SAND: fine to medium grained, pale brown, trace of shell. — D, M — DREDGED MATERIAL. Fill appears to behave as a very loose sand.
- SPT 2,2,4 N'=6
- SP — FILL: SAND: fine to medium grained, dark brown.
- U50
- U75 — Average pocket penetrometer test results on extruded sample is 40kPa. 160mm thick band of high plasticity dark grey clay observed in extruded sample.
- SP — FILL: SAND: fine to medium grained, grey. — W
- CH — FILL: CLAY: high plasticity, dark grey. — Fill appears to behave as a firm clay. PP=70kPa
- U75 — Fill appears to behave as a very soft clay. PP=20kPa
- SP — SAND: fine to medium grained, dark grey, some medium plasticity clay, some broken shell. — L — HOLOCENE
- SPT 6,4,4 N'=8

Depth markings: 6, 5, 4, 3, 2, 1, 0, -1 (RL) ; 1, 2, 3, 4, 5, 6, 7, 8 (depth metres)

NOT MEASURED

RCB — c

Legend:

method		support		notes, samples, tests		classification symbols and soil description	consistency/density index	
AS	auger screwing'	M mud	N nil	U50	undisturbed sample 50mm diameter	based on unified classification system	VS	very soft
AD	auger drilling'	C casing		U63	undisturbed sample 63mm diameter		S	soft
RR	roller/tricone	penetration		D	disturbed sample		F	firm
W	washbore	1 2 3 4		N	standard penetration test (SPT)	moisture	St	stiff
CT	cable tool		no resistance ranging to refusal	N'	SPT - sample recovered	D dry	VSt	very stiff
HA	hand auger			Nc	SPT with solid cone	M moist	H	hard
DT	dialube	water		V	vane shear (kPa)	W wet	Fb	friable
B	blank brt		10/1/98 water level on date shown	P	pressuremeter	Wp plastic limit	VL	very loose
V	V brt			Bs	bulk sample	WL liquid limit	L	loose
T	TC brt		water inflow	E	environmental sample		MD	medium dense
	'brt shown by suffix		water outflow	R	refusal		D	dense
e g	ADT						VD	very dense

Figure 11.16 A typical bore log (Courtesy of Dr. Jay Ameratunga, Coffey Geotechnics)

coffey geotechnics

Engineering Log - Borehole

Client:	**PORT OF BRISBANE CORPORATION**
Principal:	
Project:	**FUTURE RECLAMATION**
Borehole Location:	**TERMINAL 11**

Borehole No.	**XXX**
Sheet	2 of 4
Project No:	
Date started:	**13.9.2006**
Date completed:	**13.9.2006**
Logged by:	**JG**
Checked by:	

drill model and mounting:	Hydrapower Scout Track	Easting: 4935	slope: -90°	R.L. Surface: 6.7
hole diameter:	100 mm	Northing: 3695	bearing:	datum:

drilling information / material substance

method	penetration 1 2 3	support	water	notes samples, tests, etc	RL	depth metres	graphic log	classification symbol	material — soil type: plasticity or particle characteristics, colour, secondary and minor components.	moisture condition	consistency/ density index	pocket penetro-meter kPa 100 200 300 400	structure and additional observations
RCB	C							SP	**SAND:** fine to medium grained, dark grey, some medium plasticity clay, some broken shell. *(continued)*	W	L		
		M				-2 / 9							
				SPT 3,5,6 N'=11				CH	**CLAY:** high plasticity, dark grey.		VS		
						-3		SP	**SAND:** fine to medium grained, dark grey, some medium plasticity clay, some broken shell.		VL		
				SPT 2,1,1 N=2		10 / -4							
			NOT MEASURED	U₇₅		11		CH	**CLAY:** high plasticity, dark grey.		S		
						-5						×	PP=50kPa
				U₇₅		12					VS		
						-6						×	PP=20kPa
						13							
						-7							
				U₇₅		14					S		Average pocket penetrometer test results on extruded sample is 30kPa. PP=30kPa
						-8						×	
						15							
						-9 / 16							

method		support		notes, samples, tests		classification symbols and soil description	consistency/density index	
AS	auger screwing*	M mud N nil		U₅₀	undisturbed sample 50mm diameter	classification symbols and soil description based on unified classification system	VS	very soft
AD	auger drilling*	C casing		U₆₃	undisturbed sample 63mm diameter		S	soft
RR	roller/tricone	**penetration** 1 2 3 4		D	disturbed sample		F	firm
W	washbore	no resistance ranging to refusal		N	standard penetration test (SPT)		St	stiff
CT	cable tool			N*	SPT - sample recovered		VSt	very stiff
HA	hand auger	**water**		Nc	SPT with solid cone	**moisture**	H	hard
DT	diatube	10/1/98 water level on date shown		V	vane shear (kPa)	D dry	Fb	friable
B	blank bit			P	pressuremeter	M moist	VL	very loose
V	V bit	water inflow		Bs	bulk sample	W wet	L	loose
T	TC bit	water outflow		E	environmental sample	Wp plastic limit	MD	medium dense
*bit shown by suffix e.g. ADT				R	refusal	W∟ liquid limit	D	dense
							VD	very dense

Figure 11.16 (Continued)

❖ Laboratory and in situ tests are complements; one should not be carried out at the expense of the other.

❖ The standard penetration test is unreliable in cohesive soils. Still, there are a few empirical correlations that can be used to derive the approximate undrained shear strength.

❖ In cone penetration tests, clays have higher f_s and sands have higher q_c. As a result, clays have higher f_R and sands have lower f_R.

❖ 80–90% of in situ tests consist of standard penetration tests and cone penetration tests.

❖ The vane shear test is mainly for soft clays and determines c_u.

❖ The borehole shear test and the K_0 stepped-blade test are very specialized tests.

WORKED EXAMPLES

1. Show from the first principles that the undrained shear strength in a vane shear test is given by Equation 11.21.

   Solution: The vane is rotated quickly enough to ensure that the test is carried out under undrained conditions. The vane shears a cylindrical failure surface as shown in the figure on page 281, where the shear stress at failure is the same at the upper and lower horizontal circular areas and the vertical cylindrical surface. Let's calculate the torque resisted by the shear stresses along these surfaces.

   Cylindrical surface:

   $$T_1 = \pi dh\, \tau_f \frac{d}{2} = \frac{\pi d^2 h}{2}\tau_f$$

   One circular surface:

   $$T_2 = \int_{r=0}^{r=0.5d} \tau_f\, 2\pi r\, dr\, r = \tau_f\, 2\pi \int_0^{0.5d} r^2\, dr = \frac{\pi d^3}{12}\tau_f$$

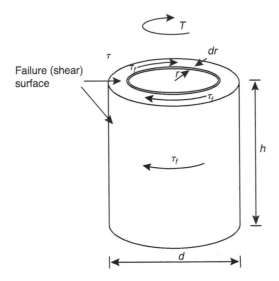

For equilibrium, the torque applied to shear the clay is given by $T = T_1 + 2\,T_2$:

$$\therefore T = \tau_f \left\{ \frac{\pi d^2 h}{2} + \frac{\pi d^3}{6} \right\}$$

The test being undrained, $\tau_f = c_u$:

$$\therefore c_u = \frac{2\,T}{\pi d^2 \left(h + \dfrac{d}{3} \right)}$$

2. In a standard penetration test in sands, the blow count measured at 10.0 m depth was 22. An automatic hammer released by a trip with an efficiency of 70% was used in the test. The unit weight of sand is 18.0 kN/m³.

   a. Find N_{60} and $(N_1)_{60}$
   b. Estimate the friction angle, relative density, and Young's modulus by all possible correlations

Solution:

   a.

$$N_{60} = 22 \times \frac{70}{60} = 25.7$$

$$C_N = 9.78 \sqrt{\frac{1}{\sigma'_{vo}(\text{kPa})}} = 9.78 \times \sqrt{\frac{1}{10 \times 18}} = 0.73$$

$$\therefore (N_1)_{60} = 0.73 \times 25.7 = 18.8$$

b. Peck et al. (1974):

$$\phi' = 27.1 + 0.3 \times 25.7 - 0.00054 \times 25.7^2 = 34.5 \text{ deg}$$

Kulhawy and Mayne (1990):

$$\phi' = \tan^{-1} \left[\frac{25.7}{12.2 + 20.3 \left(\dfrac{180}{101.3} \right)} \right]^{0.34} = 38.9 \text{ deg}$$

Hatanaka and Uchida (1996):

$$\phi' = \sqrt{20 \times 18.8} + 20 = 39.4 \text{ deg}$$

Skempton (1986):

$$\frac{(N_1)_{60}}{D_r^2} \approx 60 \rightarrow D_r^2 = \frac{18.8}{60} = 0.3133 \rightarrow D_r = 56\%$$

Leonards (1986):

$$E = 8 \times 25.7 \times 100 \text{ kPa} = 20.5 \text{ MPa}$$

Kulhawy and Mayne (1990) give similar values for E.

3. A 65 mm × 130 mm vane was pushed into a clay and rotated; the shearing occurred when the applied torque was 20.0 Nm. When the vane was further rotated to remold the clay, the torque dropped to 8.5 Nm. The plasticity index of the clay was 40. Find the undrained shear strength and the sensitivity of the clay.

What would be the maximum load that can be applied to a 50 mm diameter sample collected from this depth?

Solution: From Equation 11.21:

$$c_u = \frac{2T}{\pi d^2 \left(h + \dfrac{d}{3} \right)} = \frac{2 \times 20}{\pi \times 0.065^2 \left(0.130 + \dfrac{0.065}{3} \right)} \text{Pa} = 19870 \text{ Pa} = 19.9 \text{ kPa}$$

From Equation 11.23:

Bjerrum's correction:

$$\lambda = 1.7 - 0.54 \log \text{PI} = 1.7 - 0.54 \log 40 = 0.83$$

∴ Peak undrained shear strength = 0.83 × 19.9 = 16.5 kPa

Similarly, residual undrained shear strength = 7.0 kPa

∴ Sensitivity = 16.5/7.0 = 2.4

Unconfined compressive strength $= 2\,c_u = 33.0$ kPa

Cross-sectional area of sample $= 1963.5$ mm^2

$\therefore$ Load $= 33.0 \times 1000 \times 1963.5 \times 10^{-6}$ N $= 64.8$ N

4. A static cone penetrometer test gives the following values at 8 m depth: $q_c = 15$ MPa and $f_s = 140$ kPa. What is the soil at this depth?

Solution:

$$f_R = \frac{140}{15000} \times 100 = 0.93\%$$

From Figure 11.6, the soil is sand.

5. Estimate the friction angle and Young's modulus of the above sand in Example 4 and the equivalent blow count at this depth, assuming that the median grain size is 0.5 mm and the unit weight of the sand is 18.0 kN/m^3. The water table is deeper than 8 m.

Solution: At 8 m depth:

$$\sigma'_{v0} = 8 \times 18 = 144.0 \text{ kPa}$$

Robertson and Campanella (1983):

$$\phi' = \tan^{-1}\left[0.1 + 0.38 \log\left(\frac{15000}{144}\right)\right] = 40.9 \text{ deg}$$

Schmertmann (1970):

$$E = 2\,q_c = 2 \times 15 = 30 \text{ MPa}$$

Kulhawy and Mayne (1990):

$$\frac{\left(\dfrac{q_c}{p_a}\right)}{N_{60}} = 5.44 \times 0.5^{0.26} = 4.54$$

$$\therefore N_{60} = (15{,}000/100) \div 4.54 = 33$$

REVIEW EXERCISES

1. Compute the area ratio of a split-barrel sampler used in a standard penetration test and see if it gives good quality, undisturbed samples.
   What are the different types of hammers used in a standard penetration test? Give their approximate energy ratings.

2. What is a screw plate test? Prepare a short summary with a figure where appropriate, discussing the salient features.

3. Surf the Internet for information on seismic cone tests and write a short summary with a simple schematic diagram.

4. Carry out a literature review and discuss the advantages and disadvantages of in situ testing and laboratory testing.

5. Carry out a literature review and list five references on in situ testing of soils and two each on pressuremeter tests, dilatometer tests, borehole shear tests, and K_0 stepped-blade tests.

6. A 75 mm × 150 mm vane was pushed into a clay in a borehole and rotated. At initial shearing, the applied torque was 60 Nm. Later when the vane was rotated further, the torque was reduced to 35 Nm. The plasticity index of the clay is 35. Find the peak and residual shear strengths of the clay. What is the sensitivity of the clay?

   Answer: 33.6 kPa, 19.6 kPa; 1.7

7. A borehole shear test was carried out where the following data were measured at shear failure on the borehole walls:

Normal stress (kPa)	38.5	84.0	124.0	168.0
Shear stress (kPa)	28.0	32.0	81.0	103.0

   Find the effective cohesion and friction angle (after Handy and Spangler, 2007)

   Answer: 0 and 30°

8. A K_0 stepped-blade test was carried out in a soil and readings were obtained from all four blades at two subdepths. The readings are as follows:

Step thickness (mm)	3.0	4.5	6.0	7.5
Pressure (kPa)	110	168	190	205 at subdepth 1
	152	183	241	210 at subdepth 2

   Plot the logarithm of the measured pressure against the thickness and estimate the in situ horizontal stress.

   Answer: 60 kPa

9. The following questions are related to the bore log given in Figure 11.16.
   a. What is the predominant soil within the top 4 m?
   b. What is the predominant soil below 10 m depth?
   c. What is the reduced level at the ground level?
   d. What is the diameter of the borehole?
   e. What is the blow count from the standard penetration test at 10 m? How would you classify this sand?
   f. What is the undrained shear strength of the clay at 4.5 m depth?
   g. What is the undrained shear strength at the bottom of the borehole?
   h. How many 75 mm diameter tube samples were collected in clays?
   i. What is the difference between the N-values with and without the * sign?

10. Access http://www.gintsoftware.com and use their trial version of *gint* to prepare a bore log with as much detail as possible. What other software packages are available for this purpose? Compare them.

11. The data from a piezocone penetration test is given in the figure on page 286. Note that the cone resistance q_c is plotted to two different scales. The groundwater table lies 1 m below the ground level. Develop the soil profile for this site.

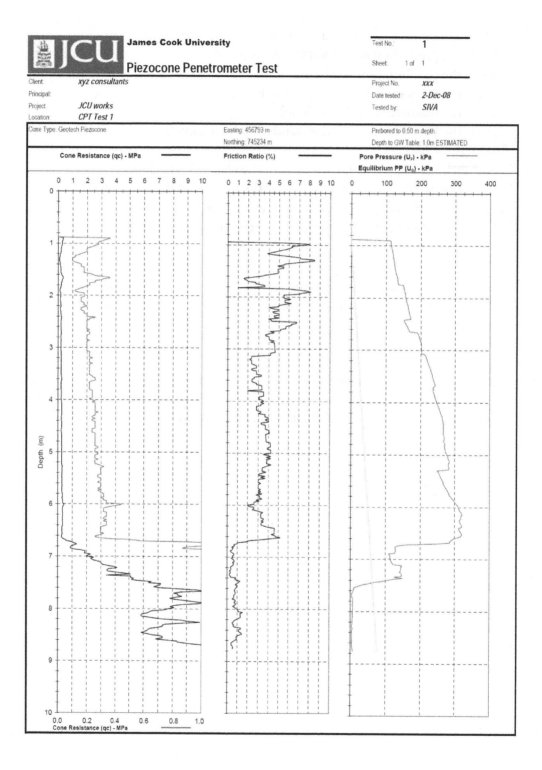

Quiz 6. Site Investigation

Duration: 20 minutes

1. State whether the following are true or false.

   a. A standard penetration test is mainly applicable to granular soils.
   b. A vane shear test is mainly applicable to soft clays.
   c. The higher the blow count, the lower the friction angle.
   d. In a cone penetration test, the friction ratio f_R is higher in granular soils than in cohesive soils.
   e. In a standard penetration test of granular soil, the higher the hammer efficiency, the higher the measured blow count.
   f. In a cone penetration test, skin friction is generally greater than the tip resistance.
   g. A clay with an unconfined compressive strength of 30 kPa will be classified as very soft clay.
   h. A sand with a relative density of 75% will be classified as dense sand (Chapter 3).

   (4 points)

2. What is a blow count or a penetration number in a standard penetration test?

   (1 point)

3. What is the parameter derived from a vane shear test?

   (1 point)

4. What parameters can be derived from the blow count N in a standard penetration test of granular soils?

   (1 point)

5. What parameters are derived from a borehole shear test?

(1 point)

6. What parameter is derived from a K_0 stepped-blade test?

(1 point)

7. What are the parameters that can be derived from a pressuremeter test?

(1 point)

Shallow Foundations 12

12.1 INTRODUCTION

Foundations are structural elements that are intended to safely transfer the loads from the structure (e.g., building, transmission tower) to the ground. The two major classes of foundations are *shallow foundations* and *deep foundations*. Shallow foundations transfer the entire load to the soil at relatively shallow depths. A common understanding is that the depth of a shallow foundation D_f must be less than the breadth B. Breadth is the shorter of the two plan dimensions. Shallow foundations include *pad* footings, *strip* (or wall) footings, and *mat* foundations as shown in Figure 12.1. Pad footings, typically 1–4 m in breadth, are placed under the columns, spreading the column loads evenly to the ground. Similarly, strip footings are placed under the walls that carry the line loads. *Combined footings* or *strap footings* carry more than one column load. Mat foundations, also known as *raft* foundations, carry multiple column and/or wall loads. When a substantial plan area of the building (e.g., more than 50%) would be occupied by isolated footings, it may be cost effective to provide a raft foundation by concreting the entire

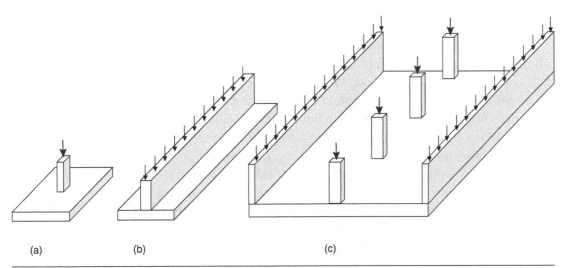

(a) (b) (c)

Figure 12.1 Types of shallow foundations: (a) pad footing (b) strip footing (c) mat or raft foundation

plan area. A typical high-rise building can apply 10–15 kPa per floor. Deep foundations have a depth greater than the breadth, and are discussed in Chapter 13.

A pad footing of plan dimensions B and L, carrying a load Q, applies a pressure of Q/BL to the underlying soil. A strip footing of width B, carrying a line load of Q kN/m, applies a pressure of Q/B to the underlying soil. The length of a strip footing is significantly greater than the breadth; hence B/L is often assumed to be zero.

Example 12.1: A 2.0 m-wide strip footing carries a wall load of 300 kN/m. What would be the pressure applied to the underlying soil?

Solution:

$$q_{applied} = \frac{300}{2} = 150 \text{ kPa}$$

12.2 DESIGN CRITERIA

Shallow foundations are generally designed to satisfy two criteria; *bearing capacity* and *settlement*. Bearing capacity criterion ensures that there is adequate protection against possible shear failure of the underlying soil; the criterion is similar to designing for the ultimate limit state, and is ensured through the provision of an adequate *factor of safety* of about three. In other words, shallow foundations are designed to carry a working load of ⅓ of the failure load. In raft foundations, a slightly lower safety factor can be recommended (Bowles 1996). Settlement criterion ensures that the settlement is within acceptable limits. For example, the pad and strip footings used in granular soils are generally designed to settle less than 25 mm. This is similar to the design for the serviceability limit state.

Why do we have to limit settlements? The building consists of a framework of slabs, beams, columns, and foundations—all of which are structural elements made of engineering materials such as concrete, steel, timber, etc. When the entire building settles equally at every location, the magnitude of settlement is of little concern. The Palace of Fine Arts, built in the early 1900s in Mexico City, settled more than 3.5 m but is still in use; it is the *differential settlement* that is a concern. When adjacent footings undergo settlements that are quite different in magnitude, the structural elements connected to these footings can undergo severe structural distress. Differential settlement is simply the difference in settlements between two nearby footings. *Angular distortion* is the ratio of the differential settlement between two adjacent columns to the span length. Limiting values of acceptable angular distortions have been reported in the literature (e.g., Lambe and Whitman 1979), with approximately 1/300 as the limit for architectural damages such as the cracking of plasters and 1/150 as the limit for structural damage. By limiting the total settlements, differential settlements and angular distortions are automatically kept in check.

Example 12.2: Two columns at a spacing of 6 m are resting on pad footings that have settled by 5 mm and 20 mm. Determine if there is excessive angular distortion.

Solution:

Differential settlement = 20 − 5 = 15 mm
Angular distortion = 15/6000 = 1/400 → within limits

12.3 BEARING CAPACITY OF A SHALLOW FOUNDATION

Prandtl (1921) modeled a narrow metal tool bearing against the surface of a block of smooth softer metal, which was later extended by Reissner (1924) to include a bearing area located *below* the surface of the softer metal. The Prandtl-Reissner plastic-limit equilibrium plane-strain analysis of a hard object that penetrates into a softer material was later extended by Terzaghi (1943) into the first rational bearing capacity equation for soil-embedded strip footings. Terzaghi assumed the soil to be a semi-infinite, isotropic, homogeneous, weightless, *rigid plastic* material and that the footing is rigid and the base of the footing is sufficiently rough to ensure there is no separation between the footing and the underlying soil. When the failure load is reached, the shear stresses are exceeded along the failure surface shown in Figure 12.2 and failure takes place.

When the foundation load is increased from zero, the settlement also increases. The applied pressure-settlement plot can take one of the three forms shown in Figure 12.3, representing three different failure mechanisms: *general shear failure* (Figure 12.3a), *local shear failure* (Figure 12.3b), and *punching shear failure* (Figure 12.3c). General shear failure is the most common mode of failure that occurs in firm ground, including dense granular soils and stiff clays, where the failure load is well-defined (see Figure 12.3a). Here, the shear resistance is fully developed along the entire failure surface that extends to the ground level as shown in Figure 12.2, and a

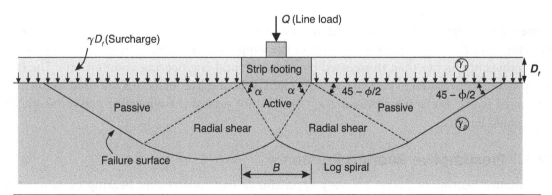

Figure 12.2 Assumed failure surface within the soil during bearing capacity failure

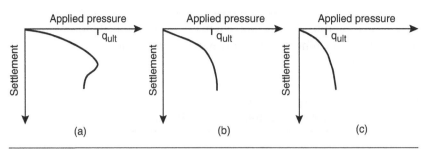

Figure 12.3 Failure modes of a shallow foundation: (a) general shear (b) local shear (c) punching shear

clearly formed heave appears at the ground level near the footing. The other extreme is punching shear failure, which occurs in weak, compressible soils such as very loose sands where the failure surface does not extend to the ground level, the failure load is not well defined, and there is no noticeable heave at the ground level (Figure 12.3c). Between these two modes, there is local shear failure (Figure 12.3b), which occurs in soils of intermediate compressibility such as medium-dense sands, where only slight heave occurs at the ground level near the footing.

In reality, the ground conditions are always improved through compaction before placing the footing. For shallow foundations in granular soils with a $D_r > 70\%$ and in stiff clays, the failure will occur in general shear mode (Vesic 1973). Therefore, it is reasonable to assume that the general shear failure mode applies in most situations.

The applied pressure at failure is known as the ultimate bearing capacity q_{ult} (Figure 12.3). This is the maximum pressure that the footing can apply to the underlying ground before failure occurs within the soil. Obviously, we want to see that the pressure applied by the footing is significantly less than the ultimate bearing capacity, thus limiting the probability of failure. The allowable bearing capacity q_{all} is defined as:

$$q_{all} = \frac{q_{ult}}{F} \tag{12.1}$$

where F is the safety factor, which is usually about 3 for shallow foundations. To account for the uncertainty in the design parameters and in the simplified theories, we use safety factors that are significantly higher than those used by our structural engineering counterparts. The high safety factor is attributed in part to the *unfactored* dead and live loads that are used to calculate the design loads. The applied pressure q_{app} should not exceed the allowable pressure—ideally, they should be equal.

12.3.1 Presumptive Bearing Pressures

Presumptive bearing pressures are *very approximate and conservative* bearing pressures that can be assumed in preliminary designs. These are given in building codes and geotechnical

Table 12.1 Presumed bearing capacity values (after BS8004:1986, Canadian Geotechnical Society 1992)

Soil type	Bearing capacity (kPa)
Rocks:	
Igneous and metamorphic rock in sound condition	10000
Hard limestone/sandstone	4000
Schist/slate	3000
Hard shale/mudstone or soft sandstone	2000
Soft shale/mudstone	600–1000
Hard sound chalk or soft limestone	600
Granular soils:	
Dense gravel or sand/gravel	> 600
Medium-dense gravel or sand/gravel	200–600
Loose gravel or sand/gravel	< 200
Dense sand	> 300
Medium-dense sand	100–300
Loose sand	< 100
Cohesive soils:	
Very stiff clays	300–600
Stiff clays	150–300
Firm clays	75–150
Soft clays and silts	< 75

textbooks (see U.S. Army 1993, Bowles 1986). Here, the specified values do not reflect the site or geologic conditions, shear strength parameters of the soil, or the foundation dimensions. Some typical values are given in Table 12.1.

Example 12.3: A square footing is required to carry a 600 kN column load in a medium-dense sand. Estimate its width.

Solution: From Table 12.1, $q_{all} = 200$ kPa

Assuming the footing width as B:

$$q_{app} = \frac{600}{B \times B} \leq 200 \text{ kPa}$$

$$\therefore B \geq 1.73 \text{ m} \rightarrow \text{Take } B \text{ as } 1.75 \text{ m}$$

12.3.2 Terzaghi's Bearing Capacity Equation

Assuming that the bearing capacity failure occurs in general shear mode, Terzaghi (1943) expressed his first bearing capacity equation for a *strip* footing as:

$$q_{ult} = c\,N_c + \gamma_1 D_f + 0.5\,B\gamma_2 N_\gamma \qquad (12.2)$$

Here, c, γ_1, and γ_2 are the cohesion and unit weights of the soil above and below the footing level respectively. N_c, N_q, and N_γ are the bearing capacity factors that are functions of the friction angle. The ultimate bearing capacity is derived from three distinct components. The first term in Equation 12.2 reflects the contribution of cohesion to the ultimate bearing capacity, and the second term reflects the frictional contribution of the overburden pressure or surcharge. The last term reflects the frictional contribution of the self-weight of the soil below the footing level in the failure zone.

For *square* and *circular footings*, the ultimate bearing capacities are given by Equations 12.3 and 12.4 respectively.

Square: $\qquad\qquad\qquad q_{ult} = 1.2\,c\,N_c + \gamma_1 D_f + 0.4\,B\gamma_2 N_\gamma \qquad (12.3)$

Circle: $\qquad\qquad\qquad q_{ult} = 1.2\,c\,N_c + \gamma_1 D_f + 0.3\,B\gamma_2 N_\gamma \qquad (12.4)$

Remember that the bearing capacity factors in Equations 12.3 and 12.4 are those of *strip* footings. In local shear failure, the failure surface is not fully developed, and thus the friction and cohesion are not fully mobilized. For this local shear failure, Terzaghi reduced the values of friction angle and cohesion to $\tan^{-1}(0.67\,\phi)$ and $0.67\,c$ respectively.

Terzaghi neglected the shear resistance provided by the overburden soil, which was simply treated as a surcharge (see Figure 12.2). Also, he assumed in Figure 12.2 that $\alpha = \phi$. Subsequent studies by several others show that $\alpha = 45 + \phi/2$ (Vesic 1973), which makes the bearing capacity factors different from what were originally proposed by Terzaghi. With $\alpha = 45 + \phi/2$, the bearing capacity factors N_q and N_c become:

$$N_q = e^{\pi \tan\phi}\,\tan^2\!\left(45 + \frac{\phi}{2}\right) \qquad (12.5)$$

$$N_c = (N_q - 1)\cot\phi \qquad (12.6)$$

The above expression for N_c is the same as the one originally proposed by Prandtl (1921), and the one for N_q is the same as the one given by Reissner (1924). While there is a consensus about Equations 12.5 and 12.6, various expressions have been proposed for N_γ in the literature, the most used being those proposed by Meyerhof (1963) and Hansen (1970). Some of these different expressions for N_γ are presented in Table 12.2. The bearing capacity equation can be applied in terms of total or effective stresses, using c' and ϕ', or c_u and ϕ_u.

Table 12.2 Expressions for N_γ

Expression	Reference
$(N_q - 1) \tan(1.4\phi)$	Meyerhof (1963)
$1.5(N_q - 1)\tan\phi$	Hansen (1970)
$2.0(N_q - 1)\tan\phi$	Eurocode 7 (EC7 1995)
$2.0(N_q + 1)$	Vesic (1973)
$1.1(N_q - 1)\tan(1.3\phi)$	Spangler & Handy (1982)
$0.1054 \exp(9.6\phi)^\#$	Davis & Booker (1971)
$0.0663 \exp(9.3\phi)^{\#\#}$	Davis & Booker (1971)

Notes: $^\#$rough footing with ϕ in radians
$^{\#\#}$smooth footing with ϕ in radians

For undrained loading in clays, when $\phi_u = 0$ it can be shown that $N_q = 1$, $N_\gamma = 0$, and $N_c = 2 + \pi (= 5.14)$. Skempton (1951) studied the variation of N_c with the shape and depth of the foundation. He showed that for strip footing, it varies from $2 + \pi$ at the surface to 7.5 at a depth greater than $4B$. For square footings, it varies between 2π at the surface and 9.0 at depth greater than $4B$. Therefore, for pile foundations, it is generally assumed that $N_c = 9$.

Most of the bearing capacity theories (e.g., Prandtl, Terzaghi) assume that the footing-soil interface is rough. Concrete footings are made by pouring concrete directly on the ground, and therefore the soil-footing interface is rough. Schultze and Horn (1967) noted that the way the concrete footings are cast in place, there is adequate friction at the base, which mobilizes friction angles equal to ϕ. Even the bottom of a metal storage tank is not smooth since the base is always treated with paint or asphalt to resist corrosion (Bowles 1996). Therefore, the assumption of a rough base is more realistic than a smooth one. Based on experimental studies, Vesic (1975) stated that foundation roughness has little effect on the ultimate bearing capacity, provided the footing load is vertical.

Meyerhof's (used predominantly in North America) and Hansen's (used in Europe) N_γ appear to be the most popular of the different expressions given for N_γ in Table 12.2. The values of N_γ, proposed by Meyerhof (1963), Hansen (1970), Vesic (1973), and Eurocode 7 (EC7 1995) are shown in Figure 12.4 along with those of N_q and N_c. For $\phi < 30°$, Meyerhof's and Hansen's values are essentially the same. For $\phi > 30°$, Meyerhof's values are larger, the difference increasing with ϕ. Indian standard recommends Vesic's N_γ (Raj 1995). *The Canadian Foundation Engineering Manual* (1992) recommends Hansen's N_γ factor.

12.3.3 Meyerhof's Bearing Capacity Equation

In spite of the various improvements to the theoretical developments proposed by Terzaghi, his original form of the bearing capacity equation is still used today because of its simplicity and practicality. Terzaghi neglected the shear resistance within the overburden soil (i.e., above the

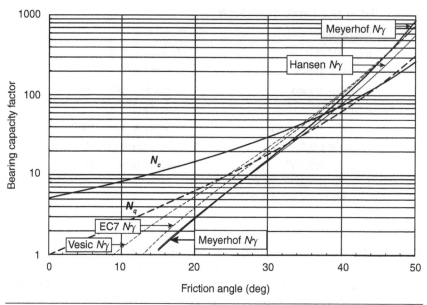

Figure 12.4 Bearing capacity factors for shallow foundations

footing level), which was included in Meyerhof's (1951) modifications, which are discussed here. Meyerhof's (1963) modifications, which are accepted worldwide, are summarized here. Meyerhof (1963) proposed the general bearing capacity equation of a rectangular footing as:

$$q_{ult} = s_c d_c i_c \, c \, N_c + s_q d_q i_q \, \gamma_1 D_f N_q + s_\gamma d_\gamma i_\gamma \, 0.5 \, B \gamma_2 N_\gamma \qquad (12.7)$$

where N_c, N_q, and N_γ are the bearing capacity factors of a *strip* footing. The shape of the footing is accounted for through the shape factors s_c, s_q, and s_γ. The depth of the footing is taken into account through the depth factors d_c, d_q, and d_γ. The inclination factors i_c, i_q, and i_γ account for the inclination in the applied load. These factors are summarized below.

Shape factors (Meyerhof 1963):

$$s_c = 1 + 0.2 \frac{B}{L} \tan^2 \left(45 + \frac{\phi}{2} \right) \qquad (12.8)$$

$$s_q = s_\gamma = 1 + 0.1 \frac{B}{L} \tan^2 \left(45 + \frac{\phi}{2} \right) \quad \text{for } \phi \geq 10° \qquad (12.9)$$

$$= 1 \quad \text{for } \phi = 0$$

Depth factors (Meyerhof 1963):

$$d_c = 1 + 0.2 \frac{D_f}{B} \tan \left(45 + \frac{\phi}{2} \right) \qquad (12.10)$$

$$d_q = d_\gamma = 1 + 0.1 \frac{D_f}{B} \tan\left(45 + \frac{\phi}{2}\right) \quad \text{for } \phi \geq 10°$$

$$= 1 \quad \text{for } \phi = 0$$

(12.11)

Inclination factors (Meyerhof 1963; Hanna and Meyerhof 1981):

$$i_c = i_q = \left(1 - \frac{\alpha°}{90}\right)^2$$

(12.12)

$$i_\gamma = \left(1 - \frac{\alpha}{\phi}\right)^2 \quad \text{for } \phi \geq 10°$$

$$= 1 \quad \text{for } \phi = 0$$

(12.13)

Here, α is the inclination (degrees) of the footing load to the vertical. Note that in spite of the load being inclined, the ultimate bearing capacity computed from Equation 12.7 provides its vertical component.

Plane-strain correction:

It has been reported by several researchers that the friction angle obtained from a plane-strain compression test ϕ_{ps} is greater than that obtained from a triaxial compression test ϕ_{tx} by about 4° to 9° in dense sands and 2° to 4° in loose sands (Ladd et al. 1977). A conservative estimate of the plane-strain friction angle may be obtained from the triaxial test by (Lade and Lee 1976):

$$\phi_{ps} = 1.5\,\phi_{tx} - 17° \quad \text{for } \phi_{tx} > 34°$$

$$= \phi_{tx} \quad \text{for } \phi_{tx} \leq 34°$$

(12.14)

Allen et al. (2004) related the peak friction angles from direct shear ϕ_{ds} and plane-strain compression tests through the following equation:

$$\phi_{ps} = \tan^{-1}(1.2 \tan \phi_{ds})$$

(12.15)

The soil element beneath the centerline of a strip footing is subjected to *plane-strain loading*, and therefore the plane-strain friction angle must be used to calculate its bearing capacity. The plane-strain friction angle can be obtained from a plane-strain compression test, which is uncommon. The loading condition of a soil element along the vertical centerline of a square or circular footing resembles more of an *axisymmetric loading* than a plane-strain one, thus requiring an axisymmetric friction angle that can be determined from a consolidated-drained or undrained-triaxial compression test.

Based on the suggestions made by Bishop (1961) and Bjerrum and Kummeneje (1961) that the plane-strain friction angle is 10% greater than that from a triaxial compression test, Meyerhof (1963) proposed the corrected friction angle for the use with rectangular footings as:

$$\phi_{rectangular\ ftg} = \left(1.1 - 0.1\frac{B}{L}\right)\phi_{triaxial} \tag{12.16}$$

Equation 12.16 simply enables interpolation between $\phi_{triaxial}$ (for $B/L=1$) and $\phi_{plane\ strain}$ (for $B/L=0$). The friction angles that are available in most geotechnical designs are derived from triaxial tests in the laboratory or in situ penetration tests. Plane-strain tests are complex and uncommon. Therefore, unless stated otherwise, it can be assumed that the friction angle is derived from axisymmetric loading conditions, and should be corrected using Equation 12.16 for rectangular or strip footings.

Eccentric loading:

When the footing is loaded with some eccentricity, the ultimate bearing capacity is reduced. Meyerhof (1963) suggested the effective footing breadth B' and length L' as $B' = B - 2\ e_B$ and $L' = L - 2\ e_L$, where e_B and e_L are the eccentricities along the breadth and length directions as shown in Figure 12.5.

For footings with eccentricities, B' and L' should be used to compute the ultimate bearing capacity (Equation 12.7) and shape factors (Equations 12.8 and 12.9). To compute the depth factors (Equations 12.10 and 12.11), B should be used. The unhatched area ($A' = B' \times L'$) in Figure 12.5 is the effective area that contributes to the bearing capacity. Therefore, the ultimate

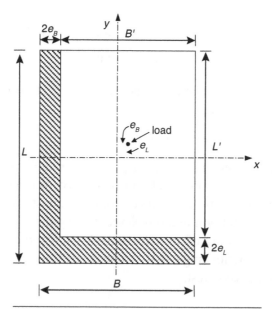

Figure 12.5 Meyerhof's eccentricity correction

footing load is computed by multiplying the ultimate bearing capacity by this area A'. When the hatched area is disregarded, the load acts at the center of the remaining area.

Meyerhof's bearing capacity equation (Equation 12.7), with the correction factors for shape, depth, and inclination, is a significant improvement from Terzaghi's equation. There are also similar approaches suggested by Hansen (1970) and Vesic (1973, 1975) where the bearing capacity equation and the correction factors are different. They have two additional sets of correction factors to account for the ground inclination (g_c, g_q, and g_γ) and base inclination (b_c, b_q, and b_γ) that cater to the footings constructed on sloping grounds and footings where the base is not horizontal.

12.3.4 Gross and Net Pressures and Bearing Capacities

The ultimate bearing capacities computed using Equations 12.2, 12.3, 12.4, and 12.7 are all *gross* ultimate bearing capacities. There is already an overburden pressure of γD_f acting at the foundation level. The *net* ultimate bearing capacity is the maximum *additional* soil pressure that can be sustained before failure. Therefore, the net ultimate bearing capacity is obtained by subtracting the overburden pressure from the gross ultimate bearing capacity. Similarly, the net applied pressure is the additional pressure applied at the foundation level in excess of the existing overburden pressure. The safety factor with respect to bearing capacity failure is generally defined in terms of the *net* values as:

$$F = \frac{q_{\text{ult,net}}}{q_{\text{app,net}}} = \frac{q_{\text{ult,gross}} - \gamma D_f}{q_{\text{app,gross}} - \gamma D_f} \tag{12.17}$$

In most spread footing designs, the gross pressures are significantly larger than the overburden pressures. In other words, the gross and net pressures are not very different as seen in most of the examples in this chapter. Only in problems involving the removal of large overburden pressures, such as buildings with basements, can gross and net pressures be quite different. The difference can be substantial when D_f is large as in the case of excavations for deep basements and rafts. In *compensated* or *floating foundations*, the net pressure applied is reduced substantially (almost to the extent of making it negligible) by increasing D_f. The safety factor for such foundations would be very high. Here, the design is governed by the settlement criterion.

In clays under undrained conditions ($\phi_u = 0$), $N_c = 5.14$, $N_q = 1$, and $N_\gamma = 0$. Therefore, the *net* ultimate bearing capacity of a shallow foundation can be written as:

$$q_{\text{ult,net}} = 5.14\, c_u \left(1 + 0.2\frac{D_f}{B}\right)\left(1 + 0.2\frac{B}{L}\right) \tag{12.18}$$

We generally use c_u and $\phi_u = 0$ for short-term stability analysis in terms of total stresses, assuming undrained conditions.

Example 12.4: In a clayey sand with $c' = 10$ kPa, $\phi' = 32°$, and $\gamma = 18$ kN/m³, a 1.5 m × 2.0 m rectangular footing is placed at a depth of 0.5 m as shown below. The unit weight of concrete is 23 kN/m³. The water table lies well below the foundation level. What is the maximum column load allowed on this footing?

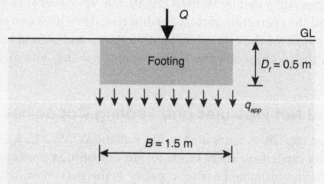

Solution:

$$\phi' = 32° \rightarrow \phi'_{rect} = \left(1.1 - 0.1\frac{1.5}{2.0}\right)32 = 32.8 \text{ deg}$$

$$N_q = 25.5, N_c = 38.0, N_{\gamma, \text{Meyerhof}} = 25.3$$

Shape factors:

$$s_c = 1 + 0.2\frac{1.5}{2.0}\tan^2(45 + 16.4) = 1.50$$

$$s_q = s_\gamma = 1.25$$

Depth factors:

$$d_c = 1 + 0.2\frac{0.5}{1.5}\tan(45 + 16.4) = 1.12$$

$$d_q = d_\gamma = 1.06$$

No inclination $\rightarrow i_c = i_q = i_\gamma = 1$

No eccentricity:

$$\therefore q_{\text{ult, gross}} = 1.50 \times 1.12 \times 10 \times 38 + 1.25 \times 1.06 \times 0.5 \times 18 \times 25.5$$
$$+ 1.25 \times 1.06 \times 0.5 \times 1.5 \times 18 \times 25.3 = 1395.0 \text{ kPa}$$

$$q_{\text{ult, net}} = 1395.0 - 18 \times 0.5 = 1386 \text{ kPa}$$

$$q_{\text{app, gross}} = \frac{Q(kN)}{1.5 \times 2.0} + 0.5 \times 23 = \frac{Q}{3} + 11.5 \text{ kPa}$$

Continues

Example 12.4: *Continued*

Applying a safety factor of 3:

$$\frac{Q}{3}+11.5\le\frac{1386}{3}\to Q\le1352\,\text{kN}$$

12.3.5 Effects of the Water Table

When computing the ultimate bearing capacity in terms of effective stress parameters, it is necessary to use the correct unit weights, depending on the location of the water table. If the water table lies at or above the ground level, γ' must be used in both bearing capacity equation terms when dealing with effective stress parameters. If the water table lies at the footing level, γ_m must be used in the second bearing capacity equation term, and γ' in the third. It can be seen from Figure 12.2 that the failure zone within the soil is confined to a depth of approximately B below the footing width. Therefore, if the water table lies at depth B or deeper beneath the footing, the bulk unit weight γ_m must be used in both bearing capacity terms. Terzaghi and Peck (1967) stated that the friction angle is reduced by 1–2° when a sand is saturated. Therefore, if a future rise in the water table is expected, the friction angle may be slightly reduced when computing the ultimate bearing capacity.

12.4 PRESSURE DISTRIBUTIONS BENEATH ECCENTRICALLY LOADED FOOTINGS

The pressure distribution beneath a *flexible* footing is often assumed to be uniform if the load is concentric, applied at the center. This is not the case when the load is applied with some eccentricity in one or both directions. Eccentricity can be introduced through moments and/or lateral loads such as wind loads. It can reduce the ultimate bearing capacity, and with the reduced effective area, the allowable load on the footing is further reduced.

On the *strip* footing shown in Figure 12.6a, a line load Q kN/m is applied with an eccentricity of e. To compute the pressure distribution beneath the footing, the eccentric line load can be replaced by a concentric line load Q kN/m and a moment Qe as shown in Figure 12.6b. The vertical pressures beneath the strip footing due to these two load components are $\frac{Q}{B}$ and $\frac{12Qe}{B^3}x$, respectively, where x is the horizontal distance to the point of interest from the centerline. Here, the moment of inertia about the longitudinal centerline for a unit length of the footing is $\frac{B^3}{12}$. Therefore, the soil pressure at any point beneath the strip footing becomes:

$$q(x)=\frac{Q}{B}\left(1+\frac{12\,e\,x}{B^2}\right) \tag{12.19}$$

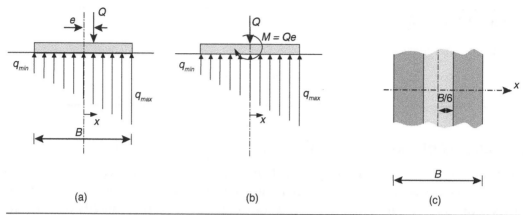

Figure 12.6 Pressure distribution beneath an eccentrically loaded strip footing: (a) eccentric load (b) equivalent concentric load with moment (c) plan view

The maximum and minimum values of the soil pressure, which occur at the two edges of the strip footing, at $x = 0.5 B$ and $x = -0.5 B$ respectively, are given by:

$$q_{max} = \frac{Q}{B}\left(1 + \frac{6e}{B}\right) \tag{12.20}$$

$$q_{min} = \frac{Q}{B}\left(1 - \frac{6e}{B}\right) \tag{12.21}$$

It can be seen from Equation 12.21 that the soil pressure beneath the footing will be compressive at all points, provided $e < B/6$. Since there cannot be tensile normal stress between the foundation and the soil when e exceeds $B/6$, one edge of the footing will lift off the ground, reducing the contact area, resulting in a redistribution of the contact pressure. It is therefore desirable to limit the eccentricity to a maximum of $B/6$, as shown by the shaded area in Figure 12.6c.

Figure 12.7a shows a rectangular footing with eccentricities of e_B and e_L in the breadth and length directions respectively. As before, the eccentric load Q can be replaced by a concentric load Q and moments $Q\,e_B$ and $Q\,e_L$ about the y and x axes respectively (see Figure 12.7b). The contact pressure at any point beneath the footing can be shown as:

$$q(x, y) = \frac{Q}{BL}\left(1 + \frac{12e_B}{B^2}x + \frac{12e_L}{L^2}y\right) \tag{12.22}$$

Here, the origin is at the center of the footing and the x and y axes are in the directions of breadth and length respectively. The shaded area at the center—a rhombus—is known as the *kern*. Provided the foundation load acts within this area, the contact stresses are compressive at all points beneath the footing.

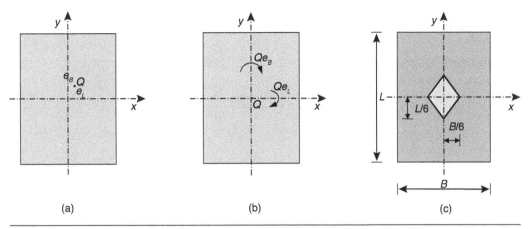

Figure 12.7 Two-way eccentricity in a rectangular footing: (a) eccentric load (b) concentric load with moments (c) kern

Example 12.5: A column load of Q is applied on a rectangular footing of dimensions B and L, with eccentricities of B/12 and L/16. Draw the pressure distribution around the perimeter and find the contact pressure at the center and the maximum and minimum pressures.

Solution: Substituting $e_B = B/12$ and $e_L = L/16$ in Equation 12.22 gives:

$$q(x, y) = \frac{Q}{BL}\left(1 + \frac{1}{B}x + \frac{0.75}{L}y\right)$$

At A, $x = 0.5B$ and $y = 0.5L \rightarrow$

$$q_A = \frac{Q}{BL}(1 + 0.5 + 0.375) = 1.875\frac{Q}{BL}$$

At B, $x = 0.5B$ and $y = -0.5L \rightarrow$

$$q_B = \frac{Q}{BL}(1 + 0.5 - 0.375) = 1.125\frac{Q}{BL}$$

At C, $x = -0.5B$ and $y = -0.5L \rightarrow$

$$q_C = \frac{Q}{BL}(1 - 0.5 - 0.375) = 0.125\frac{Q}{BL}$$

At D, $x = -0.5B$ and $y = 0.5L \rightarrow$

$$q_D = \frac{Q}{BL}(1 - 0.5 + 0.375) = 0.875\frac{Q}{BL}$$

Continues

Example 12.5: *Continued*

At the center, $x = 0, y = 0 \rightarrow q = \dfrac{Q}{BL}$

$$q_{max} = q_A = 1.875 \frac{Q}{BL}$$

and

$$q_{min} = q_C = 0.125 \frac{Q}{BL}$$

The pressure distribution around the perimeter is shown:

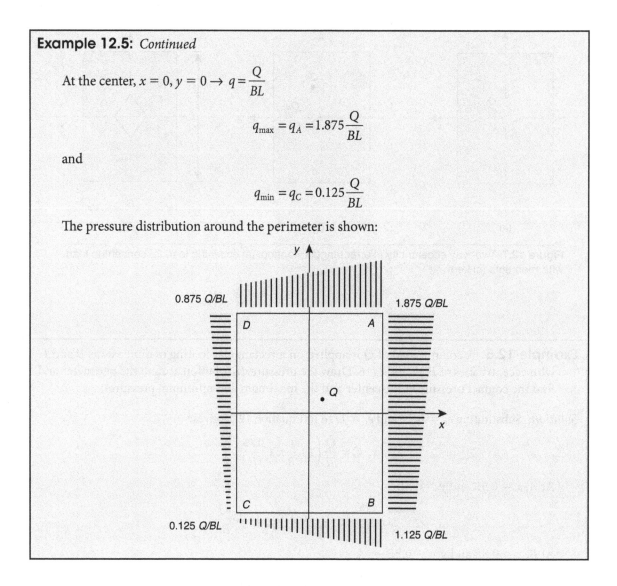

12.5 INTRODUCTION TO RAFT FOUNDATION DESIGN

A *raft* foundation, also known as a *mat* foundation, is a large, thick concrete slab supporting all or some of the columns and/or walls of a structure. Rafts can also support entire structures such as silos, storage tanks, chimneys, towers, and machinery. Hollow rafts can reduce the heavy self-weight of a large slab and still provide sufficient structural stiffness. A widely accepted practical criterion is to use rafts when more than 50% of the building plan area is covered by isolated footings. Compared to isolated footings, a raft spreads the structural load over a larger area and reduces the bearing pressure. Because of the high stiffness of the thick concrete slab, rafts can reduce differential settlements.

The bearing capacity computations for raft foundations are similar to those of the pad or strip footings discussed in previous sections of this chapter. For clays under undrained conditions, Equation 12.18 can be used to compute the net ultimate bearing capacity of a raft. Generally, due to the size of the raft, the safety factor with respect to the bearing capacity failure in sands is quite large. Extending Meyerhof's (1956) work, Bowles (1988) proposed an empirical relation for estimating the net allowable bearing capacity of shallow foundations in sands as:

$$q_{all,net}(kPa) = 12.5 N_{60} \left(\frac{B+0.3}{B} \right)^2 \left(1 + \frac{1}{3} \frac{D_f}{B} \right) \left(\frac{\text{maximum settlement (mm)}}{25} \right) \quad (12.23)$$

In rafts, total settlements as high as 50 mm can be allowed while differential settlements are still within tolerable limits. This is about twice the total settlement allowed for isolated footings in granular soils.

Example 12.6: A 10 m × 12 m raft is placed 5 m below the ground level in a clay with $c_u = 50$ kPa and $\gamma = 18.5$ kN/m³. For undrained conditions, find the net allowable bearing capacity.

How effective is it to increase the raft width and length to increase the net allowable bearing capacity?

Solution: From Equation 12.18:

$$q_{ult,net} = 5.14\, c_u \left(1 + 0.2 \frac{D_f}{B} \right) \left(1 + 0.2 \frac{B}{L} \right) = 5.14 \times 50 \left(1 + 0.2 \frac{5}{10} \right) \left(1 + 0.2 \frac{10}{12} \right) = 330 \text{ kPa}$$

With $F = 3$, $q_{all,net} = 110$ kPa

Increasing B and L has a negligible effect in increasing $q_{all,net}$ in undrained clays; it helps to reduce the net applied pressure by spreading the load over a larger area.

The structural design of a raft foundation can be carried out in two ways: the *rigid method* and the *flexible method*. These are briefly discussed below.

12.5.1 Rigid Method

The rigid method, also known as the *conventional method*, is more popular due to its simplicity. Here, the raft is assumed rigid and the settlement translational or rotational; there is no bending. For rigid, rectangular rafts with area $B \times L$, the contact pressure q at any point beneath the raft with coordinates x and y with respect to a Cartesian coordinate system passing through the centroid of the raft area (see Figure 12.8), with the axes parallel to the edges, is given by:

$$q(x, y) = \frac{Q_t}{BL} + \frac{M_x}{I_x} y + \frac{M_y}{I_y} x \qquad (12.24)$$

where $Q_t = \Sigma Q_i$ = total column loads acting on the raft; $M_x = Q_t\, e_y$ = moment of the column loads about the x-axis; $M_y = Q_t\, e_x$ = moment of the column loads about the y-axis; e_x, e_y = eccentricities about the y and x axes respectively; $I_x = BL^3/12$ = moment of inertia about the x-axis; $I_y = LB^3/12$ = moment of inertia about the y-axis. The maximum net contact pressure

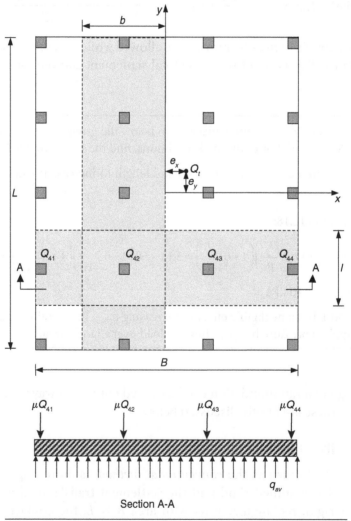

Figure 12.8 Raft foundation design as a two-way slab

computed from Equation 12.24 must be less than the net allowable bearing capacity of the raft. It can be seen from Equation 12.24 that the pressure distributions along the x and y directions are linear.

Static equilibrium in the vertical direction causes the resultant of column loads Q_t to be equal and opposite to the resultant load obtained from integration of the reactive contact pressure in Equation 12.24. For simplicity, the rigid method suggests that the raft be analyzed by tributary areas in each of the two perpendicular directions, similar to the structural design of an inverted two-way flat slab, as shown by the shaded areas in Figure 12.8.

To calculate bending moments and shear forces, each of the two perpendicular bands is assumed to be an independent, continuous beam under constant average upward pressure q_{av}, estimated by Equation 12.24.

This simplification violates equilibrium, because bending moments and shear forces at the common edge between adjacent bands are neglected. Therefore, the contact pressure obtained by dividing the sum of the column loads in each band by the total area of the band is not equal to q_{av}, as computed by Equation 12.24. Therefore, all loads are multiplied by a factor μ as shown in Figure 12.8 such that $q_{av} \times B \times l = \mu \, \Sigma Q_{4i}$, ensuring equilibrium.

12.5.2 Flexible Method

Flexible methods are based on analytical linear-elastic solutions and numerical solutions such as finite differences and finite elements, where the stiffness of both soil and structural members can be taken into account. Early flexible numerical methods are based on the numerical solution of the fourth order differential equation governing the flexural behavior of a plate by the method of finite differences. The raft is treated as a linear elastic structural element whose soil reaction is replaced by an infinite number of independent linear elastic springs following the Winkler hypothesis. The elastic constant of these springs is given by the *coefficient of subgrade reaction k_s*, also known as the *modulus of subgrade reaction* or the *subgrade modulus*, defined as the ratio of applied pressure to settlement. The pressure distribution is non-linear.

Figure 12.9a shows an infinitely long beam of width b (m) and thickness h (m) resting on the ground and is subjected to some point loads where the soil reaction is q (kN/m) at distance x from the origin. Here, the soil reaction is nonuniformly distributed along the length of the beam. From engineering mechanics principles, it can be shown that:

Bending moment at x:

$$M(x) = E_F I_F \frac{d^2 z}{dx^2} \tag{12.25}$$

Shear force at x:

$$V(x) = \frac{dM}{dx} = E_F I_F \frac{d^3 z}{dx^3} \tag{12.26}$$

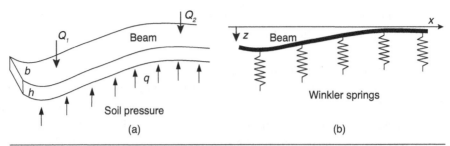

Figure 12.9 (a) flexible beam resting on soil (b) soil pressure replaced by Winkler springs

Soil reaction at x:

$$q(x) = \frac{dV}{dx} = E_F I_F \frac{d^4 z}{dx^4} = -z\,k' \qquad (12.27)$$

Here, E_F = Young's modulus of the foundation beam, $I_F = bh^3/12$ = moment of inertia of the cross section of the beam about the neutral axis, and z = vertical deflection of the beam at x and k' (kN/m^2) = subgrade reaction of the Winkler *beam* (Figure 12.9b). Note the difference between k' and k_s; k' is for the *beam*, expressed in kN/m per m, and k_s is for the loaded *area*, expressed in kPa per m.

k' (kN/m^2) and k_s (kN/m^3) are related by:

$$k' = k_s\,b \qquad (12.28)$$

Therefore, Equation 12.27 becomes:

$$E_F I_F \frac{d^4 z}{dx^4} = -z\,k_s b \qquad (12.29)$$

Solving the governing differential Equation 12.29, deflection z is given by:

$$z = e^{-ax}\,(C_1 \cos\beta x + C_2 \sin\beta x) \qquad (12.30)$$

where C_1 and C_2 are constants; and β, with the unit of length^{-1}, is an important parameter given by:

$$\beta = \sqrt[4]{\frac{b\,k_s}{4\,E_F I_F}} \qquad (12.31)$$

According to The American Concrete Institute Committee 336 (1988), the mat should be designed by the rigid method if the column spacing in a strip is less than $1.75/\beta$. If the spacing is greater than $1.75/\beta$, the flexible method may be used.

Example 12.7: A 3 m-wide and 450 mm-thick tributary strip from a raft footing applies an average contact pressure of 250 kPa to the underlying sandy soil and is expected to settle 15 mm. Find the modulus of subgrade reaction k_s. If $E_{concrete}$ = 30 GPa, up to what column spacing should this strip be designed by the rigid method?

Solution:

$$k_s = \frac{250\,(\mathrm{kN/m^2})}{0.015\,\mathrm{m}} = 16.7\,\mathrm{MN/m^3}$$

$$I_F = \frac{(3.0)(0.450)^3}{12} = 0.0228\,\mathrm{m^4}$$

$$\beta = \sqrt[4]{\frac{3.0 \times 16.7 \times 10^6}{4 \times 30 \times 10^9 \times 0.0228}} = 0.37\,\mathrm{m^{-1}} \rightarrow 1.75/\beta = 4.73\,\mathrm{m}$$

---

k_s can be determined from a plate loading test. Vesic (1961) suggested that:

$$k_s = \frac{0.65\,E_s}{B(1-v_s^2)} \sqrt[12]{\frac{E_S B^4}{E_F I_F}} \tag{12.32}$$

where E_s = Young's modulus of the soil and v_s = Poisson's ratio of the soil. For practical purposes, Equation 12.32 can be approximated as:

$$k_s = \frac{E_s}{B(1-v_s^2)} \tag{12.33}$$

Example 12.8: A 2.5 m-wide strip footing rests in a sandy soil where E_s = 25 MPa and v_s = 0.3. The thickness of the footing is 0.30 m and $E_{concrete}$ = 30 MPa. Estimate the coefficient of the subgrade reaction using Equations 12.32 and 12.33. Determine if the approximation holds.

Solution:

$$I_F = 2.5 \times 0.3^3/12 = 0.0056\,\mathrm{m^4}$$

Equation 12.32 →

$$k_s = \frac{0.65\,E_s}{B(1-v_s^2)} \sqrt[12]{\frac{E_s B^4}{E_F I_F}} = \frac{0.65 \times 25 \times 10^6}{2.5(1-0.3^2)} \sqrt[12]{\frac{25 \times 10^6 \times 2.5^4}{30 \times 10^6 \times 0.0056}}\,\mathrm{N/m^3} = 8.2\,\mathrm{MN/m^3}$$

Equation 12.33 →

$$k_s = \frac{E_s}{B(1-v_s^2)} = \frac{25 \times 10^6}{2.5(1-0.3^2)}\,\mathrm{N/m^3} = 11.0\,\mathrm{MN/m^3}$$

12.6 SETTLEMENT IN A GRANULAR SOIL

Settlements of footings in granular soils are instantaneous with the possibility for long-term creep. There are more than 40 different settlement prediction methods, but the quality of predictions is still poor as demonstrated in the Settlement 94 settlement-prediction symposium in Texas in 1994 (Briaud and Gibbens 1994). Some of the popular settlement prediction methods are discussed below.

The five most important factors that govern footing settlements are *applied pressure, soil stiffness* (or Young's modulus), *footing breadth, footing shape,* and *footing depth.* The soil stiffness is often quantified indirectly through penetration resistance such as *N*-value or blow count from a standard penetration test or the tip resistance q_c from a cone penetration test. Das and Sivakugan (2007) summarized the empirical correlations relating soil stiffness to the penetration resistance.

12.6.1 Terzaghi and Peck (1967) Method

Terzaghi and Peck (1967) proposed the first rational method for predicting the settlement of a shallow foundation in granular soils. They related the settlement of a square footing ($\delta_{footing}$) of width B (meters) to the settlement of a 300 mm square plate (δ_{plate}) under the same pressure, obtained from a plate-loading test through the following expression:

$$\delta_{footing} = \delta_{plate} \left(\frac{2B}{B+0.3} \right)^2 \left(1 - \frac{1}{4} \frac{D_f}{B} \right) \qquad (12.34)$$

The last term in Equation 12.34 is to account for the reduction in settlement with the increase in footing depth. Leonards (1986) suggested replacing ¼ with ⅓ based on additional load test data. The values of δ_{plate} can be obtained from Figure 12.10, which summarizes the plate-loading test data that is supplied by Terzaghi and Peck (1967). This method was originally proposed for square footings, but is also applicable to rectangular and strip footings, provided it is prudently applied. In the case of rectangular or strip footings, the deeper influence zone and increase in the stresses within the soil mass are compensated for by the increase in the soil stiffness.

Example 12.9: A 2 m square pad footing carrying a column load of 900 kN is placed at a depth of 1.0 m in a sand where the average N_{60} is 28. What would be the settlement?

Solution:

$$q_{app} = 900/4 = 225 \text{ kPa}; N_{60} = 28$$

From Figure 12.10, δ_{plate} = 6 mm:

$$\delta_{footing} = 6 \left(\frac{2 \times 2}{2+0.3} \right)^2 \left(1 - \frac{1}{3} \times \frac{1}{2} \right) = 15.1 \text{ mm}$$

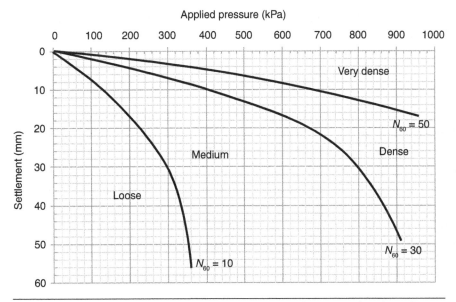

Figure 12.10 Settlements of 300 mm × 300 mm plate (load test data from the late Professor G. A. Leonards, Purdue University)

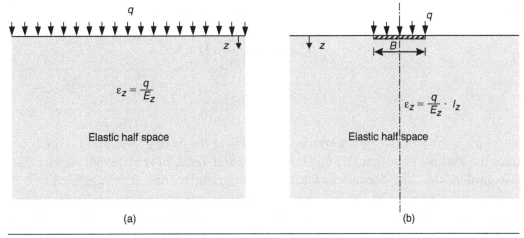

(a)   (b)

Figure 12.11 Uniform pressures on elastic half space: (a) infinite lateral extent (b) limited lateral extent

12.6.2 Schmertmann et al. (1970, 1978) Method

When an elastic half space is subject to a uniform pressure of q that is spread over a very large area as shown in Figure 12.11a, the vertical strain at a point within the material at depth z is given by q/E_z. When the same pressure is applied only over a limited width of B (see Figure

12.11b), the strains would be obviously less. The vertical strain ε_z along the centerline at depth z can be written as:

$$\varepsilon_z = \frac{q}{E_z} I_z \tag{12.35}$$

where E_z and I_z are the Young's modulus and *strain influence factor* respectively at depth z. Based on some finite element studies and load tests on model footings, Schmertmann proposed that the influence factor varies with depth, as shown in Figure 12.12a, which is known as the *2B–0.6 distribution*. This 2B–0.6 distribution does not take into account the shape of the footing.

The influence factor increases linearly from 0 at the footing level to 0.6 at a depth of $0.5B$ below the footing and then decreases linearly to 0 at a depth of $2B$ below the footing. Dividing the granular soil beneath the footing into sublayers of constant Young's modulus and integrating the above equation, the vertical settlement s can be expressed as:

$$s = C_1 C_2 \, q_{net} \sum_{z=0}^{z=2B} \frac{I_z dz}{E_z} \tag{12.36}$$

where C_1 and C_2 are two correction factors that account for the embedment and strain relief due to the removal of overburden and time-dependence of the settlement respectively, and q_{net} is the net applied pressure at the footing level. C_1 and C_2 are given by:

$$C_1 = 1 - 0.5 \left(\frac{\sigma'_{v0}}{q_{net}} \right) \geq 0.5 \tag{12.37}$$

$$C_2 = 1 + 0.2 \log \frac{t}{0.1} \tag{12.38}$$

where σ'_{vo} is the effective in situ overburden stress *at the footing level*, and t is the time in *years* since the loading. Leonards (1986), Holtz (1991) and Terzaghi et al. (1996) suggest that $C_2 = 1$, disregarding the time-dependent settlements in granular soils. It is suggested that the time-dependent settlements in the footings studied by Schmertmann are probably due to the thin layers of clays and silts interbedded within the sands in Florida, where most of Schmertmann's load test data originated. Schmertmann (1970) recommended that Young's modulus be derived from the static cone resistance as $E = 2 \, q_c$. Leonards (1986) suggested that E (kg/cm^2) = 8 N_{60} for normally consolidated sands, where N_{60} is the blow count from a standard penetration test (1 kg/cm^2 = 98.1 kPa).

Schmertmann's (1970) original method does not take into account the footing shape. Realizing the need to account for the footing shape, Schmertmann et al. (1978) made some modifications to the original method. The modified influence factor diagram is shown in Figure 12.12b where the strain influence factor extends to a depth of $2B$ for square footings and $4B$

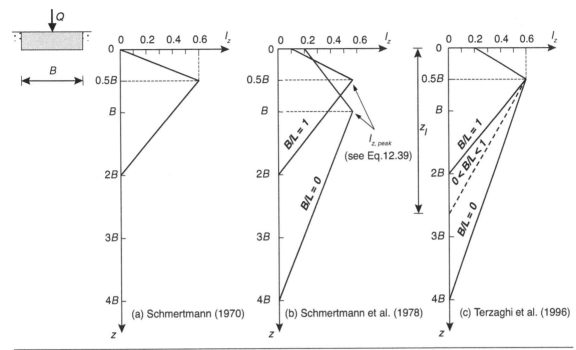

Figure 12.12 Schmertmann et al.'s influence factors

for strip footings, peaking at depths of 0.5B and B respectively. The peak value of the influence factor is given by:

$$I_{z,\text{peak}} = 0.5 + 0.1\sqrt{\frac{q_{\text{net}}}{\sigma'_{v0}}}$$ (12.39)

where σ'_{v0} is the original overburden pressure at a depth of 0.5B below the square footing and B below the strip footing, where the peak values of I_z occur. The equations for computing the settlement and the correction factors remain the same. Schmertmann et al. (1978) suggested that $E = 2.5\ q_c$ for axisymmetric loading and $E = 3.5\ q_c$ for plane-strain loading, based on the observation by Lee (1970) that the Young's modulus is about 40% greater for plane-strain loading than for axisymmetric loading. For a rectangular footing, the settlement can be calculated separately for $B/L = 0$ and 1, and interpolated on the basis of B/L.

Example 12.10: A 2.5 m-wide strip footing carrying a wall load of 550 kN/m is placed at a depth of 1.5 m below the ground as shown in part (a) in the figure on page 314. The entire soil below the ground level is granular and the average tip resistance from a cone penetration test is given for each layer. Estimate the settlement after 10 years using Schmertmann's (1970) method. The average unit weight of sand = 18 kN/m³. *Continues*

Example 12.10: *Continued*

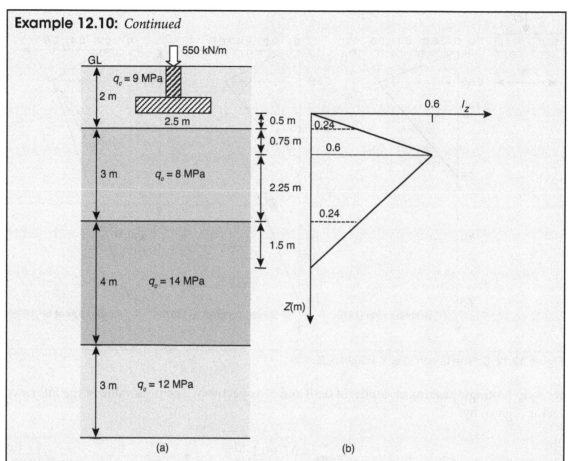

(a)   (b)

Solution: Assuming that the concrete and soil unit weights are about the same, the *net* pressure applied to the underlying soil is 550/2.5 = 220 kPa.

The first step is to draw the influence factor diagram as shown in part (b):

$$C_1 = 1 - 0.5\frac{1.5 \times 18}{220} = 0.94$$

and

$$C_2 = 1 + 0.2 \log\frac{10}{0.1} = 1.4$$

Calculate $\frac{I_z dz}{E_z}$ for each layer assuming constant $E = 2q_c$ within the layer, and then find the sum.

$$\sum_{z=0}^{z=2B} \frac{I_z dz}{E_z} = \frac{0.5 \times 0.5 \times 0.24}{2 \times 9} + \frac{0.5 \times 0.75(0.24 + 0.6) + 0.5 \times 2.25(0.6 + 0.24)}{2 \times 8} + \frac{0.5 \times 1.5 \times 0.24}{2 \times 14} = 0.0885 \text{ m/MPa}$$

∴ Settlement = 0.94 × 1.4 × 220 × 0.0885 mm = 25.6 mm

Terzaghi et al. (1996) suggested the simpler influence factor diagram shown in Figure 12.12c with the influence factors starting at the same point, reaching the same maximum of 0.6 at the same depth of 0.5B, but extending to depths of 2B and 4B for square and strip footings respectively. For rectangular footings, they suggested an interpolation function to estimate the depth of influence z_I (see Figure 12.9c) between 2B and 4B as:

$$z_I = 2B\left(1 + \log\frac{L}{B}\right) \quad \text{for } L/B \le 10 \tag{12.40}$$

Terzaghi et al. (1996) suggested taking $E = 3.5\, q_c$ for axisymmetric loading and increasing it by 40% for plane-strain loading, and suggested the following expression for E of a rectangular footing:

$$E_{\text{rectangular ftg}} = 3.5\left(1 + 0.4\log\frac{L}{B}\right)q_c \tag{12.41}$$

where L/B should be limited to 10. These modifications provide more realistic and less conservative estimates of settlements. Nevertheless, the above values of E in the range of 3.5–4.9 q_c are significantly larger than what is recommended in the literature.

12.6.3 Burland and Burbidge (1985) Method

Burland et al. (1977) collated more than 200 settlement records of shallow foundations of buildings, tanks, and embankments on granular soils, and plotted the settlement per unit pressure against the footing breadth, as shown in Figure 12.13, defining the *upper limits* for the possible settlements

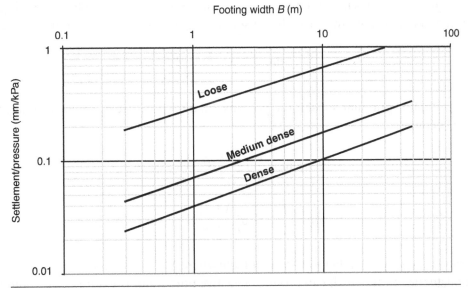

Figure 12.13 Upper limits of settlement per unit pressure (after Burland et al. 1977)

that can be expected. This figure can be used to see if the settlement predicted by a specific method falls within bounds. They suggested that the probable settlement is about 50% of the upper limit shown in the figure, and that in most cases, the maximum settlement will be unlikely to exceed 75% of the upper limit.

Burland and Burbidge (1985) reviewed the above settlement records and proposed an indirect and empirical method for estimating the settlements of shallow foundations in granular soils based on N-values from standard penetration tests that remain uncorrected for overburden pressure. The influence depth z_I was defined as:

$$z_I = B^{0.7} \tag{12.42}$$

where z_I and B are in meters. They expressed the compressibility of the soil by the compressibility index I_c, which is similar to the coefficient of volume compressibility m_v used in the consolidation of saturated clays. For normally consolidated granular soils, I_c was related to the average blow count within the influence depth $\overline{N}_{60}$ by:

$$I_c = \frac{1.71}{\overline{N}_{60}^{1.4}} \tag{12.43}$$

where I_c is in MPa^{-1}. For overconsolidated granular soils, I_c is ⅓ of what is given in Equation 12.43.

Burland and Burbidge (1985) suggested that the settlement can be estimated from:

$$\text{settlement} = q\, I_c\, z_I \tag{12.44}$$

Note that Equation 12.44 is in similar form to Equation 8.3, which is used for estimating consolidation settlements in clays. In normally consolidated granular soils, Equation 12.44 becomes:

$$\text{settlement} = q\, \frac{1.71}{\overline{N}_{60}^{1.4}} B^{0.7} \tag{12.45}$$

In overconsolidated granular soils, if the preconsolidation pressure σ'_p can be estimated, Equation 12.44 becomes:

$$\text{settlement} = \frac{1}{3} q\, \frac{1.71}{\overline{N}_{60}^{1.4}} B^{0.7} \quad \text{for } q \le \sigma'_p \tag{12.46}$$

$$\text{settlement} = \left(q - \frac{2}{3}\sigma'_p\right) \frac{1.71}{\overline{N}_{60}^{1.4}} B^{0.7} \quad \text{for } q \ge \sigma'_p \tag{12.47}$$

For fine sands and silty sands below the water table where $N_{60} > 15$, driving the split-spoon sampler can dilate the sands, which can produce negative pore water pressures that would increase the effective stresses, and hence overestimate the blow counts. Here, we should apply Terzaghi's correction as shown in Equation 12.48:

$$N_{60,\text{corrected}} = 15 + 0.5(N_{60} - 15) \tag{12.48}$$

In gravel or sandy gravel, N_{60} should be increased by 25% using Equation 12.49:

$$N_{60,\text{corrected}} = 1.25\, N_{60} \tag{12.49}$$

The settlements estimated as above apply to square footings. For rectangular or strip footings, the settlements have to be multiplied by the following factor f_s:

$$f_s = \left(\frac{1.25\, L/B}{0.25 + L/B} \right)^2 \tag{12.50}$$

The maximum value of f_s is 1.56 when $L/B = \infty$. The settlements estimated above imply that there is granular soil at least to a depth of z_I. If the thickness H_s of the granular layer below the footing is less than the influence depth, the settlements have to be multiplied by the following reduction factor f_l:

$$f_l = \frac{H_s}{z_I} \left(2 - \frac{H_s}{z_I} \right) \tag{12.51}$$

Burland and Burbidge (1985) noted some time-dependent settlements of the footings and suggested a multiplication factor f_t given by:

$$f_t = 1 + R_3 + R_t \log \frac{t}{3} \tag{12.52}$$

where R_3 takes into consideration the time-dependent settlement during the first three years of loading, and the last component accounts for the time-dependent settlement that takes place after the first three years. Suggested values for R_3 and R_t are 0.3–0.7 and 0.2–0.8 respectively. The lower end of the range is applicable for static loads and the upper end for fluctuating loads such as bridges, silos, and tall chimneys.

12.6.4 Accuracy and Reliability of the Settlement Estimates and Allowable Pressures

Das and Sivakugan (2007) reviewed the different settlement prediction methods and discussed the current state-of-the-art. The three methods above, discussed in detail, are the most popular methods for estimating settlements of shallow foundations in granular soils. These methods typically overestimate the settlements, and are thus conservative. Sivakugan et al. (1998) studied 79 settlement records where the footing width was less than 6 m and concluded that the settlements predicted by Terzaghi and Peck (1967) and Schmertmann (1970) overestimate the settlements by about 220% and 340% respectively.

Tan and Duncan (1991) introduced two parameters, *accuracy* and *reliability*, to quantify the quality of the settlement predictions, and applied these to 12 different methods using a large database of settlement records. Accuracy was defined as the average ratio of the predicted settlement to the measured settlement. Reliability is the probability that the predicted settlement is greater than the measured settlement. Therefore, an ideal settlement prediction method will have an accuracy close to 1 and a reliability approaching 100%. There is often a tradeoff between accuracy and reliability. The methods of Terzaghi and Peck (1967) and Schmertmann et al. (1978) have high reliability but poor accuracy, showing their conservativeness in the estimates. They overestimate the settlement, which leads to an underestimation of the allowable pressure. On the other hand, the Burland and Burbidge (1985) method has good accuracy and poor reliability with more realistic predictions, which can also underestimate the settlements and is therefore less conservative.

It is widely documented in the literature that the designs of shallow foundations in granular soils are usually governed more by settlement considerations than by bearing capacity considerations. Therefore, more care is required in the settlement computations. The Burland and Burbidge (1985) method gives significantly smaller settlements and higher allowable pressures compared to the more conservative Terzaghi and Peck (1967) method.

12.6.5 Probabilistic Approach

The settlements predicted by the different methods can vary widely. Therefore, the magnitude of settlement can have a different meaning depending on the method used in the settlement computations. Sivakugan and Johnson (2004) proposed a probabilistic design chart based on an extensive database of settlement records previously reported in the literature. The purpose of the chart was to quantify the probability that the settlement predicted by a certain method will exceed a specific limiting value in the field. Figure 12.14 provides three separate charts for

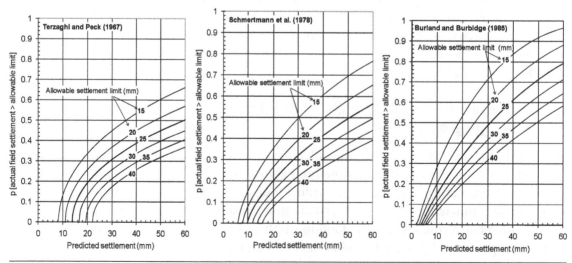

Figure 12.14 Probabilistic charts for settlement predictions (after Sivakugan and Johnson 2004)

the Terzaghi and Peck, Schmertmann et al., and Burland and Burbidge methods. For example, if the Schmertmann et al. method predicts a settlement of 20 mm, the probability that the actual settlement will exceed 25 mm is 0.2.

Example 12.11: The settlement of a 3.0 m square footing in a medium-dense sand under an applied pressure of 200 kPa was estimated to be 18 mm using the Burland and Burbidge method.

  a. Determine if the settlement is within the limit suggested by Burland and Burbidge in Figure 12.13.
  b. What is the probability that the actual settlement would exceed (i) 25 mm or (ii) 40 mm?

Solution: (a) Settlement (estimated)/applied pressure = 18/200 = 0.09 mm/kPa

With B = 3.0 m, the point lies below the limit in Figure 12.13 as expected.

(b) From Figure 12.14c:

$$p[\text{actual settlement exceeds 25 mm}] = 0.31$$

$$p[\text{actual settlement exceeds 40 mm}] = 0.18$$

12.7 SETTLEMENT IN A COHESIVE SOIL

The settlement patterns and the mechanisms in granular and cohesive soils are quite different. Let's look at the settlement of a footing shown in Figure 12.15a.

    The more porous and free-draining nature of the granular soils (Figure 12.15b) is such that the settlements are almost instantaneous, irrespective of whether they are above or below the water table. Lately, there is an expectation that there can be some time-dependent creep settlements as

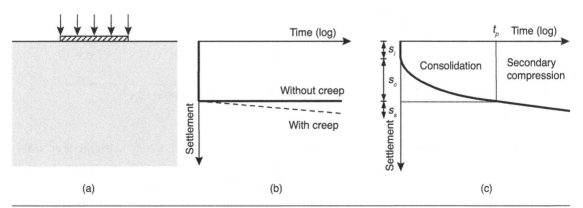

Figure 12.15 Settlement variation with time: (a) footing (b) in sands (c) in clays

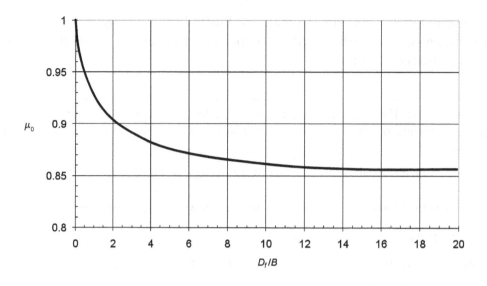

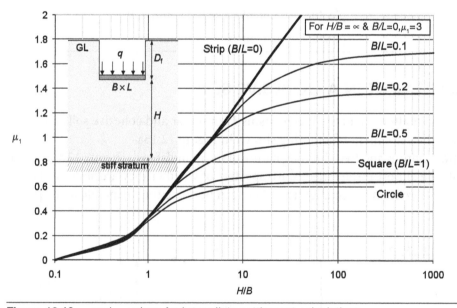

Figure 12.16 μ_0 and μ_1 values for immediate settlement calculations

suggested by Schmertmann (1970) and Burland and Burbidge (1985). Creep settlement accounts for a small fraction of the overall settlement in granular soils.

Unlike in granular soils, the settlements are not instantaneous in cohesive soils. In saturated cohesive soils (Figure 12.15c), the settlements consist of three components: immediate settlement s_i, consolidation settlement s_c, and secondary compression s_s. Immediate settlement occurs immediately after the load is applied and is instantaneous. Generally, it is only a small fraction of the total settlement that also includes consolidation and secondary compression settlements. The consolidation settlement occurs due to the expulsion of water from the saturated clay and dissipation of excess pore water pressure. This can take place over a period of several years. The secondary compression settlement (see Section 8.6), also known as creep, is assumed to occur after the consolidation is completed. Therefore, there will be no *excess* pore water pressure during the secondary compression stage.

12.7.1 Immediate Settlements

Immediate settlement, also known as *distortion settlement, initial settlement,* or *elastic settlement,* occurs immediately upon the application of the load due to lateral distortion of the soil beneath the footing. In clays where drainage is poor, it is reasonable to assume that immediate settlements take place under *undrained* conditions where there is no volume change (i.e., $v = 0.5$). The average immediate settlement under a flexible footing is generally estimated with the theory of elasticity using the following equation, originally proposed by Janbu et al. (1956):

$$s_i = \frac{q\,B}{E_u}\mu_0\mu_1 \tag{12.53}$$

The values of μ_0 and μ_1, originally suggested by Janbu et al. (1956), were later modified by Christian and Carrier III (1978) based on the work by Burland (1970) and Giroud (1972). The values of μ_0 and μ_1, assuming an undrained state with $v = 0.5$, are given in Figure 12.16.

Obtaining a reliable estimate of the undrained Young's modulus E_u of clays through laboratory or in situ tests is quite difficult. It can be estimated using Figure 12.17 proposed by Duncan and Buchignani (1976) and the U.S. Army (1994). E_u/c_u can vary from 100 for very soft clays to 1500 for very stiff clays. Typical values of elastic moduli for different types of clays are given in Table 12.3. Immediate settlement is generally a small fraction of the total settlement, and therefore a rough estimate is often adequate.

Table 12.3 Typical values of elastic moduli for clays (after U.S. Army 1994)

Clay	E_u (MPa)
Very soft clay	0.5–5
Soft clay	5–20
Medium clay	20–50
Stiff clay, silty clay	50–100
Sandy clay	25–200
Clay shale	100–200

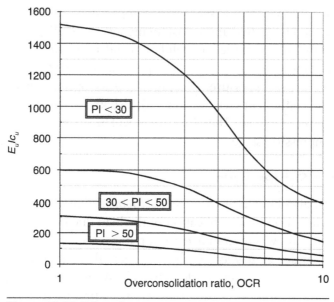

Figure 12.17 E_u/c_u values

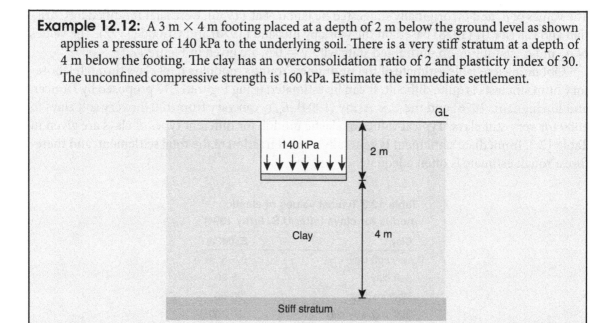

Example 12.12: A 3 m × 4 m footing placed at a depth of 2 m below the ground level as shown applies a pressure of 140 kPa to the underlying soil. There is a very stiff stratum at a depth of 4 m below the footing. The clay has an overconsolidation ratio of 2 and plasticity index of 30. The unconfined compressive strength is 160 kPa. Estimate the immediate settlement.

Continues

Example 12.12: *Continued*

Solution: Let's find the factors μ_0 and μ_1 first.

$$D_f/B = 2/3 = 0.67 \rightarrow \text{From Figure 12.16a, } \mu_0 = 0.93$$

$$B/L = 3/4 = 0.75 \text{ and } H/B = 4/3 = 1.33 \rightarrow \text{From Figure 12.16b, } \mu_1 = 0.45$$

$$\text{OCR} = 2 \text{ and PI} = 30 \rightarrow \text{From Figure 12.17, } E_u/c_u = 570$$

$$c_u = 0.5\, q_u = 80 \text{ kPa} \rightarrow E_u = 570 \times 80 \text{ kPa} = 45.6 \text{ MPa}$$

$$s_i = \frac{q\,B}{E_u}\mu_0\mu_1 = \frac{140\times3000}{45,600} \times 0.93 \times 0.45 \text{ mm} = 3.9 \text{ mm}$$

12.7.2 Consolidation Settlements

Consolidation is a time-dependent process in saturated clays where the foundation load is gradually transferred from the pore water to the soil skeleton. Immediately after loading, the entire applied normal stress is carried by the water in the voids in the form of excess pore water pressure. With time, the pore water drains out into the more porous granular soils at the boundaries, thus dissipating the excess pore water pressure and increasing the effective stresses. Depending on the thickness of the clay layer and its consolidation characteristics, this process can take anywhere from a few weeks to several years. Chapter 8 covers consolidation in good detail.

Consolidation settlement is generally computed assuming a one-dimensional consolidation, and then a correction factor is applied for three-dimensional effects (Skempton and Bjerrum 1957). In a one-dimensional consolidation, the normal strains and drainage are assumed to be taking place only in the vertical direction. This situation arises when the applied pressure at the ground level is uniform and is of a very large lateral extent, as we saw in most of the questions in Chapter 8. Subsequently, the vertical stress increase $\Delta\sigma'$ is also the same at any depth within the clay layer. In reality, when the foundations and the applied pressures are of limited lateral extent, the consolidation is not one-dimensional. The vertical stress increase $\Delta\sigma'$ will be decaying with depth, and we will use the value of $\Delta\sigma'$ at the middle of the clay layer. This value can be computed using the methods discussed in Chapter 7, and it can be significantly less than what is applied at the ground level. Leonards (1986) suggested conservatively using the maximum pressure that occurs under the center of the footing rather than the average pressure in settlement computations.

When the clay layer is thick, it is a good practice to divide it into several sublayers and to use the appropriate values of σ'_{v0}, $\Delta\sigma'$, etc. for each sublayer when computing the changes in void ratios Δe and the consolidation settlements within the layers. These settlements are then added to give the total consolidation settlement of the clay layer.

12.7.3 Secondary Compression Settlements

Secondary compression, also known as creep, can produce ongoing settlements that can continue well beyond consolidation. These were discussed in Section 8.6. Equation 8.24 can be used to compute the secondary compression settlement. In reality, consolidation and secondary settlements may occur simultaneously. For simplicity, we assume that the secondary compression begins upon completion of the consolidation.

Reminder

- ❖ Foundations must satisfy bearing capacity *and* settlement criteria.
- ❖ By limiting the total settlements, you limit both the differential settlement and the angular distortion.
- ❖ Define safety factors in terms of *net* pressures.
- ❖ Terzaghi's bearing capacity equation is too conservative; use Meyerhof's, Hansen's, or Vesic's.
- ❖ For *short-term* stability use c_u and ϕ_u and analyze in terms of *total stresses*; for long-term stability use c' and ϕ' and analyze in terms of *effective stresses*.
- ❖ Use the friction angle corrected for plane strain (Equation 12.16) in all bearing capacity calculations, including N_c, N_q, and N_γ.
- ❖ Use N from the SPT or q_c from the CPT to estimate the Young's modulus of granular soils, which is rarely measured directly; in clays, E_u is estimated based on the value of c_u using Figure 12.17.
- ❖ Most settlement calculation methods for granular soils overestimate the settlements. It is better to be conservative by overestimating settlements than by underestimating them.
- ❖ The Schmertmann et al. method works better with cone data than SPT data.

WORKED EXAMPLES

1. It is required to provide a strip footing to carry a wall load of 450 kN/m in a sandy soil with $\phi' = 32°$ and $\gamma = 18$ kN/m³. The unit weight of concrete is 23 kN/m³. What is the necessary width so that the safety factor with respect to bearing capacity is 3?

Solution: Let's assume $D_f = 0.5$ m, and that the entire 0.5 m is made of concrete.

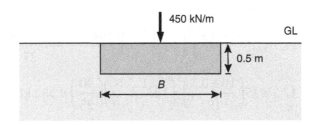

For strip footings, $L = \infty$ and hence $s_q = s_\gamma = 1$

$$d_q = d_\gamma = 1 + 0.1\frac{0.5}{B}\tan\left(45 + \frac{34}{2}\right) = 1 + 0.0940/B\,(\text{m})$$

Plane-strain correction:

$$\phi'_{strip} = 32 \times 1.1 - 0.1 \times 0 = 35.2$$
$$\phi' = 35.2° \rightarrow N_q = 34.1 \text{ and } N_\gamma = 38.5$$

$$q_{app,gross} = \frac{450}{B} + 0.5 \times 23 = \frac{450}{B} + 11.5 \text{ kPa}$$

$$q_{app,net} = \frac{450}{B} + 11.5 - 0.5 \times 18 = \frac{450}{B} + 2.5 \text{ kPa}$$

$$q_{ult,gross} = \left(1 + \frac{0.0940}{B}\right) \times 18 \times 0.5 \times 34.1 + \left(1 + \frac{0.0940}{B}\right) \times 0.5 \times B \times 18 \times 38.5$$

$$= 339.5 + 346.5\,B + \frac{28.85}{B} \text{ kPa}$$

$$q_{ult,net} = 339.5 + 346.5\,B + \frac{28.85}{B} - 0.5 \times 18 = 330.5 + 346.5B + \frac{28.85}{B} \text{ kPa}$$

$$q_{all,net} = \frac{q_{ult,net}}{3} = 110.2 + 115.5B + \frac{9.62}{B} \text{ kPa}$$

Equating the applied and allowable pressures:

$$\frac{450}{B}+2.5=110.2+115.5B+\frac{9.62}{B} \rightarrow B=1.55 \text{ m}$$

2. A 10 m × 15 m raft is placed at a depth of 6 m in a sand where the average N_{60} is 15. If the total permissible settlement is 40 mm, what would be the net allowable bearing capacity?

Solution: Applying Equation 12.23:

$$q_{\text{all, net}}(\text{kPa})=12.5N_{60}\left(\frac{B+0.3}{B}\right)^2\left(1+\frac{1}{3}\frac{D_f}{B}\right)\left(\frac{\text{maximum settlement (mm)}}{25}\right)$$

$$=12.5\times15\left(\frac{10+0.3}{10}\right)^2\left(1+\frac{1}{3}\times\frac{6}{10}\right)\left(\frac{40}{25}\right)=382 \text{ kPa}$$

3. A rectangular footing of breadth B and length L carries a column load with eccentricities of B/9 along the breadth and L/12 along the length. Identify the area beneath the footing where the contact pressure is not compressive.

Solution:

$$q(x, y)=\frac{Q}{BL}\left(1+\frac{12e_B}{B^2}x+\frac{12e_L}{L^2}y\right)=\frac{Q}{BL}\left(1+\frac{4}{3B}x+\frac{1}{L}y\right)$$

At $x=-B/2$ and $y=-L/2 \rightarrow q(-B/2, -L/2)=\frac{Q}{BL}\left(1-\frac{4}{3B}\frac{B}{2}-\frac{1}{L}\frac{L}{2}\right)=-\frac{Q}{6BL}=q_{min}$

Since $q_{min}<0$, some areas beneath the footing are not in compression. Let's identify the region where this occurs by locating the points on the bottom ($y=-L/2$) and left ($y=-B/2$) edges where $q=0$.

$$y=-L/2 \rightarrow q\left(x,-\frac{L}{2}\right)=\frac{Q}{BL}\left(1+\frac{4}{3B}x-\frac{1}{L}\frac{L}{2}\right)=\frac{Q}{BL}\left(\frac{1}{2}+\frac{4}{3B}x\right)=0$$

∴ For $q=0$, $x=-\frac{3B}{8}$

$$x=-B/2 \rightarrow q(-B/2, y)=\frac{Q}{BL}\left(1-\frac{4}{3B}\frac{B}{2}+\frac{1}{L}y\right)=\frac{Q}{BL}\left(\frac{1}{3}+\frac{1}{L}y\right)=0$$

∴ For $q=0$, $y=-\frac{L}{3}$

The area not in compression is shown by the shaded area in the figure on page 327.

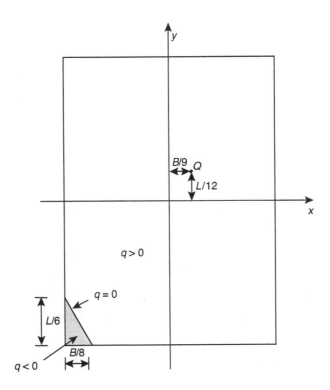

4. A 2.5 m square pad footing is constructed at a depth of 1.5 m in a sandy soil where the average N_{60} is about 30. Using the Terzaghi and Peck method, plot the expected pressure versus settlement plot if the pressure is increased from zero. What should be the maximum pressure applied so that the settlement is less than 25 mm?

Solution:

$$\delta_{footing} = \delta_{plate}\left(\frac{2B}{B+0.3}\right)^2\left(1-\frac{1}{3}\frac{D_f}{B}\right) = \delta_{plate}\left(\frac{2\times2.5}{2.5+0.3}\right)^2\left(1-\frac{1}{3}\times\frac{1.5}{2.5}\right)$$

$$= 2.55\ \delta_{plate}$$

q (kPa)	0	100	200	300	400	500	600	700
δ_{plate} (mm)	0	2.0	4.0	6.5	9.5	13.0	16.5	21.5
$\delta_{footing}$ (mm)	0	5.1	10.2	16.6	24.2	33.2	42.1	54.8

The plot is shown at the top of page 328. To limit the settlement to 25 mm, the applied pressure should not exceed 400 kPa.

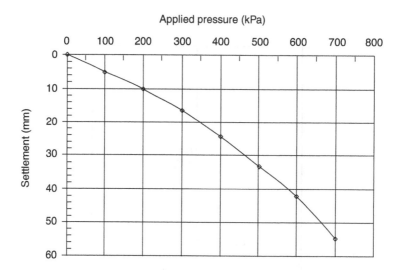

5. Repeat the problem in Example 12.10 using the Schmertmann et al. (1978) method. What is the probability that the actual settlement will exceed 25 mm?

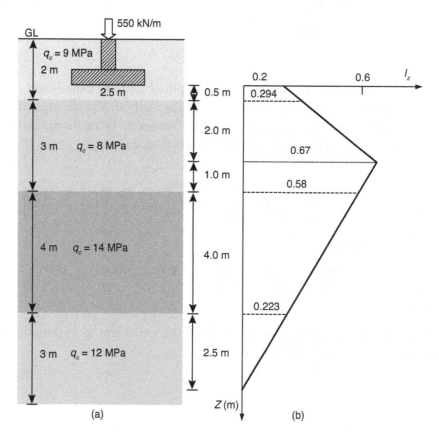

Solution: From Example 12.10, $q_{app, net} = 220$ kPa; $C_1 = 0.94$, and $C_2 = 1.4$:

$$I_{z, peak} = 0.5 + 0.1\sqrt{\frac{220}{4 \times 18}} = 0.67$$

The influence factor diagram is shown at the bottom of page 328, along with the soil profile:

$$\sum_{z=0}^{z=2B} \frac{I_z dz}{E_z} = \frac{0.5 \times 0.5(0.2 + 0.294)}{3.5 \times 9} + \frac{0.5 \times 2.0(0.294 + 0.67) + 0.5 \times 1.0(0.67 + 0.58)}{3.5 \times 8}$$

$$+ \frac{0.5 \times 4.0(0.58 + 0.223)}{3.5 \times 14} + \frac{0.5 \times 2.5 \times 0.223}{3.5 \times 12} = 0.1001 \text{ m/mPa}$$

∴ Settlement = 0.94 × 1.4 × 220 × 0.1001 = 29.0 mm

From Figure 12.14b, with a predicted settlement of 29 mm, it can be seen that there is a probability of 0.32 that the actual settlement will exceed 25 mm.

6. Assuming that the settlement criteria is the one that governs the designs of shallow foundations in granular soils, develop a design chart of allowable pressure based on a settlement of 25 mm against the footing width for different N_{60} values for (a) the Terzaghi and Peck method and (b) the Burland and Burbidge method. Assume $D_f = 0$.

Solution: (a) From the Terzaghi and Peck method:

$$\delta_{plate} = 25 \times \left(\frac{B + 0.3}{2B}\right)^2$$

Substituting values of B in the above equation, the settlement of a 0.3 m × 0.3 m plate (δ_{plate}) can be determined. The corresponding pressures (i.e., allowable pressures) can be found in Figure 12.10.

(b) From the Burland and Burbidge method, assuming the sand is normally consolidated:

$$0.025 = q \text{ (MPa)} \frac{1.71}{\overline{N}_{60}^{1.4}} B^{0.7}$$

$$q \text{(kPa)} = 25 \frac{\overline{N}_{60}^{1.4}}{1.71} \frac{1}{B^{0.7}}$$

Substituting different values for N_{60} and B in the above equation, the allowable pressures can be determined. These are shown in the figure on page 330.

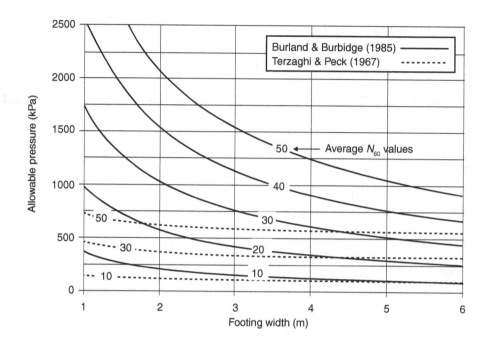

7. A 5 m × 6 m loaded area shown in part (a) in the figure on page 331 is expected to apply 120 kPa pressure to the underlying soil. The soil profile consists of a top 3 m of overconsolidated clay underlain by 8 m of normally consolidated clay with the following soil properties:

	O.C. Clay	N.C. Clay
Average OCR	5	1
Average initial void ratio	0.8	1.2
Compression index	0.60	0.65
Recompression index	0.06	0.07
Saturated unit weight (kN/m³)	19.0	17.5
Undrained modulus (MPa)	20.0	16.0

Estimate the immediate settlement and the final consolidation settlement.

Solution:

Immediate settlement: The charts to compute the immediate settlements (Figure 12.16) are valid only in homogeneous soils, not when there are two layers. Let's make some adjustments, and using the principles of superposition, break the soil profile into three separate soil profiles as shown in part (b) in the figure on page 331 and compute the immediate settlements $s_{i,1}$, $s_{i,2}$, and $s_{i,3}$. The immediate settlement is given by $s_{i,1} + s_{i,2} - s_{i,3}$.

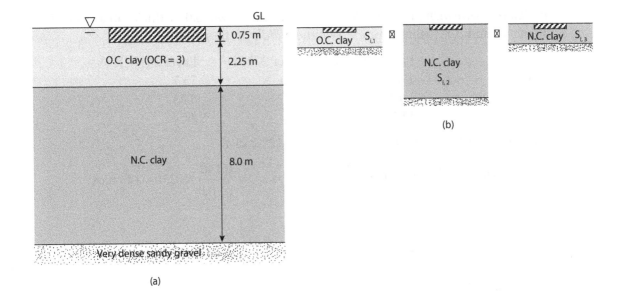

(a)

(b)

Profile 1:

$$D_f/B = 0.75/5 = 0.15 \rightarrow \mu_0 = 0.97$$

$$B/L = 5/6 = 0.833, H/B = 2.25/5 = 0.45 \rightarrow \mu_1 = 0.15$$

$$s_{i,1} = \frac{q\,B}{E_u}\,\mu_0\mu_1 = \frac{120 \times 5000}{20{,}000} \times 0.97 \times 0.15 \text{ mm} = 4.4 \text{ mm}$$

Profile 2:

$$D_f/B = 0.75/5 = 0.15 \rightarrow \mu_0 = 0.97$$

$$B/L = 5/6 = 0.833, H/B = 10.25/5 = 2.05 \rightarrow \mu_1 = 0.62$$

$$s_{i,2} = \frac{q\,B}{E_u}\,\mu_0\mu_1 = \frac{120 \times 5000}{16{,}000} \times 0.97 \times 0.62 \text{ mm} = 22.6 \text{ mm}$$

Profile 3:

$$D_f/B = 0.75/5 = 0.15 \rightarrow \mu_0 = 0.97$$

$$B/L = 5/6 = 0.833, H/B = 2.25/5 = 0.45 \rightarrow \mu_1 = 0.15$$

$$s_{i,3} = \frac{q\,B}{E_u}\,\mu_0\mu_1 = \frac{120 \times 5000}{16{,}000} \times 0.97 \times 0.15 \text{ mm} = 5.5 \text{ mm}$$

$\therefore$ Immediate settlement $= 4.4 + 22.6 - 5.5 = 21.5$ mm

Consolidation settlement: Let's divide the N.C. layer into two, 4.0 m-thick sublayers and compute σ'_{v0} and $\Delta\sigma'$ (using Figure 7.5) at the *middle* of the sublayers.

	Layer 1 (OC)	Layer 2 (NC)	Layer 3 (NC)
σ'_{v0} (kPa)	17.2	43.0	73.7
Depth below ftg (m)	1.125	4.25	8.25
$\Delta\sigma'$ (kPa)	112	55	22

Layer 1: $\sigma'_p = 5\times17.2 = 86 \text{ kPa} < \sigma'_{v0} + \Delta\sigma' = 129.2 \text{ kPa}$

$$\Delta e = 0.06 \log\frac{86}{17.2} + 0.60\log\frac{129.2}{86} = 0.148 \rightarrow s_{c,1} = \frac{0.148}{1+0.8}\times 2250 = 185 \text{ mm}$$

Layer 2: Normally consolidated

$$\Delta e = 0.60\log\frac{43+55}{43} = 0.215 \rightarrow s_{c,2} = \frac{0.215}{1+1.2}\times 4000 = 391 \text{ mm}$$

Layer 3: Normally consolidated

$$\Delta e = 0.60\log\frac{73.7+22}{73.7} = 0.068 \rightarrow s_{c,3} = \frac{0.068}{1+1.2}\times 4000 = 124 \text{ mm}$$

Final consolidation settlement $= 185 + 391 + 124 = 700 \text{ mm}$

Note: The settlement is high. Some ground improvement or alternate foundations may be proposed.

8. The soil profile at a site consists of a 3 m depth of medium dense sand underlain by 2 m of dense sand. The bedrock is at a depth of 5 m. The average soil properties of the two granular layers are:

Medium-dense sand: $\phi' = 34°$, $q_c = 14$ MPa, $N_{60} = 20$ and $\gamma = 17.5$ kN/m^3
Dense sand: $\phi' = 38°$, $q_c = 23$ MPa, $N_{60} = 38$ and $\gamma = 17.5$ kN/m^3

A 2 m $\times$ 3 m footing carrying a net column load of 1800 kN is placed at a depth of 1.5 m below the ground level. Estimate (a) the expected settlement after 10 years using the Schmertmann et al. (1978) method with the modification suggested by Terzaghi et al. (1996), and (b) the safety factor with respect to bearing capacity.

Solution: The I_z vs. depth diagram is shown on page 333.

a. Settlement:

$$z_I = 2B\left(1+\log\frac{L}{B}\right) = 2\times 2\left(1+\log\frac{3}{2}\right) = 4.70 \text{ m}$$

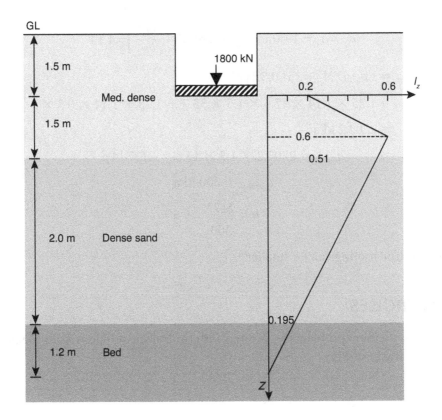

Equation 12.41 → For $B/L = 2/3$:

$$E_{\text{rectangular } ftg} = 3.5\left(1 + 0.4\log\frac{L}{B}\right)q_c \rightarrow E = 3.7\, q_c$$

$$\sum\frac{I_z dz}{E_z} = \frac{0.5(0.2+0.6)+0.5\times0.5(0.6+0.519)}{3.7\times14} + \frac{0.5(0.519+0.195)\times2.0}{3.7\times23} = 0.0215 \text{ m/MPa}$$

$$C_1 = 1 - 0.5\times\frac{0.5\times1.5\times17.5}{300} = 0.956, \text{ and } C_2 = 1 + 0.2\log\frac{10}{0.1} = 1.4$$

∴ Settlement = $300 \times 0.956 \times 1.4 \times 0.0215$ mm = 8.6 mm

b. Bearing capacity: In the bearing capacity region, let's take average ϕ' as 34°.

$$\phi'_{\text{rectangular } ftg} = 34\left(1.1 - 0.1\times\frac{2}{3}\right) = 35.1° \rightarrow N_q = 33.7 \text{ and } N_\gamma \text{ (Meyerhof)} = 37.8$$

$$s_c = 1 + 0.2\times\frac{2}{3}\tan^2\left(45 + \frac{35.1}{2}\right) = 1.49 \rightarrow s_q = s_\gamma = 1.25$$

$$d_q = d_\gamma = 1 + 0.1 \times \frac{1.5}{2} \times \tan\left(45 + \frac{35.1}{2}\right) = 1.14$$

$$q_{\text{ult,gross}} = s_q d_q \gamma D_f N_q + s_\gamma d_\gamma 0.5 B \gamma N_\gamma$$

$$= 1.25 \times 1.14 \times 17.5 \times 1.5 \times 33.7 + 1.25 \times 1.14 \times 0.5 \times 2 \times 17.5 \times 37.8$$

$$= 2203.2 \text{ kPa}$$

$$q_{\text{ult,net}} = 2203.2 - 1.5 \times 17.5 = 2177 \text{ kPa}$$

$$q_{\text{app,net}} = 300 \text{ kPa}$$

$$\therefore F = \frac{2177}{300} = 7.3$$

Note: The footing can be smaller.

REVIEW EXERCISES

1. Discuss the settlement problems associated with the Palace of Fine Arts in Mexico City and the Leaning Tower of Pisa.

2. Discuss *floating* or *compensated* foundations.

3. List the maximum tolerable values of angular distortions for various types of structures.

4. Compare the allowable pressures obtained from the Terzaghi and Peck (1967) and Burland and Burbidge (1985) methods in Worked Example 12.5 to what is reported in literature. How realistic are these values?

5. From Equations 12.5 and 12.6, show that for undrained conditions in clays, where $\phi_u = 0$, $N_c = 5.14$.

6. Bulk modulus K is defined as the ratio of volumetric stress to volumetric strain and is given by $K = E/3(1 - 2v)$. Deduce that for undrained conditions in saturated clays, $v = 0.5$.

7. A 0.5 m-thick, 3 m-square concrete footing is placed in a clayey sand with $c' = 15$ kPa and $\phi' = 29°$ at a depth of 0.5 m. The unit weight of the soil is 18 kN/m³. Estimate the maximum load that can be applied on the footing with a safety factor of 3.0. The unit weight of concrete is 23 kN/m³.

   Answer: 2900 kN

8. Design a square footing to support a column load of 1400 kN in a sandy clay soil having $c' = 40$ kPa, $\phi' = 28°$, and $\gamma = 18.5$ kN/m³. The water table lies 1.5 m below the ground level.

   Answer: 1.6 m square at 1 m depth

9. A 2 m-wide and 0.5 m-thick square footing is placed in a clay where the soil properties are as follows: $c' = 15$ kPa, $\phi' = 26°$, and $\gamma = 18.5$ kN/m³. The water table lies 6 m below the ground level. The unconfined compressive strength of the clay is 120 kPa. The unit weight of concrete is 24 kN/m³. Considering both the short-term and long-term stability, what is the allowable maximum safe load on the footing? Is the foundation's stability more critical in the short-term or in the long-term? Discuss your answer.

   Answer: 480 kN; short-term is more critical

10. A cylindrical grain storage silo is to be placed on a ring foundation as shown on page 336. The foundation rests on a medium-dense sand with $\phi'_{\text{triaxial}} = 34°$, q_c (CPT) = 10 MPa, and $\gamma = 17$ kN/m³. Considering the ring foundation as a strip footing, estimate the maximum weight that the silo can carry, considering bearing capacity and settlements.

    Answer: 19.5 MN including self-weight

11. A rectangular footing of dimensions B and L carries a column load with eccentricities in both directions. If the eccentricity in the direction of breadth is $B/8$, what is the maximum allowable eccentricity in the direction of length that ensures that the entire area beneath the footing is in compression?

    Answer: L/24

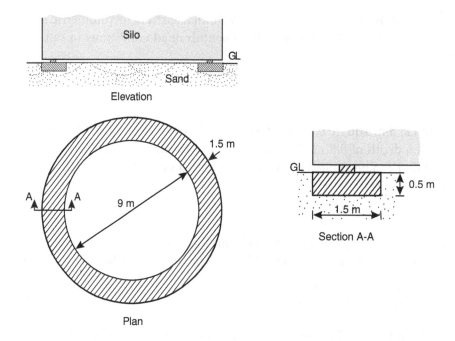

12. A circular footing of diameter B is expected to carry a column load with some eccentricity. The moment of inertia of the footing about its diameter is $\pi B^4/64$. What would be the maximum eccentricity such that the entire soil beneath the footing is in compression? For this value of eccentricity, draw the pressure distribution around the perimeter.

    Answer: B/8

13. A 2.0 m-wide strip at a depth of 1 m in a stiff clay carries a wall load of 425 kN/m. The saturated unit weight of the clay is 20 kN/m^3. Triaxial tests give the following results: $c' = 10$ kPa, $\phi' = 28°$; $c_u = 105$ kPa, and $\phi_u = 0$. Estimate the short-term and long-term safety factors.

    Answer: 2.8, 3.8

14. A 2.0 m-wide strip footing carries a 280 kN/m wall load as shown at the top of page 337. The soft clay layer is expected to be normally consolidated. The unit weight of the sand is 17.5 kN/m^3. Piezometer readings during the past few months show that the consolidation of the soft clay would be nearly completed in two years.

    a. Compute the safety factor of the footing with respect to the bearing capacity failure.
    b. Estimate the total settlement of the footing after 10 years, considering the settlements in sand and clay.

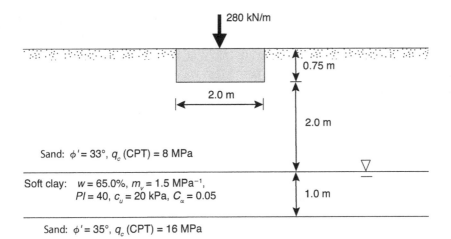

Sand: $\phi' = 33°$, q_c (CPT) = 8 MPa

Soft clay: $w = 65.0\%$, $m_v = 1.5$ MPa^{-1},
$PI = 40$, $c_u = 20$ kPa, $C_\alpha = 0.05$

Sand: $\phi' = 35°$, q_c (CPT) = 16 MPa

c. Considering the large safety factor computed in (a), would you consider reducing the footing width?

Answer: F = 10; settlement = 125 mm; No, the settlement would be too high.

15. A 4 m × 6 m pad footing is subjected to 120 kPa as shown below. The soil profile consists of 4 m of sand underlain by a *very thick* clay deposit. The water table is at a depth of 4 m below the ground level. The bulk unit weight of the sand is 17.5 kN/m³.

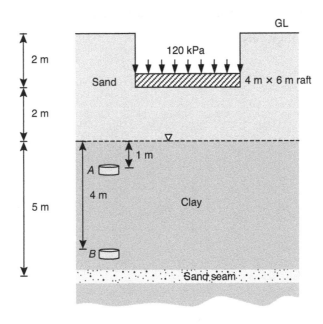

The average N_{60} of the sand is 30, and D_{50} is 1.0 mm. There is a thin sand seam present at a depth of 9 m below the ground level. Consolidation tests were carried out on two undisturbed clay samples, A and B, which were taken at depths of 1 m and 4 m respectively into the clay. A consolidated, undrained triaxial test was carried out only on sample A. The test data are summarized:

Property	Sample A	Sample B
w_n (%)	32	37
G_s	2.71	2.71
LL	85	74
PL	44	36
σ'_p (kPa)	150	115
C_c	0.65	0.59
C_r	0.07	0.06
c_v (m²/year)	2.60	0.45
C_α	0.03	0.04
q_u (kPa)	76	45
c' (kPa)	15	—
ϕ' (°)	21	—

Plot the variation of settlement with time and estimate the settlement of the footing after 15 years.

Evaluate the safety of the footing against the bearing capacity failure.

Clearly state your assumptions.

Quiz 7. Shallow Foundations

Duration: 20 minutes

1. Show that the allowable bearing capacity of a square surface footing in an undrained clay is approximately equal to the unconfined compressive strength.

(3 points)

2. A square footing of a 1.0 m-width settles by 4 mm in a sandy soil when subjected to a uniform pressure of 25 kPa. Estimate the settlement under 50 kPa using (a) the Terzaghi and Peck method and (b) the Schmertmann method. What is the modulus of subgrade reaction?

(4 points)

3. What are the values of N_q and N_γ of a footing in undrained clay?

(1 point)

4. How large is the soil stiffness (Young's modulus) in plane-strain loading compared to axisymmetric loading?

(1 point)

5. Give an estimate of the gross contact pressure beneath a raft foundation that supports a 10-story building with no basement.

(1 point)

Web
Added
Value™

This book has free material available for download from the
Web Added Value™ resource center at *www.jrosspub.com*

Deep Foundations **13**

13.1 INTRODUCTION

Deep foundations include *piles*, *pile groups*, and *piers*. Piles are slender structural elements made of timber, concrete, steel, or composites. They are commonly circular in cross section, but square, hexagonal, and octagonal sections are also common among precast concrete piles. Steel piles can simply be I-sections (i.e., H-piles) that are driven into the ground. A pile group consists of two or more piles under the same pile cap that share the column load. Piers are large diameter piles used for carrying larger loads in bridge abutments, etc. They are typically greater than 750 mm in diameter. Cast-in-place piers are also known as *drilled piers* or *drilled shafts*.

By definition, a deep foundation has a greater depth than breadth. Generally, the depth (i.e., the pile length) is significantly greater than the breadth (i.e., the pile diameter). Even for *short piers*, the length to diameter ratio is more than 5. In the case of piles, this ratio can exceed 50. Pilings require specialist contractors and skills, and consequently, they cost more than shallow foundations. For this reason, deep foundations are preferable only when shallow foundations are inadequate.

Piles are commonly used when the column loads are too large (e.g., in high-rise buildings), for a pad footing, or when the underlying soils are weak (e.g., soft clays). They carry their loads through the tip and the shaft. When most of the load is carried by the shaft, they are known as *friction piles* (Figure 13.1a). When most of the load is carried by the tip, they are known as *end-bearing piles* (Figure 13.1b). When the column loads are quite large, it may be better to provide a pile group as shown in Figure 13.1c. Piles can also be provided to resist uplift forces on the foundation (Figures 13.1d and 13.1e). The one shown in Figure 13.1d is known as an *underreamed pier* or *belled pier*. A special underreaming tool is used to make the enlarged base in clayey soils, thus increasing the load carrying capacity of the tip. It can also provide good anchorage against uplift in expansive soils. *Franki pile* also has an enlarged base, where a concrete plug at the bottom of the hole is rammed into the soil, thus forming the enlarged base. Piles are also useful for resisting lateral loads (Figures 13.1 f and 13.1g). In Figure 13.1f, *batter piles*, also known as *raker piles*, are used to provide lateral resistance to the anchor. They are also useful in wharves and jetties carrying lateral impact forces from berthing ships, offshore platforms, etc. Figure 13.1g shows a transmission tower resting on piles, which are provided to resist the uplift induced by the wind loads. Small diameter *micro piles* or *root piles* (*pali radice*), typically 50–200 mm in

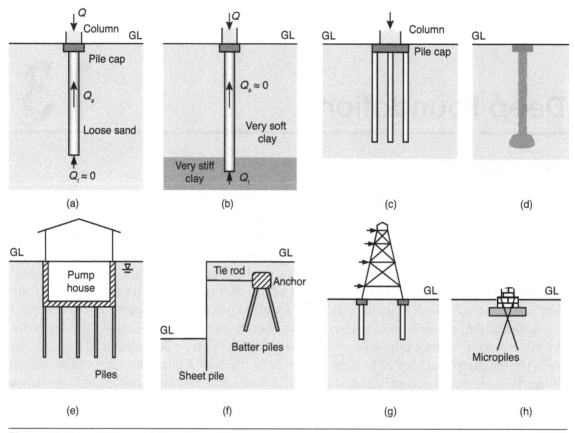

Figure 13.1 Applications of deep foundations

diameter, can be used in *underpinning work* to restore foundations that have settled excessively, as shown in Figure 13.1h. Compaction piles can be driven into loose sand in an attempt to densify them and to provide reinforcement.

13.2 PILE MATERIALS

Piles are commonly made of timber, concrete, steel, or composites. While timber, concrete, and steel are the traditional and most common types, composites also have their place. They are effective in marine environments, which can have adverse effects on concrete, timber, or steel.

13.2.1 Timber Piles

Timber is cheaper than concrete or steel, but is becoming shorter in supply. It is light, easy to handle, and readily trimmed to required lengths. Timber trunks are tree trunks that have their branches and bark removed, generally forming a circular cross section with a natural taper, with the upper end (*butt* or *head*) larger in diameter than the lower end (*tip*). The load carrying capacity of a timber pile, generally limited to about 500 kN, is significantly lower than that of a concrete or steel pile. Timber piles are available in up to 20 m lengths, but can be spliced to make longer piles. They are generally installed by driving, and are better used as friction piles than as end-bearing piles. Very high driving stresses can cause *brooming* or *splitting* at the head, and can also damage the tip. This can be minimized by using a cushion at the pile head and by providing a steel shoe at the tip.

Timber can rot when attacked by marine organisms such as marine borers and fungi. This happens only when there is oxygen and water, and is more serious under a fluctuating water table when it is exposed to both. They are pressure impregnated with creosotes and oils to make them durable. Figure 13.2a shows a *drop hammer* driving a timber pile into the ground.

13.2.2 Concrete Piles

The two major types of concrete piles are *precast* and *cast-in-place*. Precast piles can be conventionally reinforced or prestressed, cast in yards from where they were transported to the site, and generally installed by driving or *jetting* (with the help of a water jet, mostly in sands). Their load carrying capacities are greater than the cast-in-place piles; conventionally reinforced piles carry up to 3000 kN, and when prestressed, they carry as much as 8000 kN. Splicing is still possible with concrete piles, but is undesirable. Therefore, precast concrete piles have to be made to required lengths. Pile driving can be noisy and the vibrations can have adverse effects on nearby buildings. They require casting yards, storage space, transport, and special care in handling. Figure 13.2b shows a group of precast piles that have been driven into the ground, waiting for the pile cap to be cast. Precasting an octagonal concrete pile, its storage in the yard and transport to the site are shown in Figures 13.2c, 13.2d, and 13.2e.

Cast-in-place or bored piles are made by placing a reinforcement cage inside a hole and by filling it with a lower grade of concrete than what is used in precast piles (Figure 13.2f). They can be cased or uncased. When cased, a shell is driven into the ground, with or without a *mandrel*, and is filled with reinforced concrete. In the uncased type, the shell is withdrawn as the concrete is poured. The base can be underreamed to increase the load carrying capacity. Franki pile is also a cast-in-place pile that has a bell formed by hammering a concrete plug into the soil at the bottom of the pile.

13.2.3 Steel Piles

Steel piles are relatively expensive, but their higher load carrying capacity, high resistance to driving, and relative ease of splicing make them an attractive option. They are generally driven into the ground in the form of pipes or rolled H-sections. In marine environments, they are prone to corrosion and require cathodic protection.

Figure 13.2 Pile installation: (a) timber pile driven by drop hammer (b) group of precast concrete driven piles (c) casting a concrete pile (d) storage of precast piles (e) transporting precast piles (f) bored pile (g) composite pile (Photos: Courtesy of Dr. Warren Ng, Dynamic Pile Testing and Ms. Mary Balfour, Balfour Consulting)

13.2.4 Composite Piles

The traditional composite pile is a pile made of two different materials. Figure 13.2g shows a composite pile being made by placing a steel H-pile in a hole and filling it with concrete. Similarly, a driven steel pipe filled with concrete is a composite pile. To avoid exposure of timber piles to a fluctuating water table, it is common to have the upper portion of the pile above the water table in concrete. Lately, there are piles made of fiber-reinforced polymers (FRP) and other composite materials.

13.3 PILE INSTALLATION

The response of a pile to the applied load is very dependent on the method of installation. A pile can be installed into the ground by *boring*, *driving*, *jetting*, or *screwing*. Depending on the extent of lateral displacement of the surrounding soil, a pile can be of the *nondisplacement*, *low-displacement*, or *high-displacement* type. Bored piles, except for Franki piles, are nondisplacement piles. H-piles, open-ended pipe piles, or screw piles cause little lateral displacements and are therefore low-displacement piles. A large diameter concrete pile and a closed-ended pipe pile become high-displacement piles due to the large lateral displacement. A displacement pile will generally have a higher load carrying capacity than a nondisplacement pile.

Timber, steel, and precast concrete piles can be driven into the ground by a *pile hammer*. Pile-driving hammers can be one of the two major types, *impact* or *vibratory*. The impact type can be a *gravity* or *drop hammer*, *single-acting hammer*, or a *double-acting hammer*. A drop hammer (Figure 13.3a) is lifted mechanically by a hoist or crane and dropped under gravity. It is the oldest and simplest type of impact hammer. In a single-acting hammer, compressed air or steam is used to raise the hammer on the *upstroke*, which is then allowed to fall freely. In a double-acting hammer, compressed air or steam is also used to push the hammer down during the *downstroke* to accelerate its downward movement. In a single- or double-acting diesel hammer (Figure 13.3b and 13.3c), a fuel mixture injected into the combustion chamber is ignited at the end of the downstroke, with the combustion forcing the anvil down and the hammer up in preparation for the next stroke. Figure 13.3c shows a diesel hammer driving a raker pile into the ground.

Vibratory hammers (Figure 13.3d) can come in low, medium, and high frequencies. Rotating, eccentric weights are used for producing vertical vibration. The horizontal components of the vibrations get cancelled. They are very effective for driving nondisplacement piles in sand. In clays, low frequency hammers work well.

Jetting is an effective technique for installing piles in granular soils, but is quite ineffective in clays. A high-pressure water jet through a nozzle at the pile tip loosens the soil beneath, enabling the pile to advance further. A screw pile consists of a cylindrical concrete or steel cylinder with helical blades at the tip. The tip is screwed into the ground and can provide good uplift resistance.

Figure 13.3 Pile installation: (a) pile driving by drop hammer (b) pile driving by diesel hammer (c) driving a raker pile (d) vibratory hammer (Photos: Courtesy of Dr. Warren Ng, Dynamic Pile Testing and Ms. Mary Balfour, Balfour Consulting)

13.4 LOAD CARRYING CAPACITY OF A PILE—STATIC ANALYSIS

In shallow foundations, the entire column load is transferred through the base of the footing. In deep foundations, the load is transferred to the soil partly through the tip and the rest through the shaft.

Figure 13.4a shows a column load Q being applied to the head of a pile, which is transferred to the soil through the shaft and the tip, with the loads carried being Q_s and Q_t respectively. The length of the pile is l and the diameter is d. For equilibrium, $Q = Q_s + Q_t$. The stress distributions along the shaft and the tip are shown in Figure 13.4b, where $f_s(z)$ is the skin friction/adhesion along the shaft at depth z, and q_{ult} is the ultimate bearing capacity at the pile tip. When the load Q is increased gradually from 0, the variations of Q_s, Q_t, and Q with the settlement of the pile are shown in Figure 13.4c. It can be seen that the shaft resistance is *mobilized* well before the tip resistance. The full capacity of the shaft is reached when the pile settlement is only about 0.5–1.0% of the shaft diameter, whereas only a small fraction of the tip capacity is reached at this time. Full tip resistance is mobilized only at very large settlements of about 10% of the pile diameter for driven piles and as much as 30% for bored piles. The *gross* ultimate load carrying capacity Q_{ult} is determined by computing the ultimate load carrying capacities of the shaft $Q_{s,ult}$ and tip $Q_{t,ult}$ separately, and adding them up. As in the case of shallow foundations, a

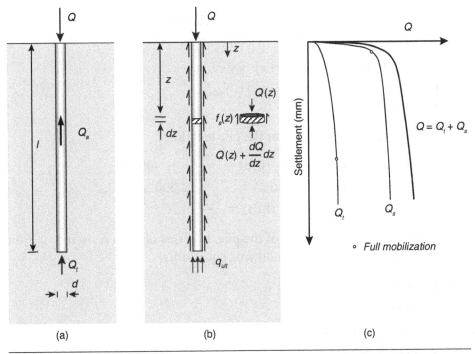

Figure 13.4 Load transfer mechanism: (a) pile loads (b) stress distribution (c) load-settlement plot

safety factor of 2–3 is applied on the ultimate load to calculate the allowable load. For piles in compression, the U.S. Army (1993) suggests a safety factor of 2.0 when verified by pile load test, 2.5 when verified by pile driving analyzer, and 3.0 when not verified. In tension (i.e., uplift), a slightly higher safety factor may be appropriate.

When the ultimate bearing capacity q_{ult} of the pile tip is known, the ultimate load carrying capacity of the pile tip can be estimated as:

$$Q_{t,ult} = q_{ult} A_{tip} \tag{13.1}$$

where A_{tip} is the cross-sectional area of the tip, which is $\pi d^2/4$ for circular cross sections. The shear resistance along the shaft f_s, often known as skin friction or adhesion, varies with the depth. Once its variation with depth is known, the ultimate load carrying capacity of the pile shaft can be estimated as:

$$Q_{s,ult} = \int_{z=0}^{z=1} f_s(z) p_z(z) dz \tag{13.2}$$

where $p_z(z)$ is the perimeter at depth z, which is πd in the case of circular piles with no taper. Now, let's see how we can determine q_{ult} and $f_s(z)$. What is discussed below is applicable mainly to driven piles. In bored piles where the installation technique is quite different, such estimates are very approximate. For bored piles, the ultimate load carrying capacity can be conservatively estimated as 50% of what is obtained for driven piles.

Meyerhof (1976) suggested the following empirical formulae for estimating f_s and q_{ult} in terms of standard penetration test results in cohesionless soils.

Large displacement driven piles: f_s (kPa) $= 2 \overline{N}_{60}$
Small displacement driven piles: f_s (kPa) $= \overline{N}_{60}$
Bored piles: f_s (kPa) $= 0.67 \overline{N}_{60}$

where $\overline{N}_{60}$ is the average value of N_{60} over the pile length.

Driven piles in sands and gravels: q_{ult} (kPa) $= \frac{40 N_{60} l}{d} \leq 400 N_{60}$
Driven piles in nonplastic silts: q_{ult} (kPa) $= \frac{40 N_{60} l}{d} \leq 300 N_{60}$
Bored piles in any granular soils: q_{ult} (kPa) $= \frac{14 N_{60} l}{d}$

where the N_{60} values are those at the tip of the pile. In terms of cone resistance q_c from a static cone penetration test, he suggested the following equations:

Driven piles in dense sands: $f_s = \frac{\overline{q}_c}{200}$
Driven piles in loose sands: $f_s = \frac{q_c}{400}$
Driven piles in nonplastic silts: $f_s = \frac{\overline{q}_c}{150}$

where $\overline{q}_c$ is the average cone resistance along the pile. The ultimate bearing capacity at the tip q_{ult} is assumed equal to the cone resistance at the tip.

13.4.1 Ultimate Bearing Capacity at the Tip q_{ult}

The ultimate bearing capacity at the tip can be computed by treating the pile as a footing of width d at depth l. In the case of granular soils, the bearing capacity equation (Equation 12.2) becomes:

$$q_{ult} = \sigma'_{v,tip} N_q + 0.5 \, d \, \gamma N_\gamma \qquad (13.3)$$

where $\sigma'_{v,tip}$ replaces the term $\gamma_1 D_f$ in Equation 12.2. The bearing capacity factors N_q and N_γ are different from those we used for shallow foundations due to different failure surfaces within the surrounding soil. Since $\sigma'_{v,tip} >> 0.5 \, d\gamma$, and N_q and N_γ are of the same order, the second term in Equation 13.3 can be neglected. Therefore:

$$q_{ult} \approx \sigma'_{v,tip} N_q \qquad (13.4)$$

A wide range of values have been proposed by Terzaghi (1943), Meyerhof (1976), Berezantzev et al. (1961), and several others based on the different failure surfaces assumed. These are shown in Figure 13.5 along with those suggested by the U.S. Army (1993). The values proposed by Berezantzev et al. (1961) appear to be more popular. The values suggested by *The Canadian Foundation Engineering Manual* (2006) for driven and cast-in-place piles are summarized in Table 13.1.

Right next to *driven* piles in granular soils, including silts, the vertical stress does not increase indefinitely with depth. Beyond a certain depth, known as critical depth d_c, the vertical stress remains constant. The critical depth is approximately 10–20 times the pile diameter with the lower and upper ends of the range for loose and dense granular soils respectively (U.S. Army 1993).

For driven and bored piles in clays, the ultimate load carrying capacity of the pile Q_{ult} is generally calculated in terms of *total stresses*, using undrained shear strength (i.e., short-term analysis). Assuming undrained conditions, $\phi_u = 0$ and $N_c = 9$ for piles with $l/d > 4$ (see Section 12.3 and Figure 13.9b), the net ultimate bearing capacity at the tip is given by:

$$q_{ult} = c_u N_c = 9 \, c_u \qquad (13.5)$$

13.4.2 Ultimate Shear Resistance along the Shaft f_s

In granular soils at depth z, the effective overburden stress is σ'_v. The effective horizontal stress next to the pile can be written as $K_s \sigma'_v$, where K_s is an earth pressure coefficient, which depends on the degree of lateral displacement taking place during the pile installation. Bored piles are nondisplacement piles, as they cause no lateral displacement to the surrounding soils. Driven piles can be of low- or high-displacement type. H-pile and open-ended pipes are low-displacement piles and other driven piles including closed-ended pipe piles are high-displacement piles. The skin friction is therefore $K_s \sigma'_v \tan \delta$, where δ is the angle of friction

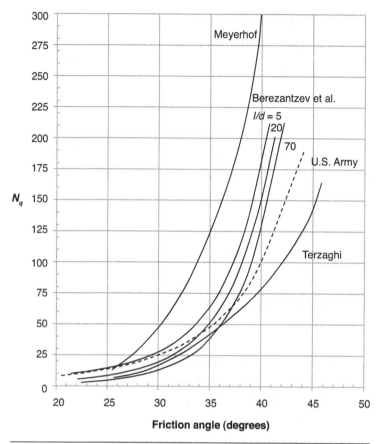

Figure 13.5 N_q values

Table 13.1 N_q **Values (after** *The Canadian Foundation Engineering Manual* **2006)**

Soil type	Driven piles	Cast-in-place piles
Silt	20–40	10–30
Loose sand	30–80	20–30
Medium sand	50–120	30–60
Dense sand	100–120	50–100
Gravel	150–300	80–150

between the soil and the pile material, similar to what was used in Section 10.4. Suggested values for δ and K_s are given in Tables 13.2 and 13.3 respectively. K_0 can be estimated from Equation 10.2. K_s tanδ is sometimes referred to as β-coefficient in the literature and is used in effective stress analysis in clays too. The typical ranges for K_s tanδ for driven and bored piles in granular soils are summarized in Table 13.4.

Table 13.2 δ/ϕ' values (after Kulhawy 1984 and U.S. Army 1994)

Pile material–soil	Kulhawy (1984)	the U.S. Army (1994)
Timber–sand	0.8–0.9	0.8–1.0
Smooth steel–sand	0.5–0.7	0.67–0.80
Rough (corrugated) steel–sand	0.7–0.9	
Precast concrete–sand	0.8–1.0	0.8–1.0
Cast-in-place concrete–sand	1.0	

Table 13.3 K_s/K_o values (after Kulhawy 1984)

Installation method	K_s/K_o
Jetted piles	0.5–0.7
Bored and cast-in-place piles	0.7–1.0
Low displacement driven piles	0.75–1.75
High displacement driven piles	1.0–2.0

Table 13.4 Typical ranges for K_s tanδ or β-coefficient (after *The Canadian Foundation Engineering Manual* 2006)

Soil	Driven piles	Cast in place piles
Silts	0.3–0.5	0.2–0.3
Loose sands	0.3–0.8	0.2–0.4
Medium sands	0.6–1.0	0.3–0.5
Dense sands	0.8–1.2	0.4–0.6
Gravels	0.8–1.5	0.4–0.7

Again, the critical depth must be considered in the case of driven piles in granular soils. The effective vertical stress σ'_v does not increase beyond this depth. This fact should be remembered in computing the skin friction.

When computing the load carrying capacity of a driven or bored pile in clays in terms of total stresses, f_s is taken as the *adhesion* between the pile and the clay. Adhesion is a fraction of the undrained shear strength, typically in the range of 0.35–1.0. Defining adhesion as $\alpha\, c_u$, for driven piles in clays, *The Canadian Foundation Engineering Manual* (2006) suggests:

$$\alpha = 0.21 + 0.26\frac{p_a}{c_u} < 1.0 \tag{13.6}$$

where p_a is the atmospheric pressure ($=101.3$ kPa). For driven piles, API (1984) suggests an adhesion factor α of 1.0 for $c_u \le 25$ kPa and 0.5 for $c_u \ge 70$ kPa, with linear interpolation in between these values. This is also supported by the U.S. Army (1993). For bored piles, 0.5 can be used as the adhesion factor.

For *heavily loaded piles driven to deep penetration* such as in offshore structures, Tomlinson (1995) noted that adhesion depends on the overconsolidation ratio and slenderness ratio of the pile and suggested that:

$$f_s = \alpha\, c_u = F_1\, F_2\, c_u \tag{13.7}$$

where F_1 accounts for the overconsolidation ratio and F_2 for the slenderness ratio. They are given by:

$$F_1 = 1.0 \quad \text{for} \quad \frac{c_u}{\sigma'_{v0}} \leq 0.35 \tag{13.8a}$$

$$F_1 = 0.3649 - 1.393 \log \frac{c_u}{\sigma'_{v0}} \quad \text{for} \quad 0.35 \leq \frac{c_u}{\sigma'_{v0}} \leq 0.8 \tag{13.8b}$$

$$F_1 = 0.5 \quad \text{for} \quad \frac{c_u}{\sigma'_{v0}} \geq 0.8 \tag{13.8c}$$

$$F_2 = 1.0 \quad \text{for} \quad \frac{l}{d} \leq 50 \tag{13.9a}$$

$$F_2 = 2.3405 - 0.789 \log \frac{l}{d} \quad \text{for} \quad 50 \leq \frac{l}{d} \leq 120 \tag{13.9b}$$

$$F_2 = 0.7 \quad \text{for} \quad \frac{l}{d} \geq 120 \tag{13.9c}$$

Example 13.1: A 12 m-long and 400 mm-diameter concrete pile is driven into sands where $\phi' = 34°$. The unit weight of the sand is 18.0 kN/m³. Estimate the maximum load allowed on this pile.

Solution: For $\phi' = 34°$, $d_c = 13d = 5.2$ m:

$$\delta = 0.9\phi' = 30.6°$$

From Equation 10.2:

$$K_0 = 1 - \sin \phi' = 0.44 \rightarrow K_s = 1.5 \, K_0 = 0.66$$

$\beta = K_s \tan \delta = 0.39$, which is at the low end of the range given in Table 13.4.

Let's increase it slightly, and use $K_s \tan\delta$ of 0.5.

At depth d_c, $\sigma'_v = 18 \times 5.2 = 93.6$ kPa. The variation of σ'_v with depth is shown on page 353.

Let's compute the tip load first: $l/d = 12/0.4 = 30$.

From Figure 13.5: $N_q = 40$, which is lower than the values from Table 13.1.

Continues

Example 13.1: *Continued*

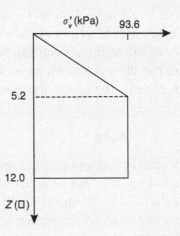

Let's increase N_q to 50:

$$q_{ult} \approx \sigma'_{v,tip} N_q \approx 93.6 \times 50 = 4680 \text{ kPa}$$

$$Q_{t,ult} = q_{ult} A_{tip} = 4680 \times \left(\frac{\pi}{4} \times 0.4^2 \right) = 588 \text{ kN}$$

Let's compute the shaft load, using $K_s \tan\delta$ of 0.5.

Area of the $\sigma'_v - z$ diagram $= 0.5 \times 93.6 \times 5.2 + 93.6 \times 6.8 = 879.8 \text{ kN/m}$:

$$Q_{s,ult} = \int_{z=0}^{z=l} f_s(z) p_z(z) dz = 879.8 \times 0.5 \times \pi \times 0.4 = 553 \text{ kN}$$

$$Q_{ult} = Q_{t,ult} + Q_{s,ult} = 588 + 553 = 1141 \text{ kN}$$

With a safety factor of 2.5, the maximum allowable load on the pile is 456 kN.

13.4.3 Negative Skin Friction

Throughout this chapter, we have assumed that the pile load is carried partly by the shaft in the form of skin friction or adhesion. This is true if the pile moves downward relative to the surrounding soil. Sometimes, in compressible materials such as soft clays or fills that are still undergoing settlements, the soil can move down relative to the pile and induce a down drag on the pile, thus reversing the direction of the skin friction. This is known as negative skin friction, which can have adverse effects on the load carrying capacity of the pile. In addition to reducing the shaft capacity, the tip has to carry a load greater than what is applied on top of the pile. Lowering the water table can also induce negative skin friction.

13.5 PILE-DRIVING FORMULAE

Pile-driving formulae can be used to estimate, however crudely, the ultimate load carrying capacities of driven piles. When a hammer of weight W_R falls over a distance h, the work done is $W_R h \eta$, where η is the efficiency of the setup, accounting for any energy loss. The penetration of a driven pile into the soil per blow is known as set s, a very important term in pile driving. Assuming that the work done by the pile is also given by $Q_{ult} s$, these two terms can be equated as:

$$W_R h \eta = Q_{ult} \, s \qquad (13.10)$$

Equation 13.10 is the basis for all pile-driving equations. Typical values of η for different pile-driving hammers are given in Table 13.5. $W_R h$ can be replaced by the hammer's energy rating. The allowable load on the pile is estimated by dividing the ultimate load computed from Equation 13.10 by a large safety factor, often of about six, reflecting the crudeness of the method. Is it a factor of safety or a factor of ignorance?

The Engineering News Record (ENR) formula is one of the oldest and simplest, and was developed for timber piles. It suggests that:

$$Q_{ult} = \frac{W_R \, h}{s + c} \qquad (13.11)$$

where the constant c accounts for the energy loss (or efficiency), and is 25.4 mm for drop hammers and 2.54 for steam hammers. With 5–10 blows per minute, drop hammers are very slow in operation and are used only for small jobs.

The *modified ENR formula* suggests that:

$$Q_{ult} = \frac{W_R \, h \, \eta}{s + c} \times \frac{W_R + n^2 W_P}{W_R + W_P} \qquad (13.12)$$

where $c = 2.54$ mm, W_R = weight of the ram, W_P = weight of the pile, and n = coefficient of restitution between the ram and the pile cap that varies from 0.25 for timber to 0.50 for steel (Bowles 1988).

The Michigan State Highway Commission formula is of the same form as Equation 13.12, where $W_R h$ is replaced by the hammer's energy rating multiplied by 1.25. The Navy-McKay formula is given by:

$$Q_{ult} = \frac{\eta \times \text{Energy rating}}{s\left(1 + 0.3\dfrac{W_P}{W_R}\right)} \qquad (13.13)$$

Table 13.5 Typical values of η (after Bowles 1988)

Hammer	Efficiency, η
Drop hammer	0.75–1.00
Single-acting hammer	0.75–0.85
Double-acting hammer	0.85
Diesel hammer	0.85–1.00

The Danish formula is given by:

$$Q_{ult} = \frac{\eta \times \text{Energy rating}}{s + \sqrt{\dfrac{\eta \times \text{Energy rating} \times l}{2A_p E_p}}} \tag{13.14}$$

where A_p = cross-sectional area of the pile and E_p = Young's modulus of the pile.

A safety factor of six must be applied to the above equations to estimate the allowable load on the pile.

Example 13.2: A 300 mm × 300 mm-square precast concrete pile, 25 m in length, is driven into the ground with a hammer having an efficiency of 0.8 and maximum energy rating of 40 kJ. The weight of the ram is 35 kN. The coefficient of restitution = 0.35. $E_{concrete}$ = 25 GPa. The unit weight of concrete = 24 kN/m³. Estimate the maximum allowable load on the pile using the modified ENR and Danish formulae when the set is 3.5 mm.

Solution: W_p = 0.3 × 0.3 × 25 × 24 = 54 kN; A_p = 0.3 × 0.3 = 0.09 m²:

Modified ENR: $Q_{ult} = \dfrac{W_R h \eta}{s+c} \times \dfrac{W_R + n^2 W_P}{W_R + W_P} = \dfrac{40{,}000 \times 0.8}{3.5 + 2.54} \times \dfrac{35 + 0.35^2 \times 54}{35 + 54}$ kN = 2477 kN

∴ Allowable load, $Q_{all} = Q_{ult}/6$ = 413 kN

Danish: $Q_{ult} = \dfrac{\eta \times \text{Energy rating}}{s + \sqrt{\dfrac{\eta \times \text{Energy rating} \times l}{2A_p E_p}}} = \dfrac{0.8 \times 40000}{\dfrac{3.5}{1000} + \sqrt{\dfrac{0.8 \times 40000 \times 25}{2 \times 0.09 \times 25 \times 10^9}}} N = 1901$ kN

∴ Allowable load, $Q_{all} = Q_{ult}/6$ = 317 kN

13.6 PILE LOAD TEST

The proof of the pudding is in the eating. The same is true for piling. Now that we have so many dodgy assumptions, how reliable is our estimate of the load carrying capacity of a pile? The best way to assess this is to carry out a *pile load test* (ASTM D1143). The pile load test loads the pile and develops the load-settlement plot, thus determining the allowable load. A

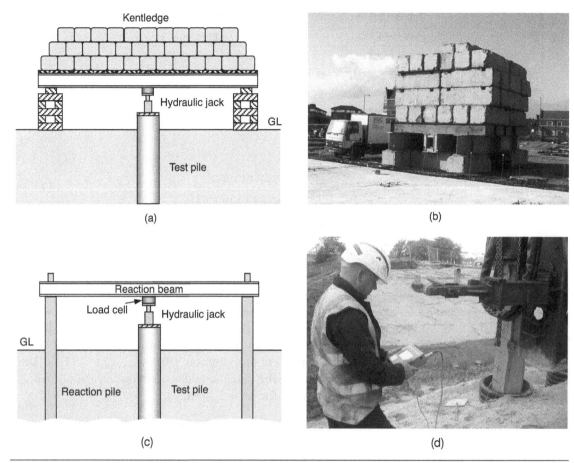

Figure 13.6 Pile load test: (a and b) using kentledge (c) using reaction pile (d) using pile-driving analyzer (Courtesy of Mr. Martyn Ellis, PMC, UK)

hydraulic jack is jacked against a *kentledge* shown in Figures 13.6a and 13.6b, or a reaction beam tied to a reaction pile shown in Figure 13.6c. During the loading, a load cell measures the applied load and a dial gauge measures the settlement. Kentledges can be used to apply loads as high as 5000 kN, provided a safe, stable arrangement can be made. For higher loads, the reaction piles would be better. Settlement of the pile head is plotted against the applied load. Pile load tests are expensive and are therefore carried out on a few randomly selected piles. A pile-driving analyzer (PDA) is a quick and economical alternative to static pile load tests (Figure 13.6d). Here, sensors attached to the pile are wired to the computer. During driving, the pulse is monitored and analyzed using a wave equation program (Smith 1960) such as *CAPWAP*. It gives hammer energy, driving stresses, pile integrity, and static bearing capacity. PDAs can also be used on nondriven piles such as bored piles—a process that uses

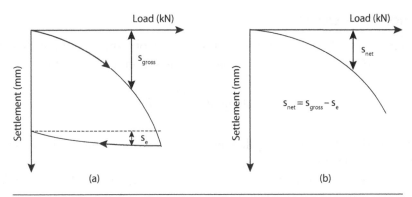

Figure 13.7 Load-settlement plot from a pile load test: (a) gross settlement (b) net settlement

a drop weight to cause an impact and create a pulse which is then analyzed. The lower safety factor that is allowed with the pile load test justifies its cost.

A typical load test plot from a loading-unloading cycle is shown in Figure 13.7a. The settlement measured at the head of the pile s_{gross} is the *gross settlement* of the pile, which includes the settlement of the tip or net settlement s_{net} and the *elastic shortening* of the pile s_e. When the pile is unloaded there is rebound, which is equal to the elastic shortening. The net settlement plot shown in Figure 13.7b is obtained by subtracting the elastic settlements from the gross settlements. The allowable pile load is determined from Figure 13.7a or 13.7b based on relevant design standards, some of which are discussed below.

Common criteria based on the gross settlements define the ultimate pile load as the one that corresponds to a limiting value: 10% of d (UK), 25 mm (Holland), or $\frac{Ql}{AE} + \frac{d}{30}$ mm (Canada), the first one being more popular. De Beer (1968) suggested plotting both load and gross settlement on logarithmic scales, which defines the ultimate load by the point of maximum curvature. In log-log space, the plot often consists of two straight-line segments. Davisson's (1973) method is quite popular. It defines ultimate pile load as the one that corresponds to a gross settlement of 4 mm $+ \frac{d}{120} + \frac{Ql}{A_{Pile}E_{Pile}}$.

The U.S. Army (1993) recommends computing the ultimate load using three different methods and averaging them. The suggested three loads are: (a) the load when the net settlement is 6 mm, (b) the load obtained by the intersection of the two tangents drawn at the start and end of the *net* settlement plot, and (c) the load where the slope of the *net* settlement plot becomes 0.0254 mm/kN (0.01 in/ton).

13.7 SETTLEMENT OF A PILE

The settlement under a pile is less than what is seen in a pad or strip footing. The settlement of a pile can be crudely estimated as about 1% of the pile diameter under the working load.

13.7.1 Poulos and Davis Method

A more rational estimate based on elastic analysis can be obtained from the chart developed by Poulos and Davis (1974) for an *incompressible pile*, the settlement given by:

$$\text{settlement} = \frac{Q}{l \, E_{\text{soil}}} I_\rho \tag{13.15}$$

where Q = pile load at the head, l = pile length, E_{soil} = Young's modulus of the soil, and I_ρ = influence factor from Figure 13.8. h is the depth of the soil above the bedrock and ν is Poisson's ratio. The elastic compression of the pile can be estimated separately as $0.5 \, Ql/A_{\text{Pile}}E_{\text{Pile}}$, and added to the above settlement to give the settlement of the pile head. This method was further refined by Poulos and Davis (1980), where they account for several factors including the stiffness of the pile relative to the soil, etc.

13.7.2 Vesic Method

The load transfer mechanism of a pile is quite complex. The shaft resistance is mobilized well before the tip. Vesic (1977) proposed a method that breaks the settlement into three components: the elastic shortening of the pile s_1, the settlement due to the tip load s_2, and the settlement due to the shaft load s_3. Let's see how we can determine these three components.

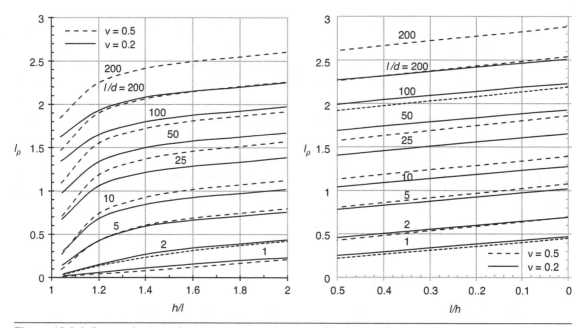

Figure 13.8 Influence factor I_ρ for *incompressible* piles (after Poulos and Davis 1974)

Elastic shortening (s_1):

When there is no shaft friction/adhesion, the pile load is the same at any cross section, and hence the elastic compression would simply be $Ql/A_{Pile}E_{Pile}$. When $f_s(z) > 0$, the load acting on the cross section of the pile decreases with depth. Considering the element shown in Figure 13.4b, it can be seen that:

$$\frac{dQ}{dz} = -f_s(z)\,\pi d \qquad (13.16)$$

Therefore, the variation of the normal load $Q(z)$ acting at the cross section of the pile with depth depends on how $f_s(z)$ varies with depth. This is illustrated in the following example.

Example 13.3: For the four different scenarios of the $Q(z) - z$ variations shown below, deduce the variation of $f_s(z)$ with depth using Equation 13.16.

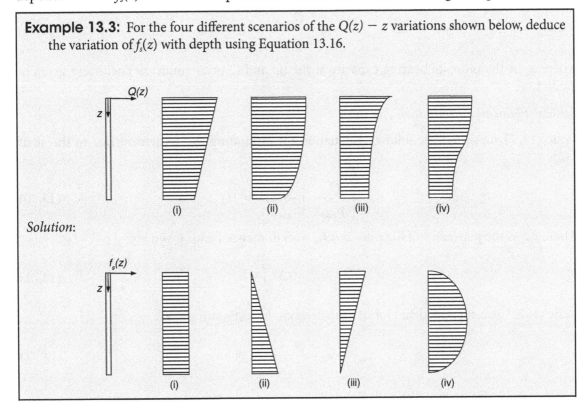

Solution:

Let Q_{wt} and Q_{ws} be the *working loads* carried by the tip and shaft respectively (i.e., $Q = Q_{wt} + Q_{ws}$). Vesic (1977) suggested that the elastic shortening can be written as:

$$s_1 = \frac{(Q_{wt} + \xi Q_{ws})l}{A_{Pile}E_{Pile}} \qquad (13.17)$$

where ξ is a constant that depends on the skin friction distribution. If the variation of f_s with depth is uniform or parabolic as shown in (i) or (iv) in Example 13.3, $\xi = \frac{1}{2}$; if f_s varies linearly with depth as shown in (ii) and (iii) in Example 13.3, $\xi = \frac{2}{3}$. See Worked Example 10 for the proof for scenario (i).

Settlement due to the tip load (s_2):

Based on elastic analysis, Vesic expressed the settlement due to the tip load as:

$$s_2 = \frac{Q_{wt}\, d}{A_{Pile}E_{Soil}}(1-v^2)I_{wt} \tag{13.18}$$

where I_{wt} is the influence factor that can be assumed as 0.85. Vesic (1977) also suggested a semi-empirical expression for estimating s_2 as:

$$s_2 = \frac{Q_{wt}\, C_t}{d\, q_{ult}} \tag{13.19}$$

where q_{ult} is the ultimate bearing capacity at the tip and C_t is an empirical coefficient given in Table 13.6.

Settlement due to the shaft load (s_3):

Vesic (1977) suggested the following equation for estimating the settlement due to the shaft load:

$$s_3 = \left(\frac{Q_{ws}}{p_{Pile}\, l}\right)\frac{d}{E_{soil}}(1-v^2)I_{ws} \tag{13.20}$$

where p_{Pile} is the perimeter of the pile, and I_{ws} is an influence factor given by:

$$I_{ws} = 2 + 0.35\sqrt{\frac{l}{d}} \tag{13.21}$$

Vesic (1977) also suggested an empirical expression for estimating s_3 as:

$$s_3 = \frac{Q_{ws}C_s}{l\, q_{ult}} \tag{13.22}$$

where C_s is an empirical coefficient, given by:

$$C_s = \left(0.93 + 0.16\sqrt{\frac{l}{d}}\right)C_t \tag{13.23}$$

Now that we have the three components, the settlement of the pile head is given by:

$$s = s_1 + s_2 + s_3 \tag{13.24}$$

Table 13.6 C_t values for driven and bored piles (after Vesic 1977)

Soil type	Driven piles	Bored piles
Sand (dense to loose)	0.02–0.04	0.09–0.18
Clay (stiff to soft)	0.02–0.03	0.03–0.06
Silt (dense to loose)	0.03–0.05	0.09–0.12

The U.S. Army (1993) and *The Canadian Foundation Engineering Manual* (2006) recommend the empirical Equations 13.19 and 13.22 for s_2 and s_3 respectively.

13.8 PILE GROUP

A pile group consists of more than one pile, which are connected at the head by a reinforced concrete pile cap, often at the ground level as in Figure 13.1c. In the case of off-shore platforms, the pile cap would be well above the ground—in fact, above the sea level. Typically, the piles in a group are spaced at a minimum of $2.5d$ (see Figure 13.2b). When piles are spaced closely, there is overlap in the stresses induced by the adjacent piles, leading to a reduction in the bearing capacity.

The efficiency η of a pile group is defined as:

$$\eta = \frac{\text{Ultimate load carrying capacity of the group}}{\text{No. of piles} \times \text{Ultimate load carrying capacity of a pile}} \qquad (13.25)$$

Computing the load carrying capacity of a pile group can be difficult. A simpler approach is to compute the capacity of a single pile and use an assumed value of efficiency to estimate the capacity of the group. In loose sands, pile driving further densifies the sand and may lead to η exceeding 1. Nevertheless, it is not recommended to use an efficiency greater than 1. In dense sand, pile driving loosens the sand; hence η can be less than 1. Generally η is taken as 1 for driven piles in sands and is slightly reduced for bored piles, where it can be as low as ⅔.

In *friction* pile groups, assuming that the entire load is carried by the sides, η is simply the ratio of the perimeter of the pile group to the sum of the perimeter of all piles. It is given by:

$$\eta = \frac{2(m+n-2)s+4d}{mn\pi d} \qquad (13.26)$$

where m = no. of piles in a row, n = no. of rows of piles, s = center-center spacing, and d = pile diameter (see Figure 13.9a). One of the earliest equations used to calculate the group efficiency was proposed by Converse and Labarre as (Bolin 1941):

$$\eta = 1 - \frac{\varepsilon}{90}\frac{(m-1)n+(n-1)m}{mn} \qquad (13.27)$$

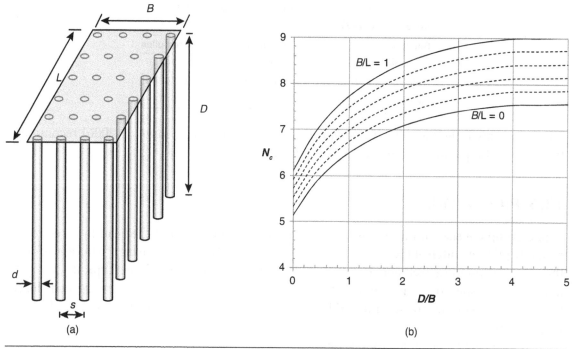

Figure 13.9 Pile group: (a) dimensions (b) N_c values

where $\varepsilon = \tan^{-1}(d/s)$, expressed in degrees. In clays, η is generally less than 1. Here, the piles and the soil in between them can act as a monolithic block, and the ultimate group capacity can be estimated as:

$$Q_{\text{ult,group}} = 2(B+L)D\,f_s + c_u N_c BL \tag{13.28}$$

where B, L, and D are the breadth, length, and depth of the pile group (see Figure 13.9a). The first component is the contribution from the sides due to adhesion and the second component is the contribution from the base. The N_c values as suggested by Skempton (1951) are shown in Figure 13.9b. The skin friction f_s can be assumed as αc_u (Equation 13.7). The ultimate group capacity can also be estimated as the single pile capacity multiplied by the number of piles, assuming $\eta = 1$. It is recommended to use the lesser of the two values as the ultimate group capacity.

Example 13.4: A 3×4 pile group consists of 12 piles of a 300 mm-diameter and a 15 m-length, spaced at 900 mm, center-to-center. The top 6 m consists of clay with $c_u = 50$ kPa, followed by 6 m of clay with $c_u = 65$ kPa, which was underlain by stiff clay with $c_u = 90$ kPa. Estimate the allowable load carrying capacity of the pile group.

Continues

Example 13.4: *Continued*

Solution: $B = 2.1$ m, $L = 3.0$ m, and $D = 15.0$ m.

Adhesion factors (Equation 13.6):

$$c_u = 50 \text{ kPa} \rightarrow \alpha = 0.74; c_u = 65 \text{ kPa} \rightarrow \alpha = 0.62; c_u = 90 \text{ kPa} \rightarrow \alpha = 0.50$$

Single pile: $Q_{s,ult} = \pi \times 0.3 \times [6 \times 0.74 \times 50 + 6 \times 0.62 \times 65 + 3 \times 0.50 \times 90] = 564.4$ kN

$$Q_{t,ult} = (\pi/4) \times 0.3^2 \times 90 \times 9 = 57.3 \text{ kN}$$

$$\therefore Q_{ult} = 564.4 + 57.3 = 621.7 \text{ kN}$$

$$\therefore Q_{ult, group} = 12 \times 621.7 = 7460 \text{ kN}$$

Block: $Q_{sides, ult} = 2 \times (2.1 + 3.0) \times [6 \times 0.74 \times 50 + 6 \times 0.62 \times 65 + 3 \times 0.50 \times 90] = 6108$ kN

$$D/B = 15.0/2.1 = 7.1, B/L = 2.1/3.0 = 0.7 \rightarrow N_c = 8.6 \text{ (Figure 13.9b)}$$

$$Q_{base,ult} = 90 \times 8.6 \times 2.1 \times 3 = 4876 \text{ kN}$$

$$\therefore Q_{ult, group} = 6108 + 4876 = 10{,}984 \text{ kN}$$

We will take the lowest of the two values (i.e., 7460 kN and 10984 kN).

The allowable load on the group is $7460/2.5 = 2984$ kN

In all soils, the settlement of a pile group can be significantly greater than that of a single pile. In sands, Vesic (1970) suggested estimating the settlement of a pile group by:

$$\text{settlement}_{group} = \text{settlement}_{single\ pile} \sqrt{\frac{B}{d}} \tag{13.29}$$

A simplified approach is adapted for computing the settlement of a pile group in general. Here, it is assumed that the column load is actually acting on an imaginary *equivalent raft* well below the pile cap, and the settlement is computed using the methods discussed for shallow foundations. In the case of friction piles as shown in Figure 13.10a, the equivalent raft is at ⅔ *l* depth. In end-bearing piles as shown in Figure 13.10b, the equivalent raft is at the bottom of the piles. Once the equivalent raft is defined, it is a common practice to assume that the loads spread 1 (horizontal):2 (vertical) for computing the vertical stresses at various depths, using Equation 7.6.

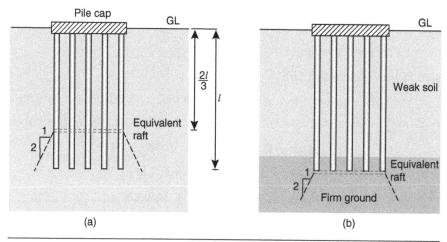

Figure 13.10 Equivalent raft for settlement calculations: (a) friction piles (b) end-bearing piles

❖ In piles, shaft resistance is mobilized well before the tip resistance.

❖ The adhesion factor is close to 1.0 in soft clays and reduces with increasing undrained shear strength.

❖ The ultimate load carrying capacity of a bored pile can be conservatively estimated as half of that of a driven pile.

❖ There is significant judgment involved in selecting the values of N_q, K_s, δ, and α. Therefore, the estimated load carrying capacity of the pile can vary substantially depending on the factors used.

❖ Pile-driving formulae are very approximate, and hence we use high safety factors.

❖ The efficiency of a pile group in granular soils is taken as 1 for driven piles and reduced to as low as ⅔ for bored piles.

❖ Treat the pile group as an equivalent raft in computing settlement.

WORKED EXAMPLES

1. An 8 m-long and 300 mm-diameter precast concrete pile is driven into a sand with $\phi' = 37°$ and $\gamma = 19.0$ kN/m³. Estimate the maximum load allowed on the pile.

Solution: For $\phi' = 37°$, $d_c = 15d = 4.5$ m:

$$\delta = 0.9\phi' = 33.3°; K_0 = 1 - \sin\phi' = 0.40 \rightarrow K_s = 1.5 K_0 = 0.60$$

$\beta = K_s \tan \delta = 0.60 \times \tan 33.3 = 0.39$ less than the range recommended in Table 13.4. Let's increase it to 0.6.

At depth d_c, $\sigma'_v = 19 \times 4.5 = 85.5$ kPa. The variation of σ'_v with z is shown.

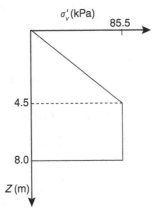

Tip load: $l/d = 8/0.3 = 27 \rightarrow$ From Figure 13.5, $N_q = 85$, which is slightly less than the values in Table 13.1. Let's increase it to 100:

$$q_{ult} \approx \sigma'_{v, tip} \, N_q = 85.5 \times 100 = 8550 \text{ kPa}$$

$$\therefore Q_{t, ult} = q_{ult} \times A_{tip} = 8550 \times \left(\frac{\pi}{4} \times 0.3^2 \right) = 604 \text{ kN}$$

Shaft load: The area of σ'_v-z diagram $= 0.5 \times 4.5 \times 85.5 + 3.5 \times 85.5 = 491.6$ kN/m:

$$\therefore Q_{s, ult} = 491.6 \times \pi \times 0.3 \times 0.6 = 278 \text{ kN}$$

$$\therefore Q_{ult} = 604 + 278 = 882 \text{ kN}$$

$$\therefore Q_{all} = 882/2.5 = 353 \text{ kN}$$

2. A 12 m-long and 300 mm-diameter precast concrete pile is driven into a clay where the unconfined compressive strength is 70 kPa and the unit weight is 19 kN/m³. Estimate its load carrying capacity. What fraction of the load is being carried by the shaft?

Solution:

Tip load:

$$Q_{t,\,ult} = c_u N_c A_{tip} = 35 \times 9 \times \frac{\pi}{4} \times 0.3^2 \text{ kN} = 22.3 \text{ kN}$$

Shaft load: From *The Canadian Foundation Engineering Manual* (2006) and API (1984), the adhesion factor $\alpha = 0.9$:

$$Q_{s,\,ult} = 0.9 \times 35 \times \pi \times 0.3 \times 12 = 356.3 \text{ kN}$$

$$Q_{ult} = 22.3 + 356.3 = 378.6 \text{ kN}$$

$$\therefore Q_{all} = 378.6/2.5 = 151.4 \text{ kN}$$

94% of the pile load is carried by the shaft → Friction pile

3. An 8 m-long and 300 mm-diameter precast concrete pile is driven into a sand with ϕ' = 37° and $\gamma_{sat} = 19.5$ kN/m^3, and $\gamma_m = 17.0$ kN/m^3. The water table is at 2 m below the ground level. Estimate the maximum load allowed on the pile. How does it compare with the load carrying capacity estimated in Worked Example 1?

Solution: As in Worked Example 13.1, $d_c = 4.5$ m, $\delta = 33.3°$, and $K_s \tan\delta = 0.6$. The variation of σ'_v with z is shown.

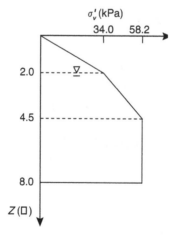

Tip load: $N_q = 100$ as before:

$$q_{ult} \approx \sigma'_{v,\,tip} N_q = 58.2 \times 100 = 5820 \text{ kPa}$$

$$\therefore Q_{t,\,ult} = q_{ult} \times A_{tip} = 5820 \times \left(\frac{\pi}{4} \times 0.3^2 \right) = 411.4 \text{ kN}$$

Shaft load:

The area of σ'_v-z diagram $= 0.5 \times 2 \times 34 + 0.5 \times (34.0 + 58.2) \times 2.5 + 3.5 \times 58.2 = 353.0$ kN/m:

$$\therefore Q_{s,\,ult} = 353.0 \times \pi \times 0.3 \times 0.6 = 199.6 \text{ kN}$$

$$\therefore Q_{ult} = 411.4 + 199.6 = 611.0 \text{ kN}$$

$$\therefore Q_{all} = 611.0 / 2.5 = 244$$

Due to the presence of the water table, the effective stresses are less. Therefore, the shaft load as well as the tip load is reduced. The allowable load is reduced from 353 kN to 244 kN.

4. A 350 mm-diameter and 12 m-long concrete pile is driven into the ground where the top 5 m has $c_u = 30$ kPa, which was underlain by clay with $c_u = 100$ kPa. Estimate the maximum load allowed safely on the pile.

Solution:

Tip load: $Q_{t,\,ult} = c_u N_c A_{tip} = 100 \times 9 \times \dfrac{\pi}{4} \times 0.35^2$ kN $= 86.6$ kN

Shaft load: Equation 13.6 $\rightarrow$ For $c_u = 30$ kPa, $\alpha = 1.0$; for $c_u = 100$ kPa, $\alpha = 0.47$

$$\therefore Q_{s,\,ult} = \pi \times 0.35 \times 5.0 \times 1.0 \times 30 + \pi \times 0.35 \times 7.0 \times 0.47 \times 100 = 526.7 \text{ kN}$$

$$\therefore Q_{ult} = 86.6 + 526.7 = 613.3 \text{ kN}$$

Allowable load $= \dfrac{613.3}{2.5} = 245$ kN

5. The undrained shear strength varies linearly from 20 kPa at the ground level to 60 kPa at a depth of 10 m. Estimate the load carrying capacity of a 600 mm-diameter and 10 m-long bored pile.

Solution: Let's assume $\alpha = 0.5$:

$$Q_{s,\,ult} = \pi \times 0.6 \times 10 \times 0.5 \times 35 = 329.9 \text{ kN}$$

$$Q_{t,\,ult} = \dfrac{\pi}{4} \times 0.6^2 \times 60 \times 9 = 152.7 \text{ kN}$$

$$\therefore Q_{ult} = 329.9 + 152.7 = 482.6 \text{ kN}$$

Allowable load $= \dfrac{482.6}{2.5} = 193$ kN

6. A 900 mm-diameter bored pile with a 1.75 m underream at the base is constructed in a clayey soil as shown. Estimate its load carrying capacity.

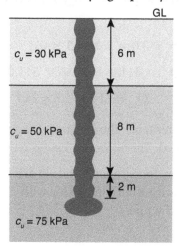

Solution: Let's assume $\alpha = 0.5$ in all layers:

$$Q_{s,\,ult} = \pi \times 0.9 \times [6 \times 0.5 \times 30 + 8 \times 0.5 \times 50 + 2 \times 0.5 \times 75] = 1032 \text{ kN}$$

$$Q_{t,\,ult} = \frac{\pi}{4} \times 1.75^2 \times 75 \times 9 = 1624 \text{ kN}$$

$$\therefore Q_{ult} = 1624 + 1032 = 2656 \text{ kN} \rightarrow Q_{all} = 2656/2.5 = 1062 \text{ kN}$$

7. A 20 m-long 350 mm bored concrete pile is load tested and the data are given below:

Load (kN)	0	250	500	750	1000	1250	1000	750	500	250	0
Settlement (mm)	0	2.0	5.0	8.2	15.0	39.6	39.5	38.2	37.5	36.1	33.5

$E_{concrete} = 25$ GPa. Estimate the allowable pile load using Davisson's (1973) method.

Solution: The load vs. gross settlement plot is shown on the top of page 369:

$$4 + \frac{d}{120} = 4 + \frac{350}{120} = 6.9 \text{ mm}$$

$$\frac{l}{AE} = \frac{20}{\frac{\pi}{4} \times 0.35^2 \times 25 \times 10^9} \text{ m/N} = 0.0083 \text{ mm/kN}$$

Let's draw a straight line with a slope of 0.0083 mm/kN with an intercept of 6.9 mm on the settlement axis. The intersection of this line (dashed) with the settlement curve gives the ultimate load as 1000 kN.

The allowable pile load is 500 kN, with safety factor of 2.0.

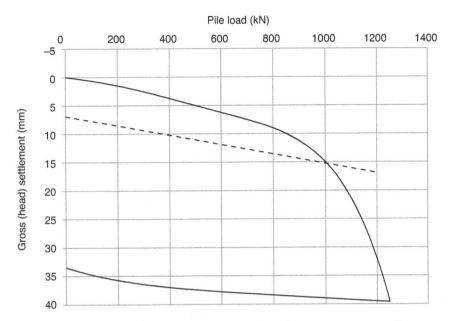

Pile load (kN)

8. A 400 mm-diameter and 15 m-long concrete pile is driven into 12 m of loose sand that was underlain by very stiff clay. Estimate its load carrying capacity. The geotechnical parameters of the sand and stiff clay are given below:

Sand: $\phi' = 29°$; $\gamma = 18$ kN/m³

Stiff clay: $c_u = 90$ kN/m²; $\gamma = 19$ kN/m³

Solution: Critical depth in loose sand, $d_c = 10d = 4.0$ m:

For $\phi' = 29°$, $\delta = 0.9\phi' = 26°$

$$K_0 = 1 - \sin 29 = 0.52 \rightarrow K_s = 1.8K_0 = 0.94 \rightarrow \beta = K_s \tan \delta = 0.46$$

Let's take β as 0.5 (see Table 13.4).

At a depth of 4 m, $\sigma'_v = 4 \times 18 = 72$ kPa $\rightarrow f_s = \beta \sigma'_v = 0.46 \times 72 = 33.1$ kPa.

Skin friction increases linearly from 0 at GL to 33.1 kPa at a depth of 4.0 m, and remains the same for 4–12 m depth.

In the stiff clay, from Equation 13.6, $\alpha = 0.50 \rightarrow f_s = 0.5 \times 90 = 45$ kPa:

$$Q_{s,\text{ult}} = \pi \times 0.4 \times [0.5 \times (12 + 8) \times 33.1 + 3 \times 45] = 585.6 \text{ kN}$$

$$Q_{t,\text{ult}} = \frac{\pi}{4} \times 0.4^2 \times 90 \times 9 = 101.8 \text{ kN}$$

$$\therefore Q_{\text{ult}} = 585.6 + 101.8 = 687.4 \text{ kN} \rightarrow Q_{\text{all}} = \frac{687.4}{2.5} = 275 \text{ kN}$$

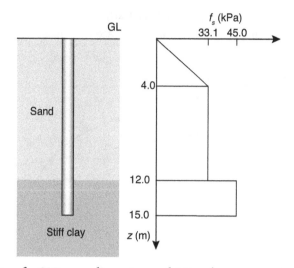

9. The design capacity of a 300 mm-diameter and 15 m-long concrete pile is 400 kN. A Vulcan 08 single-acting steam hammer (maximum energy rating = 35.2 kN-m; ram weight = 35.6 kN; stroke height = 991 mm) is used to drive the pile into the ground. $E_{concrete}$ = 30 GPa; $\gamma_{concrete}$ = 24 kN/m³. What is the set that would ensure the above design capacity? Estimate this by all possible pile-driving formulae.

Solution: Let's assume $\eta = 0.8$ from Table 13.5, and $n = 0.35$:

$$W_P = \frac{\pi}{4} \times 0.3^2 \times 15 \times 24 = 25.4 \; kN; \; A_P = \frac{\pi}{4} \times 0.3^2 = 0.0707 \; \text{m}^2$$

With a safety factor of 6, $Q_{ult} = 6 \times 400 = 2400$ kN.

ENR:

$$Q_{ult} = \frac{W_R \, h}{s + c} = \frac{35.6 \times 991}{s + 2.54} = 2400 \rightarrow s = 12.2 \text{ mm}$$

Modified ENR:

$$Q_{ult} = \frac{W_R \, h \, \eta}{s + c} \times \frac{W_R + n^2 W_P}{W_R + W_P} = \frac{35.6 \times 991 \times 0.8}{s + 2.54} \times \frac{35.6 + 0.35^2 \times 25.4}{35.6 + 25.4} = 2400 \text{ kN}$$

$$\therefore s = 4.9 \text{ mm}$$

Danish:

$$Q_{ult} = \frac{\eta \times \text{Energy rating}}{s + \sqrt{\dfrac{\eta \times \text{Energy rating} \times l}{2 A_P E_P}}} = \frac{0.8 \times 35200}{\dfrac{s}{1000} + \sqrt{\dfrac{0.8 \times 35200 \times 15}{2 \times 0.0707 \times 30 \times 10^9}}} = 2400 \times 10^3 \text{ kN}$$

$$\therefore s = 1.8 \text{ mm}$$

Navy-McKay:

$$Q_{ult} = \frac{\eta \times \text{Energy rating}}{s\left(1 + 0.3\dfrac{W_P}{W_R}\right)} = \frac{0.8 \times 35200}{s\left(1 + 0.3 \times \dfrac{25.4}{35.6}\right)} = 2400 \text{ kN}$$

$$\therefore s = 9.7 \text{ mm}$$

10. If the skin friction f_s is uniform with depth as in Example 13.3 (i), show that ξ in Equation 13.17 should be 0.5.

Solution: Let $Q_s = k\,Q$ and $Q_t = (1 - k)Q$ where Q is the pile load at the head.

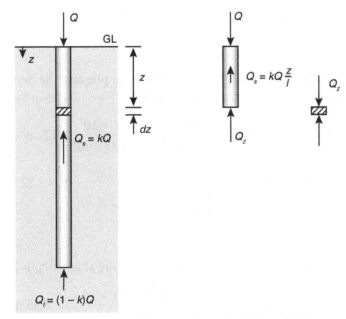

From equilibrium considerations of the top z length of the pile, the normal load on the cross section at depth z can be written as:

$$Q_z = Q - kQ\frac{z}{l} = Q\left(1 - \frac{kz}{l}\right)$$

The elastic shortening of the element of thickness dz is given by:

$$ds_1 = \frac{Q_z\,dz}{AE} = \frac{Q}{AE}\left(1 - \frac{kz}{l}\right)dz$$

$$\therefore s_1 = \frac{Q}{AE}\int_0^1\left(1 - \frac{kz}{l}\right)dz = \frac{Ql}{AE}\left(1 - \frac{k}{2}\right)$$

Applying Equation 13.17:

$$s_1 = \frac{(Q_t + \xi Q_s)l}{A_{Pile} E_{Pile}} = \frac{[(1-k)Q + \xi kQ]l}{AE} = \frac{Ql}{AE}[(1-k) + \xi k]$$

Equating the two expressions for s_1:

$$1 - \frac{k}{2} = 1 - k + \xi k \rightarrow \xi = 0.5$$

11. A 450 mm-diameter and 18 m-long concrete pile driven into medium-dense sand carries 1200 kN. 400 kN is carried by the tip and 800 kN by the shaft. Assuming $E_{soil} = 35$ MPa and $E_{concrete} = 27$ GPa, estimate the settlement of the head using the Poulos and Davis (1974) and Vesic (1977) methods.

Solution:

Poulos and Davis (1974): Assuming an incompressible pile and the depth of sand to be infinite, assuming $v = 0.2$, $l/h = 0$ and $l/d = 40$ gives $I_\rho = 1.8$ (from Figure 13.8):

$$\text{Settlement (if incompressible)} = \frac{Q}{l\, E_{soil}} I_\rho = \frac{1200 \times 10^3}{18 \times 35 \times 10^6} \times 1.8 \text{ m} = 3.43 \text{ mm}$$

$$\text{Elastic shortening} = 0.5 \times \frac{1200 \times 10^3 \times 18}{\left(\frac{\pi}{4} \times 0.45^2\right) \times 27 \times 10^9} \text{ m} = 2.52 \text{ mm}$$

$\therefore$ Settlement = 3.43 + 2.52 = 5.95 mm

Vesic (1977): Let's assume skin friction remains the same at any depth (i.e., $\xi = \frac{1}{2}$):

$$s_1 = \frac{(Q_{wt} + \xi Q_{ws})l}{A_{Pile} E_{Pile}} = \frac{(400 + 0.5 \times 800) \times 10^3 \times 18}{\left(\frac{\pi}{4} \times 0.45^2\right) \times 27 \times 10^9} \text{ m} = 3.35 \text{ mm}$$

$$s_2 = \frac{Q_{wt}\, d}{A_{Pile} E_{Soil}}(1 - v^2)I_{wt} = \frac{400 \times 10^3 \times 0.450}{\left(\frac{\pi}{4} \times 0.45^2\right) \times 35 \times 10^6} \times (1 - 0.2^2) \times 0.85 \text{ m} = 26.39 \text{ mm}$$

$$I_{ws} = 2 + 0.35\sqrt{\frac{l}{d}} = 2 + 0.35 \times \sqrt{40} = 4.2$$

$$s_3 = \left(\frac{Q_{ws}}{p_{Pile}\, l}\right)\frac{d}{E_{Soil}}(1 - v^2)I_{ws} = \left(\frac{800 \times 10^3}{\pi \times 0.45 \times 18}\right) \times \frac{0.45}{35 \times 10^6} \times (1 - 0.2^2) \times 4.2 \text{ m} = 1.63 \text{ mm}$$

$\therefore$ Settlement of pile head = 3.35 + 26.39 + 1.63 = 31.4 mm

REVIEW EXERCISES

1. Write a 500-word essay on root piles, also known as *pali radice*. Include diagrams or figures as appropriate.

2. Discuss how timber, concrete, and steel piles are being spliced.

3. Discuss the difference between the construction of an underreamed pile and a Franki pile.

4. Discuss three pile-driving formulae not included in this chapter, clearly identifying the variables, recommended safety factors, and any limitations.

5. Surf the Internet and collect some videos and images on pile installation. There are a few on YouTube.

6. An MKT-S10 single-acting steam hammer has a maximum energy rating of 44.1 kNm. The weight of the striking ram is 44.5 kN. What is the drop height?

   Answer: 991 mm

7. What would be the load carrying capacity of a 250 mm-diameter and 15 m-long timber pile driven into soft clay with an unconfined compressive strength of 40 kPa? What percentage of the total load is carried by the shaft? Use a safety factor of 2.5.

   Answer: 98 kN, 96%

8. It is required to drive a 350 mm × 350 mm square precast concrete pile into a clay where the undrained shear strength is 35 kPa. With a safety factor of 2.5, what pile length can support a column load of 250 kN?

   Answer: 15 m

9. A 12 m-long and 800 mm-diameter drilled pier is constructed in a clayey soil with a 1.5 m-diameter bell at the base. The top 7 m of the clay has an average c_u of 40 kPa and the clay underneath has a c_u of 90 kPa. Estimate its load carrying capacity, assuming a safety factor of 2.5.

Answer: 940 kN

10. A 250 mm-diameter and 16 m-long timber pile is driven into a sandy soil where $\gamma = 18$ kN/m^3 and $\phi' = 34°$. Estimate the maximum load allowed on the pile with a safety factor of 2.5.

Answer: 270 kN

11. A Delmag D-22 diesel hammer with a maximum energy rating of 53.8 kNm and ram weight of 21.6 kN is used to drive a 300 mm-diameter and 20 m-long concrete pile into the ground. Calculate the blows per 100 mm that would ensure that the pile could carry 400 kN based on the Danish formula. $E_{concrete} = 25$ GPa; $\gamma_{concrete} = 24$ kN/m^3.

Answer: 27

12. A 500 mm-diameter and 16 m-long driven concrete pile carries a 1100 kN load. A load cell placed at the bottom of the pile shows that the load carried by the tip is 300 kN. Assuming that the pile load (i.e., load at the pile cross section) decreases linearly with depth, estimate the skin friction at the middle of the pile.

If the underlying soil is sand with a friction angle of 36° and a unit weight of 18 kN/m^3, what is the safety factor of the pile with respect to bearing capacity failure?

What is the elastic shortening of the pile? $E_{concrete} = 30$ GPa.

Answer: 32 kPa, 2.4, 1.9 mm

13. The soil profile at a site consists of 10 m of sand with $\phi' = 32°$ and $\gamma = 18$ kN/m^3, followed by a thick deposit of very stiff clay with $c_u = 120$ kPa and $\gamma = 19$ kN/m^3. The water table lies 3 m below the ground level. It is required to design a driven pile foundation to carry a column load of 350 kN.
   a. Propose two alternate designs
   b. Estimate the settlement in each case
   c. Suggest a pile hammer and the appropriate set value

14. A 600 mm-diameter and 18 m-long concrete bored pile is constructed using the continuous flight auger technique. The pile passes through 8 m of gravelly sand followed by 8 m of stiff clay, with the bottom two meters founded in very dense sand. A pile load test was carried out on this pile and the load test data are summarized on page 375.

Load (kN)	0	330	660	990	1321	1650	1982	660	330	0
Settlement (mm)	0	0.33	0.93	1.64	2.40	2.92	3.98	3.19	2.69	2.34

[Data: Courtesy of Mr. Martyn Ellis, PMC, UK]

a. Plot the load vs. gross settlement.
b. Plot the load vs. net settlement.
c. Estimate the allowable load on the pile with a safety factor of 2.0.
d. For the load estimated in (c), estimate the gross settlement.
e. How does your estimate in (d) compare with the load test value?

15. If the skin friction f_s is increasing linearly with depth as in Example 13.3 (ii), show that ξ in Equation 13.17 should be 0.67.

Quiz 8. Pile Foundations

Duration: 20 minutes

1. State whether the following are true or false.
   a. Skin friction is greater for displacement piles than for nondisplacement piles
   b. Skin friction is greater for bored piles than for driven piles
   c. In a driven pile, tip resistance is fully mobilized before the skin friction
   d. Bored pile is a nondisplacement pile

   (2 points)

2. List five different applications of piles.

   (2½ points)

3. Describe each of the following terms in your own words, within *three* lines.
   (a) underreamed pile (b) kentledge (c) friction pile (d) negative skin friction

   (2 points)

4. What is another name for *batter pile*?

   (½ point)

5. What is the difference between a single-acting and a double-acting hammer?

   (1 point)

6. What is known as *set* in pile driving?

   (1 point)

7. A 600 mm-diameter pile carries 1200 kN with a safety factor of 2.5. Give an estimate of the settlement you would expect.

   (1 point)

Earth Retaining Structures

14

14.1 INTRODUCTION

Earth retaining structures retain soil and resist lateral earth pressures. They ensure stability to an area where the ground level is quite different on both sides of the structure (see Figure 10.1), even to height differences in excess of 10 m. Some major types of earth retaining structures are:

- Gravity retaining walls
- Cantilever retaining walls
- Sheet pile walls
- Gabion walls
- Diaphragm walls
- Crib walls
- Reinforced earth walls

Gravity retaining walls are bulky and are made of plain concrete or masonry. They rely on their self-weight for their stability. They are often unreinforced or nominally reinforced, and become semigravity walls. A *cantilever retaining wall* (Figure 10.1a) has a smaller cross section and is made of reinforced concrete. With a vertical wall, often with a slight batter, fixed to a horizontal base, the retaining wall acts like a vertical cantilever fixed at the bottom. The base width is generally about 50–70% of the height. Figure 14.1a shows a steel sheet pile wall being driven into the ground, isolating an area for excavation. Figure 14.1b shows a close-up view of interlocking segments of a steel sheet pile.

A *crib wall* (Figure 14.1c) can be constructed relatively fast and can tolerate large differential settlements due to its flexible nature. Centuries ago, crib walls were made from tree trunks and branches in the Alpine areas of Austria. Today, they are made of interlocking precast concrete, timber, or steel elements that are filled with free-draining, granular soils such as gravels or crushed rocks. The self-weight of the granular fill within the interior of the crib wall contributes to the stability of the wall. The width of the wall is typically 50–100% of the height, reaching as high as 6 m. They are often tilted toward the backfill with a slope of 6:1 or less. They are generally designed as rigid retaining walls, using Rankine's or Coulomb's theories.

Figure 14.1d shows a riverbank slope being stabilized through *soil nailing*. Here, 75–150 mm-diameter holes are made in the soil, where a 20–30 mm-diameter steel bar is placed and the annular area filled with cement grout, injected under pressure. This is repeated at other locations, with the soil

Figure 14.1 Earth retaining structures: (a) steel sheet pile walls (b) interlocking sections of sheet piles (c) crib wall (d) soil nailing (e) diaphragm wall (f) MSE wall

nails spaced at certain intervals. On completion of the nail installation, the face of the slope can be shotcreted. The early applications of soil nailing were limited to temporary earth-retaining systems, but today they are also used as permanent structures. *Tiebacks* are similar to soil nails, but are prestressed.

A *diaphragm wall* is a relatively thin, reinforced concrete wall, cast in place in a trench to a depth as high as 50 m. Figure 14.1e shows a diaphragm wall built several meters below the ground during the construction of an underground transport facility. A trench is excavated and filled with bentonite slurry as the excavation proceeds. During excavation, the slurry provides lateral support to the walls of the excavation. On completion of excavation, the trench is full of slurry, which has thixotropic properties (it hardens when undisturbed and liquefies when agitated). A reinforcement cage is placed into the slurry-filled trench and then concreted using tremie pipes. The tremie pipe has a funnel at the top and is used to place concrete under water. Diaphragm walls are a structural element that can also become a part of the permanent structure. They are generally constructed in short, alternating panel widths, giving sufficient time for the concrete between the panels to cure. Without the reinforcement and concrete, the *slurry wall* can be used as an impervious barrier in landfills and excavations.

Gabions are large steel wire cages filled with gravels or cobbles. They are assembled like building blocks and are tied to the adjacent cages (Figure 1.1e). They are analyzed like gravity walls. They are used as retaining walls along highways, and for erosion protection along riverbanks.

Figure 14.1f shows a *mechanically stabilized earth wall*, often known as an *MSE wall*. This is a special case of reinforced earth walls where tensile elements such as metal strips, geofabrics, or geogrids are placed within the soil at certain intervals to improve stability. Relatively thin precast concrete panels are used in the exterior of the wall.

The designs of retaining walls, sheet piles, and braced excavations are discussed in detail in the sections that follow. The lateral earth pressure theories covered in Chapter 10 will be used to compute the lateral loads on the retaining structures. Due to its simplicity, Rankine's earth pressure theory is preferred over Coulomb's in the designs of retaining walls and sheet piles. It is assumed that the active and passive states are fully mobilized on both sides of the walls.

14.2 DESIGN OF RETAINING WALLS

The two major types of retaining walls, the gravity retaining wall and the cantilever retaining wall, are shown retaining granular backfills in Figure 14.2. To ensure free draining, thus minimizing problems due to the buildup of pore water pressures, the backfills behind retaining walls are generally granular. The retaining wall can fail in three different modes:

a. *Sliding*—By sliding along the base of the wall
b. *Overturning*—By overturning or toppling about the toe
c. *Bearing capacity*—By failing beneath the base of the wall (within the soil)

Of the three failure modes, the first two are the most critical, and it is often assumed that the bearing capacity failure does not occur. Nevertheless, it is a good practice to check.

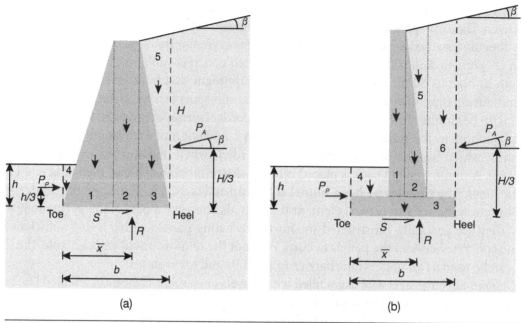

Figure 14.2 Free body diagram for equilibrium considerations of gravity and cantilever retaining walls

The equations developed in Chapter 10 under Rankine's earth pressure theory assume a *smooth vertical wall*. This is not the case with the retaining walls in Figure 14.2 where the walls are inclined. Let's make a simplification here by considering the two dashed vertical lines passing through the toe and the heel as the smooth vertical walls of heights h and H, and treating the soil and concrete enclosed within them to act as a monolithic rigid body. The soil and concrete within the rigid block are broken into triangular or rectangular zones (numbered 1, 2, . . . , 6) for ease in computing their weights W_i and horizontal distances x_i of the centroids from the toe.

Assuming that the active and passive states are fully mobilized, P_A and P_P can be computed using Rankine's earth pressure theory. They are given by:

$$P_A = \frac{1}{2} K_A \gamma H^2 \qquad (10.10)$$

$$P_P = \frac{1}{2} K_P \gamma h^2 \qquad (10.11)$$

The ground level being horizontal on the passive side, K_P can be computed as:

$$K_P = \left(\frac{1 + \sin \phi'}{1 - \sin \phi'} \right) = \tan^2(45 + \phi'/2)$$

In the active side, since the ground is inclined at an angle of β to horizontal, K_A must be calculated from Equation 10.12 given below:

$$K_A = \cos\beta \frac{\cos\beta - \sqrt{\cos^2\beta - \cos^2\phi'}}{\cos\beta + \sqrt{\cos^2\beta - \cos^2\phi'}} \qquad (10.12)$$

The retaining wall is in equilibrium under the following forces:

- $W_1, W_2, ..., W_n$
- Active thrust P_A
- Passive thrust P_P
- Vertical reaction at the base R
- Shear resistance to sliding S

a. Stability with respect to sliding:

When P_A is substantially larger than P_P, there is a possibility that the wall may slide along the base and become unstable. The stability is threatened by P_A, which is the *driving force*. Any attempt to make the wall slide is resisted by P_P and S. The maximum possible value for the shear resistance S is given by:

$$S_{max} = \left(P_A \sin\beta + \sum_{i=1}^{n} W_i \right) \tan\delta \qquad (14.1)$$

where δ is the wall friction angle discussed in Section 10.4. For a soil-concrete interface, δ can be taken as 0.5–$0.8\ \phi'$, with $\frac{2}{3}\ \phi'$ being a popular choice. The safety factor with respect to sliding is defined as:

$$F_{sliding} = \frac{\text{Maximum resistance available}}{\text{Driving force}} = \frac{P_P + S_{max}}{P_A \cos\beta} \qquad (14.2)$$

which has to be greater than 1.5. If the soil is cohesive (e.g., clayey sand), the adhesion term $b\ c'_a$ has to be included in the numerator; its contribution improves the stability, and hence increases the safety factor. Here, b is the width of the base and c'_a is the adhesion, which is about 0.5–0.7 times the cohesion c' (see Section 10.4).

b. Stability with respect to overturning:

When P_A is substantially larger than P_P, there is the possibility that the wall can overturn about the *toe*. For equilibrium, the moment about the toe has to be zero. Therefore:

$$\left(\sum_{i=1}^{n} W_i x_i \right) + P_P \frac{h}{3} + P_A \sin\beta b - P_A \cos\beta \frac{H}{3} - R \bar{x} = 0 \qquad (14.3)$$

$$R\bar{x} = \left(\sum_{i=1}^{n} W_i x_i \right) + P_P \frac{h}{3} + P_A \sin\beta\, b - P_A \cos\beta \frac{H}{3} \qquad (14.4)$$

For no overturning to occur, there has to be contact between the base and the soil, implying that $R\bar{x}$ is positive:

$$\therefore \left(\sum_{i=1}^{n} W_i x_i \right) + P_P \frac{h}{3} + P_A \sin\beta\, b > P_A \cos\beta \frac{H}{3}$$

The right side of the above inequality is the *driving moment* (counterclockwise) that attempts to cause the overturning. The left side of the inequality is the *resisting moment* (clockwise) that resists any attempt for instability. Therefore, the safety factor with respect to overturning is defined as:

$$F_{\text{overturning}} = \frac{\text{Resisting moment}}{\text{Driving moment}} = \frac{\left(\sum_{i=1}^{n} W_i x_i \right) + P_P \dfrac{h}{3} + P_A \sin\beta\, b}{P_A \cos\beta \dfrac{H}{3}} \qquad (14.5)$$

which has to be greater than 2.0.

c. Stability with respect to bearing capacity:

The walls are designed such that the resultant vertical force R acts within the middle third of the base so that the entire soil below the base is in compression. The base can be treated as a strip footing of width b in computing the ultimate bearing capacity q_{ult}. The eccentricity and inclination of the applied load should be incorporated in the bearing capacity equation (Equation 12.7) in computing q_{ult}. The footing depth D_f in the bearing capacity equation can be taken as h.

For equilibrium:

$$R = \left(\sum_{i=1}^{n} W_i \right) + P_A \sin\beta \qquad (14.6)$$

$$S = P_A \cos\beta - P_P \qquad (14.7)$$

What happens if the computed value of P_P exceeds $P_A \cos\beta$? Obviously, S is not going to reverse its direction. It just means that the passive resistance is not fully mobilized and the design is conservative. Remember that active resistance has to be fully mobilized first. The inclination α of the applied load to vertical is given by:

$$\alpha = \tan^{-1} \frac{S}{R} \qquad (14.8)$$

which should be used in Equations 12.12 and 12.13 in computing the load inclination factors. A conservative approach is to ignore the passive resistance and overestimate both S and α. Otherwise, $P_A \cos\beta$ has to be distributed between P_P and S in some sensible way (e.g., the same level of mobilization of passive and sliding resistances).

From Equation 14.4:

$$\bar{x} = \frac{\left(\sum_{i=1}^{n} W_i x_i\right) + P_P \dfrac{h}{3} + P_A \sin\beta\, b - P_A \cos\beta \dfrac{H}{3}}{R} \tag{14.9}$$

The eccentricity of the applied load is given by:

$$e = |\bar{x} - 0.5b| \tag{14.10}$$

It is a common practice to ensure that the load acts within the middle third of the base width (i.e., $e < b/6$) so that the soil beneath the entire base is in compression. Computation of the pressure distribution beneath the base can be carried out as shown in Section 12.4, where the maximum applied pressure is given by:

$$q_{max} = \frac{Q}{B}\left(1 + \frac{6\,e}{B}\right) \tag{12.20}$$

The safety factor with respect to bearing capacity can be determined as:

$$F_{bearing\ capacity} = \frac{q_{ult,net}}{q_{max}} \tag{14.11}$$

which has to be greater than 3.0.

If there is a likelihood that all or part of the soil in front of the wall on the passive side may be removed, the designer should not rely on the passive resistance and assume $P_P = 0$ or use a reduced value of P_P.

Example 14.1: Evaluate the stability of the retaining wall shown on the top of page 384 with respect to sliding, overturning, and bearing capacity.

Solution: $\phi' = 33° \rightarrow K_A = \tan^2\left(45 - \dfrac{\phi}{2}\right) = 0.295$ and $K_P = \tan^2\left(45 + \dfrac{\phi}{2}\right) = 3.392$

$$P_A = 0.5\, K_A \gamma H^2 = 0.5 \times 0.295 \times 18 \times 5^2 = 66.4 \text{ kN per m width}$$

$$P_P = 0.5\, K_P \gamma h^2 = 0.5 \times 3.392 \times 18 \times 1^2 = 30.5 \text{ kN per m width}$$

$$W = 1.5 \times 5.0 \times 1.0 \times 24 = 180.0 \text{ kN per m width}$$

Continues

Example 14.1: *Continued*

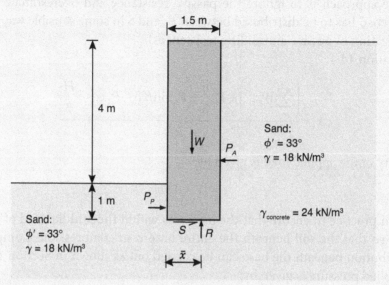

Let's assume $\delta = \frac{2}{3}\,\phi' = 22°$:

$$\therefore F_{\text{sliding}} = \frac{P_P + S_{\max}}{P_A} = \frac{30.5 + (180 \times \tan 22)}{66.4} = 1.55 > 1.5$$

$$F_{\text{overturning}} = \frac{\text{Resisting moment about toe}}{\text{Driving moment about toe}} = \frac{(180 \times 0.75) + (30.5 \times 0.33)}{66.4 \times 1.667} = 1.31 < 2.0$$

The wall is safe with respect to sliding, but unsafe with respect to overturning.

Bearing capacity: $S = P_A - P_P = 66.4 - 30.5 = 35.9$ kN; $R = W = 180.0$ kN

From Equation 14.8:

$$\alpha = \tan^{-1}\frac{S}{R} = \tan^{-1}\frac{35.9}{180} = 11.3°$$

From Equation 14.9:

$$\bar{x} = \frac{\left(\sum_{i=1}^{n} W_i x_i\right) + P_P \dfrac{h}{3} - P_A \dfrac{H}{3}}{R} = \frac{(180 \times 0.75) + (30.5 \times 0.33) - (66.4 \times 1.667)}{180} = 0.192 \text{ m}$$

The resultant acts outside the middle third, which is not acceptable. The wall must be modified.

When one of the three safety factors is less than the minimum suggested values, the section of the retaining wall has to be modified. When F_{sliding} is low, it can be improved by providing a *key*

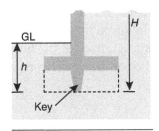

Figure 14.3 Key at the base of a retaining wall

at the base of the wall. A key is simply an extended wall that protrudes into the soil beneath the base as shown in Figure 14.3.

The soil enclosed within the dashed lines is assumed to act as a rigid body along with the key and the rest of the wall. This increases the values of h and H, and hence P_P and P_A. Since K_P is an order of magnitude greater than K_A, the increase in P_P is very much greater than that in P_A. This significantly increases the safety factor with respect to sliding.

14.3 CANTILEVER SHEET PILES

When it is required to carry out wide and deep excavations, it is required to support the sides against any possible instability. For up to about 6 m of excavations, cantilever sheet piles are quite effective. For larger depths, they become uneconomical, and it becomes necessary to use anchored sheet piles, which are discussed in Section 14.4. Sheet piles are made of interlocking sheets of timber, steel (Figure 14.1a and 14.1b), or concrete, making a continuous flexible wall. A typical cantilever sheet pile arrangement is shown in Figure 14.4, where the depth of

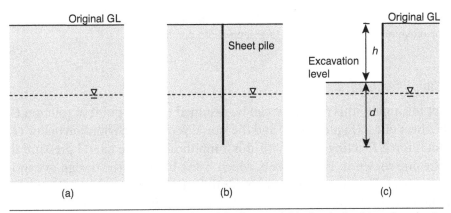

Figure 14.4 Cantilever sheet pile: (a) original ground (b) sheet pile driven in prior to excavation (c) after excavation

excavation is h and the depth of embedment is d. Here, a sheet pile is driven into the in situ soil (Figures 14.4 a and 14.4b), which is followed by excavation to the desired level (Figure 14.4c). The sheet pile acts like a vertical cantilever, fixed at the bottom and loaded horizontally; hence the name. A cantilever sheet pile relies on the passive resistance developed in the embedded portion for its stability.

14.3.1 In Granular Soils

Let's consider the situation in *granular soils*, and assume that the water table is below the tip of the sheet pile. When the sheet pile wall deflects left as a result of the excavation, it rotates about a point O near the tip (Figure 14.5a), which is at a depth of d_0 below the excavation level. The top of the sheet pile moves from A to A', and the bottom tip moves from B to B'. Assuming that there is enough movement to mobilize active and passive resistance in the surrounding soil, it is possible to define these zones as shown in Figure 14.5b based on the directions of wall movements. The lateral pressure distribution on both sides of the sheet pile is shown in Figure 14.5c.

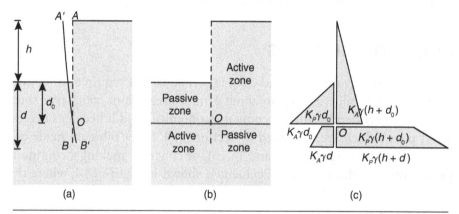

(a)   (b)   (c)

Figure 14.5 Analysis of cantilever sheet pile: (a) original and deflected positions (b) active and passive zones (c) lateral pressure distribution

Method 1: Simplified analysis

In an attempt to simplify this further, it can be assumed that the point of rotation O is close to the tip of the sheet pile B (Figure 14.6a) and the lateral pressure distribution below O is replaced by a horizontal force R acting at O. With this simplification, the lateral pressure distribution reduces to the one shown in Figure 14.6b, which is the basis for the design of cantilever sheet piles. Here, the sheet pile is in equilibrium under three forces: active thrust P_A, passive thrust P_P, and horizontal reaction R where $P_A = \frac{1}{2} K_A \gamma (h + d_0)^2$ and $P_P = \frac{1}{2} K_P \gamma d_0^2$.

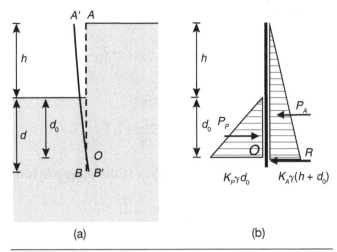

Figure 14.6 Simplified analysis: (a) original and deflected positions (b) approximate lateral pressure distribution

Taking moment about O:

$$P_A \times \frac{h+d_0}{3} = P_P \times \frac{d_0}{3}$$

$$\frac{1}{2}K_A\gamma(h+d_0)^2 \times \frac{h+d_0}{3} = \frac{1}{2}K_P\gamma d_0^2 \times \frac{d_0}{3} \qquad (14.12)$$

$$d_0 = \frac{h}{\sqrt[3]{\dfrac{K_P}{K_A}} - 1}$$

From the friction angle of the granular soil and h, d_0 can be determined. The maximum bending moment occurs below the excavation line where the shear force is zero. This depth z_* below the excavation line can be computed as follows, by equilibrium consideration of horizontal forces:

$$\frac{1}{2}K_P\gamma z_*^2 = \frac{1}{2}K_A\gamma(h+z_*)^2$$

$$z_* = \frac{h}{\sqrt{\dfrac{K_P}{K_A}} - 1} \qquad (14.13)$$

The maximum bending moment can be computed as:

$$M_{max} = \frac{1}{6}\gamma[K_A(h+z_*)^3 - K_P z_*^3]$$
(14.14)

From the theory of bending:

$$\sigma_{allowable} = \frac{M_{max}}{I}y$$

The section modulus S required for the cross section of the sheet pile is defined as:

$$S = \frac{M_{max}}{\sigma_{allowable}} = \frac{I}{y}$$
(14.15)

where I is the moment of inertia about the axis of bending and y is the distance to the edge from the neutral axis. The required sheet pile can be selected on the basis of the section modulus, which is generally provided in the sheet-piling catalogues.

With all the approximations made, we have not yet incorporated any safety factor in the analysis. It can be done in two ways:

a. Increase the value of d_0 computed by 20–40%; or
b. Provide a safety factor F of 1.5–2.0 on the passive resistance and use K_P/F. Here we assume that only a fraction of passive resistance is mobilized, and hence do not rely on the full passive resistance for stability.

Example 14.2: Develop an expression for d_0 similar to Equation 14.12 with a safety factor F on the passive resistance.

Solution:

$$P_A \times \frac{h+d_0}{3} = P_P \times \frac{d_0}{3}$$

$$\frac{1}{2}K_A\gamma(h+d_0)^2 \times \frac{h+d_0}{3} = \frac{1}{2}\frac{K_P}{F}\gamma d_0^2 \times \frac{d_0}{3}$$

$$\left(\frac{h+d_0}{d_0}\right)^3 = \frac{K_P}{K_A F}$$

$$d_0 = \frac{h}{\sqrt[3]{\dfrac{K_P}{K_A F}} - 1}$$

Method 2: Using the net lateral pressure diagram

A better and more realistic method, but one that is a little more complex, is described below. Here we will draw the *net* horizontal pressure diagram as shown in Figure 14.7. To the right of C, $\sigma'_v = \gamma h$, and therefore, $\sigma'_h = K_A \gamma h = \sigma'_1$.

Let's measure z downward *from the excavation level*. At a depth of z below the excavation level, there is active pressure on the right and passive pressure on the left, given by:

$$\sigma'_{ha} = K_A \sigma'_v = K_A \gamma(h + z)$$

$$\sigma'_{hp} = K_P \sigma'_v = K_P \gamma z$$

The net pressure, from right to left, is given by:

$$\sigma'_{hn} = \sigma'_{ha} - \sigma'_{hp} = K_A \gamma h + (K_A - K_P)\gamma z = \sigma'_1 - (K_P - K_A)\gamma z \qquad (14.16)$$

At a depth of z_0 below the excavation, the net pressure becomes zero. This depth is given by:

$$z_0 = \frac{\sigma'_1}{(K_P - K_A)\gamma} = \frac{K_A}{(K_P - K_A)}h \qquad (14.17)$$

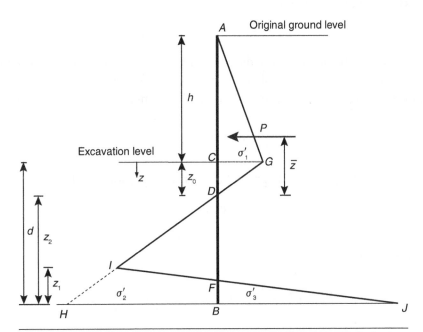

Figure 14.7 Net lateral pressure diagram for a cantilever sheet pile in dry granular soils

From Equation 14.16, it can be seen that for every unit depth increase below the excavation level, the net pressure σ'_{hn} decreases by $(K_P - K_A)\gamma$. Therefore, the slope of the line GH is 1 vertical to $(K_P - K_A)\gamma$ horizontal:

$$\therefore \sigma'_2 = \overline{HB} = z_2(K_P - K_A)\gamma = K_P\gamma d - K_A\gamma(h+d) \tag{14.18}$$

At the bottom of the sheet pile, there is active earth pressure on the left and passive earth pressure on the right. They are:

$$\sigma'_{ha} = K_A\gamma d$$

$$\sigma'_{hp} = K_P\gamma(h + d)$$

Therefore, the net lateral earth pressure from *right to left* is given by:

$$\sigma'_{hn} = \sigma'_{hp} - \sigma'_{ha} = K_P\gamma(h + d) - K_A\gamma d = \sigma'_3$$

i.e., $\sigma'_3 = K_P\gamma h + (K_P - K_A)\gamma d$.
    Substituting $d = z_0 + z_2$:

$$\sigma'_3 = K_P\gamma h + (K_P - K_A)\gamma z_0 + (K_P - K_A)\gamma z_2 = \sigma'_4 + (K_P - K_A)\gamma z_2 \tag{14.19}$$

where σ'_4 is a known quantity, given by:

$$\sigma'_4 = K_P\gamma h + (K_P - K_A)\gamma z_0 \tag{14.20}$$

Let's have a close look at the net pressure diagram in Figure 14.7. There are two unknowns, z_1 and z_2. These can be determined from equilibrium equations. Let's include the area $IHBF$ on both sides so that the computations are simpler. Adding up the horizontal forces for equilibrium:

$$P + \frac{1}{2}z_1(\sigma'_2 + \sigma'_3) - \frac{1}{2}z_2\sigma'_2 = 0 \tag{14.21}$$

where P is the area of the pressure diagram AGD. From Equation 14.21:

$$z_1 = \frac{\sigma'_2 z_2 - 2P}{\sigma'_2 + \sigma'_3} \tag{14.22}$$

Taking moment about B:

$$P(z_2 + \overline{z}) + \frac{1}{2}(\sigma'_2 + \sigma'_3)z_1\frac{z_1}{3} - \frac{1}{2}\sigma'_2 z_2\frac{z_2}{3} = 0 \tag{14.23}$$

From Equations 14.18, 14.19, 14.20, 14.22, and 14.23, it can be shown that:

$$z_2^4 + A_1 z_2^3 - A_2 z_2^2 - A_3 z_2 - A_4 = 0 \tag{14.24}$$

where:

$$A_1 = \frac{\sigma'_4}{\gamma(K_P - K_A)}; A_2 = \frac{8P}{\gamma(K_P - K_A)}; A_3 = \frac{6P[2\bar{z}\gamma(K_P - K_A) + \sigma'_4]}{\gamma^2(K_P - K_A)^2}; \text{ and } A_4 = \frac{P(6\bar{z}\sigma'_4 + 4P)}{\gamma^2(K_P - K_A)^2}$$

Equation 14.24 can be solved by a trial-and-error iterative process, and z_2 can be found. To incorporate the safety factor, the penetration d $(= z_0 + z_2)$ can be increased by 20–40%, or a safety factor F of 1.5–2.0 can be provided on passive resistance (i.e., use K_P/F). The maximum bending moment occurs at the point of zero shear that can be easily located.

Example 14.3: A 4 m-deep excavation is to be carried out in dry sands where $\phi' = 34°$ and $\gamma = 18$ kN/m³. Determine the sheet pile's required depth of penetration using (a) a net lateral pressure diagram, and (b) Equation 14.12.

Solution: $\phi' = 34° \rightarrow K_A = 0.283$ and $K_P = 3.537$

a. *Net pressure diagram approach:*

$$\sigma'_1 = K_A\gamma h = 0.283 \times 18 \times 4 = 20.4 \text{ kPa}$$

$$(K_P - K_A)\gamma = (3.537 - 0.283) \times 18 = 58.6 \text{ kPa per m depth}$$

$$z_0 = \frac{K_A}{K_P - K_A}h = \frac{0.283}{3.537 - 0.283} \times 4 = 0.35 \text{ m}$$

$$P = 0.5 \times 20.4 \times 4 + 0.5 \times 20.4 \times 0.35 = 40.8 + 3.57 = 44.37 \text{ kN per m}$$

$$\bar{z} = \frac{40.8 \times 1.683 + 3.57 \times 0.233}{44.37} = 1.57 \text{ m}$$

$$\sigma'_4 = K_P\gamma h + (K_P - K_A)\gamma z_0 = 3.537 \times 18 \times 4 + (3.537 - 0.283) \times 18 \times 0.35 = 275.16 \text{ kPa}$$

$$A_1 = \frac{\sigma'_4}{\gamma(K_P - K_A)} = \frac{275.16}{18(3.537 - 0.283)} = 4.70 \text{ m}$$

$$A_2 = \frac{8P}{\gamma(K_P - K_A)} = \frac{8 \times 44.37}{18(3.537 - 0.283)} = 6.06 \text{ m}^2$$

$$A_3 = \frac{6P[2\bar{z}\gamma(K_P - K_A) + \sigma'_4]}{\gamma^2(K_P - K_A)^2} = \frac{6 \times 44.37[2 \times 1.57 \times 18(3.537 - 0.283) + 275.16]}{18^2 \times (3.537 - 0.283)^2} = 35.62 \text{ m}^3$$

$$A_4 = \frac{P(6\bar{z}\sigma'_4 + 4P)}{\gamma^2(K_P - K_A)^2} = \frac{44.37(6 \times 1.57 \times 275.16 + 4 \times 44.37)}{18^2(3.537 - 0.283)^2} = 35.82 \text{ m}^4$$

Continues

Example 14.3: *Continued*

Equation 14.24 becomes:

$$f(z_2) = z_2^4 + 4.70z_2^3 - 6.06z_2^2 - 35.62z_2 - 35.82 = 0$$

By trial and error, $f(2.93) = 0 \rightarrow z_2 = 2.93$ m $\quad d = z_2 + z_0 = 2.93 + 0.35 = 3.28$ m

b. Using Equation 14.12:

$$d = d_0 = \frac{h}{\sqrt[3]{\dfrac{K_P}{K_A}} - 1} = \frac{4}{\sqrt[3]{\dfrac{3.537}{0.283}} - 1} = 3.03 \text{ m}$$

The above d from both methods must be increased by 20–40%.

When either the water table or more than one soil layer is present, there will be breaks in the lateral pressure diagram, but the concepts remain the same. In dredging operations where excavation takes place below the water table, sheet piles can be used to support the walls of the excavation as shown in Figure 14.8. The depth of excavation is h where the water table is at a depth of h_1 ($< h$). The water pressure is the same on both sides and will not be considered in the analysis.

At the water table level to the right of the wall, $\sigma'_v = \gamma_m h_1$, hence $\sigma'_h = K_A \gamma_m h_1$.

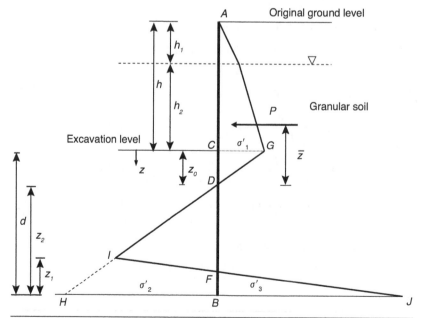

Figure 14.8 Net lateral pressure diagram for a cantilever sheet pile in partially submerged granular soils

At the excavation level to the right of C, $\sigma'_v = \gamma_m h_1 + \gamma' h_2$; hence $\sigma'_h = K_A(\gamma_m h_1 + \gamma' h_2) = \sigma'_1$.
At a depth of z below the excavation level:

$$\sigma'_{ha} = K_A(\gamma_m h_1 + \gamma' h_2 + \gamma' z) \text{ and } \sigma'_{hp} = K_P \gamma' z$$

The net pressure from right to left is given by: $\sigma'_{hn} = \sigma'_{ha} - \sigma'_{hp} = K_A(\gamma_m h_1 + \gamma' h_2) + (K_A - K_P)$
$\gamma' z = \sigma'_1 - (K_P - K_A)\gamma' z$.
The depth z_0 where the net pressure becomes zero (point D) is given by:

$$z_0 = \frac{\sigma'_1}{(K_P - K_A)\gamma'} \tag{14.25}$$

For every unit depth increase below the excavation level, the net pressure σ'_{hn} decreases by $(K_P - K_A)\gamma$. Therefore, the slope of the line GH is 1 vertical to $(K_P - K_A)\gamma'$ horizontal:

$$\therefore \sigma'_2 = \overline{HB} = z_2(K_P - K_A)\gamma' \tag{14.26}$$

At the bottom of the sheet pile, there is passive earth pressure on the right and active earth pressure on the left. They are:

$$\sigma'_{ha} = K_A \gamma' d$$

$$\sigma'_{hp} = K_P(\gamma_m h_1 + \gamma' h_2 + \gamma' d)$$

The net pressure $\sigma'_{hn} = \sigma'_{hp} - \sigma'_{ha} = K_P(\gamma_m h_1 + \gamma' h_2 + \gamma' d) - K_A \gamma' d = \sigma'_3 = \overline{BJ}$.
Substituting $d = z_0 + z_2$:

$$\sigma'_3 = K_P(\gamma_m h_1 + \gamma' h_2) + (K_P - K_A)\gamma' z_0 + (K_P - K_A)\gamma' z_2 = \sigma'_4 + (K_P - K_A)\gamma' z_2 \tag{14.27}$$

where σ'_4 is a known quantity, given by:

$$\sigma'_4 = K_P(\gamma_m h_1 + \gamma' h_2) + (K_P - K_A)\gamma' z_0 \tag{14.28}$$

From equilibrium considerations, Equations 14.21 through 14.24 still hold. The values of A_1 through A_4 are slightly different, replacing *the bulk unit weight with the submerged unit weight* (see Worked Example 5).

14.3.2 In Cohesive Soils

Let's consider a situation where the water level is above the excavation line, and the soil beneath the excavation line is cohesive, as shown in Figure 14.9. Immediately after the installation of the sheet piles, we will treat the clay as undrained with $\phi_u = 0$ (i.e., $K_A = K_P = 1$), use γ_{sat}, and work in terms of total stresses in the clay. As before, we will neglect the water pressure, which is the same on both sides.

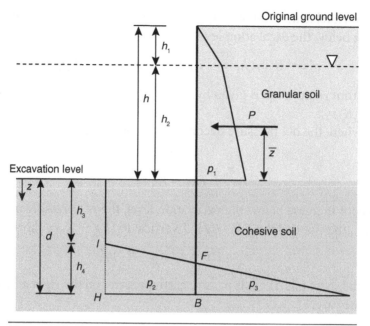

Figure 14.9 Net lateral pressure diagram in cohesive soils

In the granular soil layer to the right of the sheet pile, the active pressures can be computed as before and P and $\bar{z}$ can be determined. For example, at the bottom of the granular soil:

$$p_1 = K_{A,g}(\gamma_{m,g}h_1 + \gamma'_g h_2)$$

At a depth of z within the clay and above the point of rotation to the right of the sheet pile:

$$\sigma_{ha} = (\gamma_{m,g}h_1 + \gamma'_g h_2 + \gamma_{sat,c}z) - 2c_u$$

The subscripts g and c represent granular and cohesive soils respectively.

At a depth of z within the clay and above the point of rotation to the left of the sheet pile:

$$\sigma_{hp} = \gamma_{sat,c}z + 2c_u$$

The net lateral pressure in the clay above the point of rotation, acting from *left to right*, is given by:

$$\sigma_{hn} = \sigma_{hp} - \sigma_{ha} = 4c_u - (\gamma_{m,g}h_1 + \gamma'_g h_2) = p_2 = \overline{HB} \qquad (14.29)$$

At the bottom of the sheet pile, the passive pressure from *right to left* is:

$$\sigma_{hp} = (\gamma_{m,g}h_1 + \gamma'_g h_2 + \gamma_{sat,c}d) + 2c_u$$

and the active pressure from *left to right* is:

$$\sigma_{ha} = \gamma_{sat,c}d - 2c_u$$

Therefore, the net pressure from *right to left* is given by:

$$\sigma_{hn} = \sigma_{hp} - \sigma_{ha} = 4c_u + (\gamma_{m,g}h_1 + \gamma'_g h_2) = p_3 \qquad (14.30)$$

As before, let's include the area *IHBF* on both sides of the pressure diagram to make the solution simpler.

By equating the horizontal forces to zero:

$$P + \frac{1}{2}(p_2 + p_3)h_4 - p_2 d = 0$$

$$P + \frac{1}{2}[4c_u - (\gamma_{m,g}h_1 + \gamma'_g h_2) + 4c_u + (\gamma_{m,g}h_1 + \gamma'_g h_2)]h_4 - [4c_u - (\gamma_{m,g}h_1 + \gamma'_g h_2)]d = 0$$

$$\therefore h_4 = \frac{[4c_u - (\gamma_{m,g}h_1 + \gamma'_g h_2)]d - P}{4c_u} \qquad (14.31)$$

where h_4 is obtained in terms of d. Taking moment about the bottom of the sheet pile and equating this to zero:

$$P(d+\overline{z}) + \frac{1}{2}8c_u \frac{h_4^2}{3} - [4c_u - (\gamma_{m,g}h_1 + \gamma'_g h_2)]\frac{d^2}{2} = 0 \qquad (14.32)$$

From Equations 14.31 and 14.32:

$$[4c_u - (\gamma_{m,g}h_1 + \gamma'_g h_2)]d^2 - 2P d - \frac{P(P + 12c_u \overline{z})}{(\gamma_{m,g}h_1 + \gamma'_g h_2) + 2c_u} = 0 \qquad (14.33)$$

Solving Equation 14.33 by trial and error, d can be determined. In clays, it is required to increase the penetration depth by 40–60%.

14.4 ANCHORED SHEET PILES

When the excavations get deeper (i.e., $h > 6$ m), the loadings on the sheet piles increase significantly, resulting in larger depths of embedment d and larger bending moments, making it necessary to go for thicker sections. Both the depth of the embedment and the section can be reduced by anchoring the sheet pile as shown in Figure 14.10a. Such *anchored sheet piles* or *anchored bulkheads* are commonly used in waterfront structures. Here, the tie rod is attached to the sheet pile and anchored at the other end using a *deadman*, braced piles, sheet piles, etc. A deadman is simply a concrete block that provides anchorage to a tie rod. It can also be in the form of a *continuous beam* to which all tie rods are connected.

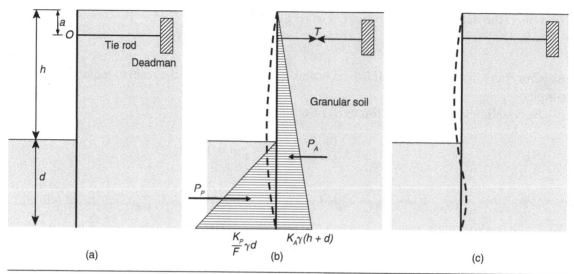

Figure 14.10 Anchored sheet pile: (a) anchored by a deadman (b) free earth support (c) fixed earth support

There are two different methods to design an anchored sheet pile: (a) the *free earth support* method, and (b) the *fixed earth support* method. The free earth support method assumes that the sheet pile is not deep enough to provide fixity at the bottom and allows rotation at the bottom tip of the sheet pile. It acts as a simply supported beam in equilibrium under P_A, P_P, and T. The analysis is quite straightforward and is discussed below. The deflected shape of a sheet pile in the free earth support method is shown by a dashed line in Figure 14.10b. The fixed earth support method assumes that the sheet pile is driven deep enough to provide some fixity at the bottom of the sheet pile, which introduces a reverse bend as shown by the dashed line in Figure 14.10c. The analysis is more complex. The depth of penetration is more for the fixed earth support method, but the maximum bending moment may be less; hence the cross section of the sheet pile can be smaller.

14.4.1 Free Earth Support Method

Let's consider a simple situation where the anchored sheet pile is in dry granular soils, where the lateral pressure distribution is as shown in Figure 14.10b. The active state is fully mobilized on the right side and the passive state is only partially mobilized on the left. The resultant active and passive thrusts are: $P_A = \frac{1}{2}K_A\gamma(h+d)^2$ and $P_P = \frac{1}{2}\frac{K_P}{F}\gamma d^2$. Equating the horizontal forces to zero:

$$T + P_P = P_A \tag{14.34}$$

Taking moment about O:

$$P_A\left(\frac{2}{3}(h+d)-a\right)=P_P\left(h-a+\frac{2}{3}d\right) \tag{14.35}$$

d can be determined from Equation 13.35. T can be obtained by substituting for d in Equation 13.34. The safety factor F on passive resistance is generally 1.5–2.0, as in the case of a cantilever sheet pile. As before, an alternate approach is not to use F and simply increase d by 20–40%. It is also possible to use net pressure diagrams as before.

Example 14.4: Find the depth of embedment d for the anchored sheet pile in sands ($\phi' = 32°$, $\gamma_m = 16.0$ kN/m³, $\gamma_{sat} = 19.5$ kN/m³) as shown below, with a safety factor of 2.0 on passive resistance. Also, find the force on the tie rod, placed at 3 m horizontal intervals.

Solution:

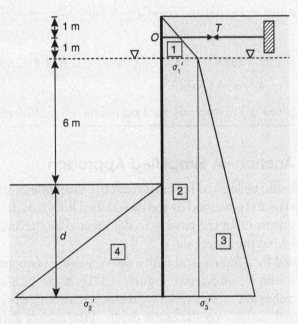

$\phi' = 32° \rightarrow K_A = 0.307$ and $K_P = 3.255$

The horizontal pressures σ'_1, σ'_2, and σ'_3 are given by:

$$\sigma'_1 = 0.307[2 \times 16.0] = 9.82 \text{ kPa}$$

$$\sigma'_2 = \frac{3.255}{2.0}[9.69d] = 15.77d$$

$$\sigma'_3 = 0.307[2 \times 16.0 + (6 + d) \times 9.69] = 2.98d + 27.67 \qquad \text{*Continues*}$$

Example 14.4: *Continued*

Let's divide the pressure diagram into rectangles and triangles as shown, and number them from 1 to 4.

Block	Hor. force (kN/m)	Depth below O (m)	Moment about O (kN-m/m)
1	$0.5 \times 9.82 \times 2 = 9.82$	0.33	3.24
2	$9.82(6 + d) = 58.92 + 9.82d$	$4.0 + 0.5d$	$235.7 + 68.74d + 4.91d^2$
3	$0.5(2.98d + 17.85)(6 + d)$ $= 1.49d^2 + 17.87d + 53.55$	$5.0 + 0.67d$	$d^3 + 19.42d^2 + 125.23d$ $+ 267.75$
4	$0.5 \times 15.77d \times d = 7.89d^2$	$7.0 + 0.67d$	$55.23d^2 + 5.29d^3$

Taking moment about O:

$$3.24 + (235.7 + 68.74d + 4.91d^2) + (d^3 + 19.42d^2 + 125.23d + 267.75) = 55.23d^2 + 5.29d^3$$

$$4.29d^3 + 30.9\ d^2 - 193.97d - 506.69 = 0$$

Solving the above equation by trial and error, $d = 5.3$ m.

For equilibrium,

$$T = P_A - P_P = 9.82 + (58.92 + 9.82d) + (1.49d^2 + 17.87d + 53.55) - (7.89d^2)$$

Substituting $d = 5.3$ m, $T = 89.3$ kN/m.

If the tie rods are spaced at 3 m intervals, the load per tie rod is 267.9 kN.

14.4.2 Deadman Anchor—A Simplified Approach

The deadman anchor should be located far away from the sheet pile as shown in Figure 14.11a, where the minimum distance is governed by the two dashed lines that define the active and passive failure zones. This ensures that the passive wedge created by the anchor does not interfere with the active wedge behind the sheet pile.

The anchor is designed for a higher load with a safety factor F of about 1.5–2. If the anchor is near the ground surface with $b > 0.5d_a$ (see Figure 14.11b), it can be assumed that the anchor and the soil above the anchor act together as a rigid block, with active pressure on the right and passive pressure on the left, acting over the entire depth of the anchor d_a (on DF and AC). In the case of a continuous beam deadman, from equilibrium considerations:

$$F \times T = \left(\frac{1}{2}K_P \gamma d_a^2 - \frac{1}{2}K_A \gamma d_a^2 \right)s = \frac{1}{2}(K_P - K_A)\gamma d_a^2 s \tag{14.36}$$

where T is the tie rod force and s is the horizontal spacing of the tie rods. The depth of the anchor can be determined from Equation 14.36. The same steps apply to isolated anchors as well.

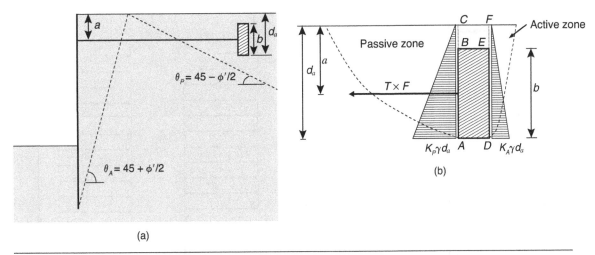

Figure 14.11 Anchor: (a) location (b) force equilibrium

14.5 BRACED EXCAVATIONS

When narrow and deep trenches are excavated for the installation of pipelines or other services, it is necessary to protect the walls against any potential failure. Here, sheet piles are driven into the ground prior to the excavation. As excavation proceeds, *wales* and *struts* are placed from top to bottom. Wales are the beams placed longitudinally along the length of the excavation. Struts are placed between the wales on the opposite sides of the wall to carry the earth pressure in compression. To design the bracing system, it is necessary to know the lateral pressure distribution along the walls of the excavation. Based on the in situ strut load measurements of several excavations under different soil conditions in Chicago and in other areas, Peck (1969) proposed *pressure envelopes* and suggested using them in designs. A schematic diagram of a braced excavation and the pressure envelopes for three different soil conditions are shown in Figure 14.12.

The analysis of the bracing systems to determine the strut loads is a straightforward exercise. It is assumed that all the wall-strut joints, except for those of the top and the bottom struts, act as hinges. In Figure 14.12, joints B, C, D, and E act like hinges that do not carry any moments. The pressure diagrams can be broken along each hinge into several blocks, and equilibrium equations can be written for each block to solve for the unknown strut loads. At any hinge where the pressure diagram is divided, the strut force is broken into two components (e.g., F_2' and F_2''), one acting on each adjacent block. After these components are computed separately, they are added together (i.e., $F_2 = F_2' + F_2''$) to give the strut load.

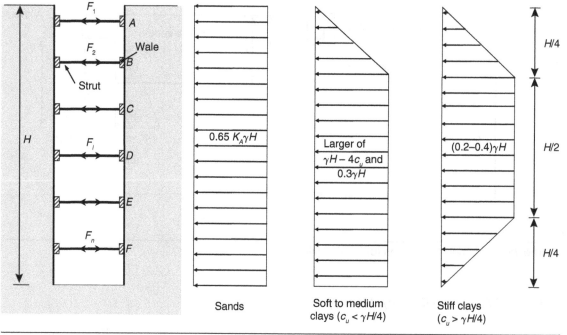

Figure 14.12 Pressure envelopes

Example 14.5: The braced excavation system shown in figure (a) at the top of page 401 is proposed for a 12 m-deep excavation in clays where the unconfined compressive strength is 90 kPa and saturated unit weight is 18.9 kN/m³. Estimate the strut loads if the struts are spaced at 3.5 m intervals horizontally.

Solution: $H = 12$ m, $c_u = 45$ kPa, $\gamma = 18.9$ kN/m³ $\rightarrow c_u < \dfrac{\gamma H}{4} \rightarrow$ soft-medium clay

The pressure diagram is shown in figures b and c on page 401:

$\gamma H - 4 c_u = 18.9 \times 12 - 4 \times 45 = 46.8$ kPa; $0.3\, \gamma H = 0.3 \times 18.9 \times 12 = 68.0$ kPa (larger)

The wall-strut joints for struts 2 and 3 at B and C are taken as hinges, and the lateral pressure diagram is divided (represented by dashed lines) through these hinges into blocks 1, 2, and 3 (see figure c on page 401). The strut loads F_2 and F_3 are split into two components.

For equilibrium of block 1:

$$\sum \text{Moment}_B = 0 \rightarrow F_1 \times 3.0 = \left(\frac{1}{2} \times 68 \times 3.0 \times 3.5\right) \times 3.0 + (68 \times 2 \times 3.5) \times 1.0$$

$$F_1 = 515.7 \text{ kN}$$

Continues

Example 14.5: *Continued*

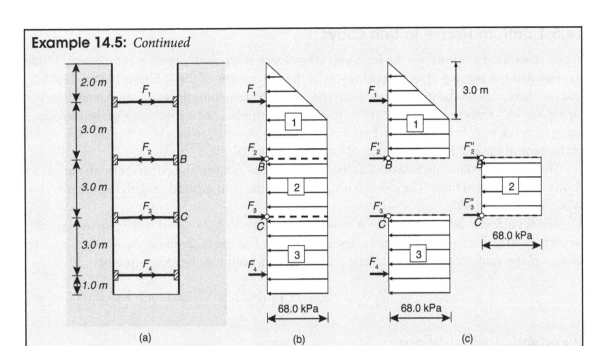

$$\sum \text{Hor.forces} = 0 \rightarrow \quad F_1 + F_2' = \left(\frac{1}{2} \times 68 \times 3.0 \times 3.5\right) + (68 \times 2 \times 3.5) = 833.0 \text{ kN}$$

$$\therefore F_2' = 317.3 \text{ kN}$$

For equilibrium of block 2, and by symmetry:

$$F_2'' = F_3' = \frac{1}{2} \times 68 \times 3.0 \times 3.5 = 357.0 \text{ kN}$$

For equilibrium of block 3:

$$\sum \text{Moment}_C = 0 \rightarrow \quad F_4 \times 3.0 = (68 \times 4 \times 3.5) \times 2.0 \rightarrow F_4 = 634.7 \text{ kN}$$

$$\sum \text{Hor.forces} = 0 \rightarrow \quad F_4 + F_3'' = 68 \times 4 \times 3.5 = 952.0 \text{ kN} \rightarrow F_3'' = 317.3 \text{ kN}$$

Strut load summary:

$$F_1 = 515.7 \text{ kN}$$

$$F_2 = 317.3 + 357.0 = 674.3 \text{ kN}$$

$$F_3 = 357.0 + 317.3 = 674.3 \text{ kN}$$

$$F_4 = 634.7 \text{ kN}$$

14.5.1 Bottom Heave in Soft Clays

When braced excavations are made in soft clays, there is a possibility of *bottom heave*. While the wall and the bracing system remain stable, the self-weight of the soil next to the sheet pile and the surcharge on the ground can push the soil at the base into the excavation, endangering safety. We will analyze this problem by treating *eb* as the base of a footing in undrained clays using c_u and $\phi_u = 0$. In the case of a long cut, the base is assumed to be a strip footing. The width of the excavation is *B* and the depth is *H* (see Figure 14.13).

The assumed failure surface *abcd* consists of straight lines (*ab* and *cd*) and circular arc *bc* as shown by the dashed lines. The circular arc extends to the firm ground underlying the soft clay. When $\phi_u = 0, \alpha = 45°$.

The net ultimate bearing capacity at *eb* is $c_u N_{c,\text{strip}}$ where $N_{c,\text{strip}}$ (= 5.14) is the bearing capacity factor for a strip footing in undrained clays. If the length *L* of the excavation is not long enough to assume plane-strain conditions (i.e., strip footing), the net ultimate bearing capacity is:

$$q_{\text{ult, net}} = c_u N_{c,\text{strip}} \left(1 + 0.2 \frac{B'}{L} \right) \tag{14.37}$$

The net applied pressure at *eb* is:

$$q_{\text{app,net}} = \frac{\text{Load}}{\text{Area}} = \frac{B'LH\gamma + qB'L - c_u HL}{B'L} = \gamma H + q - c_u \frac{H}{B'} \tag{14.38}$$

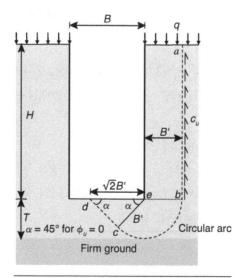

Figure 14.13 Bottom heave in soft clays

Therefore, the safety factor against bottom heave is given by:

$$F_{\text{bottom heave}} = \frac{c_u N_{c,\text{strip}}\left(1+0.2\dfrac{B'}{L}\right)}{\gamma H + q - c_u \dfrac{H}{B'}} \tag{14.39}$$

When the firm ground is near the bottom of the excavation with $T \le B/\sqrt{2}$, de is less than B and $B' = T$. When T exceeds $B/\sqrt{2}$, de extends to the full width of the excavation, and the circular arc failure surface bc would not be tangent to the firm ground underneath. Here, $B' = B/\sqrt{2}$. In other words, B' should be taken as the smaller of T and $B/\sqrt{2}$ in Equation 14.39. $F_{\text{bottom heave}}$ should be greater than 1.5. When the safety factor is less, the sheet pile is driven further into the ground.

Example 14.6: Is there a possibility of bottom heave in the braced excavation from Example 14.5? Assume that the width of the excavation is 4.0 m.

Solution: Assuming the firm ground is not in the vicinity (i.e., T is large):

$$B' = \frac{B}{\sqrt{2}} = \frac{4}{\sqrt{2}} = 2.83 \text{ m}$$

$$F_{\text{bottom heave}} = \frac{c_u N_{c,\text{strip}}\left(1+0.2\dfrac{B'}{L}\right)}{\gamma H + q - c_u \dfrac{H}{B'}} = \frac{45 \times 5.14}{18.9 \times 12 - 45 \times \dfrac{12}{2.83}} = 6.4$$

Therefore the braced excavation is quite safe against any possible bottom heave.

Reminder

❖ Rankine's earth pressure theories (rather than Coulomb's) are used in the designs of retaining walls, sheet piles, and braced excavations.

❖ In designing retaining walls, always check for sliding, overturning, and bearing capacity failures.

❖ For cantilever sheet piles, the method using the net pressure diagram (Method 2) is slightly better, although it requires a little more work. In calculating the constants A_1 through A_4, replace γ with γ' when the water level is above the excavation line. For any other situation, go from the first principles or use the simplified method (Method 1).

❖ In clays under undrained conditions, $\phi_u = 0$, and hence $K_A = K_P = 1$. Use γ_{sat} and analyze in terms of total stresses within the clays.

WORKED EXAMPLES

1. The cantilever retaining wall shown on the left of the figure at the top of page 405 retains a sandy backfill with $\phi' = 36°$ and $\gamma = 18$ kN/m³. Evaluate the stability of the wall against sliding, overturning, and bearing capacity failure. Assume that the same sand exists on the passive side and below the wall. $\gamma_{concrete} = 24$ kN/m³. Draw the pressure distribution beneath the retaining wall.

Solution: $\phi' = 36° \rightarrow K_A = 0.260$ and $K_P = 3.852$

The free body diagram is shown on the right of the figure on page 405:

$$P_A = 0.5\, K_A \gamma H^2 = 0.5 \times 0.260 \times 18 \times 5.5^2 = 70.8 \text{ kN per m width}$$

$$P_P = 0.5\, K_P \gamma h^2 = 0.5 \times 3.852 \times 18 \times 0.5^2 = 8.7 \text{ kN per m width}$$

Let's tabulate the values of W_i, x_i, and $W_i x_i$ as below.

Block No.	Weight W_i (kN per m)	Hor. Distance x_i (m)	$W_i x_i$ (kN-m per m)
1	$0.5 \times 5 \times 24 = 60.0$	1.5	90.0
2	$3.25 \times 0.5 \times 24 = 39.0$	1.625	63.4
3	$1.5 \times 5 \times 18 = 135.0$	2.5	337.5
	$\Sigma W_i = 234.0$ kN per m		$\Sigma W_i x_i = 490.9$ kNm per m

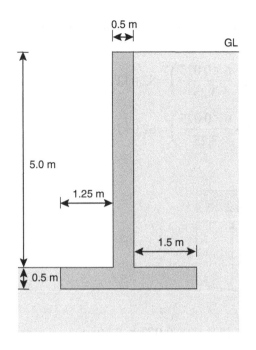

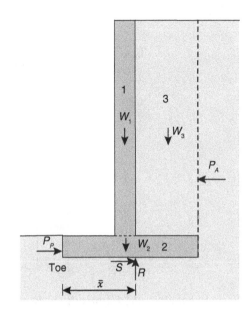

Let's take δ as ⅔ $\phi' \rightarrow \delta = 24°$:

$$F_{sliding} = \frac{\text{Maximum resistance available}}{\text{Driving force}} = \frac{P_P + S_{max}}{P_A} = \frac{8.7 + 234.0\tan 24}{70.8} = 1.59 > 1.5$$

$$F_{overturning} = \frac{\text{Resisting moment}}{\text{Driving force}} = \frac{\left(\sum_{i=1}^{n} W_i x_i\right) + P_P\dfrac{h}{3}}{P_A\dfrac{H}{3}} = \frac{490.9 + 8.7 \times 0.17}{70.8 \times 1.83} = 3.80 > 2$$

Bearing pressures:

$$R = \Sigma W_i = 234.0 \text{ kN per m}; S = P_A - P_P = 70.8 - 8.7 = 62.1 \text{ kN per m}$$

Note: We are assuming that the passive resistance is fully mobilized while the sliding resistance is only partially mobilized.

Inclination of the load to vertical:

$$\alpha = \tan^{-1}\frac{S}{R} = \tan^{-1}\frac{62.1}{234} = 14.9°$$

$$\bar{x} = \frac{\left(\sum_{i=1}^{n} W_i x_i\right) + P_P\dfrac{h}{3} - P_A\dfrac{H}{3}}{R} = \frac{490.9 + 8.7 \times 0.167 - 70.8 \times 1.83}{234} = 1.550 \text{ m}$$

∴ eccentricity, $e = 1.625 - 1.550 = 0.075$ m or 75 mm ← well within the middle third of the base

$$q_{max} = \frac{Q}{B}\left(1 + \frac{6e}{B}\right) = \frac{234}{3.25}\left(1 + \frac{6 \times 0.075}{3.25}\right) = 82.0 \text{ kPa}$$

$$q_{min} = \frac{Q}{B}\left(1 - \frac{6e}{B}\right) = \frac{234}{3.25}\left(1 - \frac{6 \times 0.075}{3.25}\right) = 62.0 \text{ kPa}$$

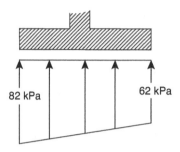

82 kPa   62 kPa

Bearing capacity calculations: $B' = B - 2e = 3.25 - 2 \times 0.075 = 3.10$ m

Plane-strain correction: $\phi' = 1.1 \times 36 = 39.6° \rightarrow N_q = 60$, $N_{\gamma, \text{Meyerhof}} = 86$

Shape factors: $s_q = s_\gamma = 1$

Depth factors:

$$d_q = d_\gamma = 1 + 0.1\frac{D_f}{B}\tan\left(45 + \frac{\phi}{2}\right) = 1 + 0.1 \times \frac{0.5}{3.25} \times \tan 64.8 = 1.04$$

Inclination factors:

$$i_q = \left(1 - \frac{\alpha°}{90}\right)^2 = \left(1 - \frac{14.9}{90}\right)^2 = 0.70$$

$$i_\gamma = \left(1 - \frac{\alpha}{\phi}\right)^2 = \left(1 - \frac{14.9}{39.6}\right)^2 = 0.39$$

Applying Equation 12.7:

$$q_{ult} = s_q d_q i_q \gamma_1 D_f N_q + s_\gamma d_\gamma i_\gamma 0.5 \, B \, \gamma_2 N_\gamma$$

$$q_{ult,gross} = 1.0 \times 1.04 \times 0.70 \times 18 \times 0.5 \times 60 + 1.0 \times 1.04 \times 0.39 \times 0.5 \\ \times 3.10 \times 18 \times 86 = 1366.3 \text{ kPa}$$

$$q_{ult,net} = 1366.3 - 0.5 \times 18 = 1357.3 \text{ kPa}$$

From Equation 14.11:

$$F_{bearing\ capacity} = \frac{1357.3}{82.0} = 16.6 > 3$$

The retaining wall is very safe with respect to sliding, overturning, and bearing capacity.

2. A gravity retaining wall shown on the left retains a sandy backfill with $\phi' = 34°$ and $\gamma = 18\ kN/m^3$. Analyze the stability of the wall. $\gamma_{concrete} = 24\ kN/m^3$.

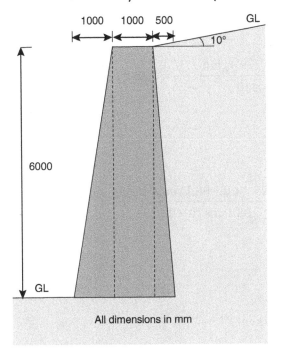

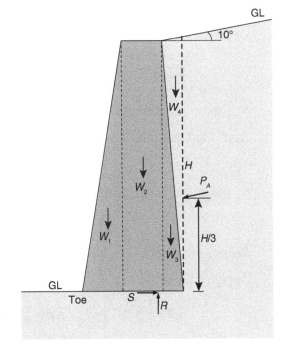

Solution: The free body diagram is shown on the right of the above figure.

Since the backfill is inclined, Equation 10.12 will be used for computing K_A.

$$K_A = \cos\beta \frac{\cos\beta - \sqrt{\cos^2\beta - \cos^2\phi'}}{\cos\beta + \sqrt{\cos^2\beta - \cos^2\phi'}} = 0.9848 \times \frac{0.9848 - \sqrt{0.9848^2 - 0.8290^2}}{0.9848 = \sqrt{0.9848^2 - 0.8290^2}} = 0.294$$

$$H = 6000 + 500\tan 10 = 6088\ mm$$

$$P_A = 0.5\ K_A\gamma H^2 = 0.5 \times 0.294 \times 18 \times 6.088^2 = 98.1\ kN\ per\ m\ width$$

$$P_P = 0$$

Let's tabulate the values of W_i, x_i, and $W_i x_i$ as shown on page 408.

Block No.	Weight W_i (kN per m)	x_i (m)	$W_i x_i$ (kN-m per m)
1	$0.5 \times 1 \times 6 \times 24 = 72.0$	0.667	48.0
2	$1.0 \times 6 \times 24 = 144.0$	1.500	216.0
3	$0.5 \times 0.5 \times 6.0 \times 24 = 36.0$	2.167	78.0
4	$0.5 \times 0.5 \times 6.088 \times 18 = 27.4$	2.333	63.9
	$\Sigma W_i = 279.4$ kN per m		$\Sigma W_i x_i = 405.9$ kN-m per m

Let's take δ as $\frac{2}{3}\ \phi' \rightarrow \delta = 22.7°$.

Applying Equation 14.2:

$$F_{\text{sliding}} = \frac{P_P + S_{\max}}{P_A \cos\beta} = \frac{0 + (98.1 \sin 10 + 279.4) \tan 22.7}{98.1 \cos 10} = 1.28 < 1.5$$

$S_{\max}$ above was obtained from Equation 14.1.

Applying Equation 14.5:

$$F_{\text{overturning}} = \frac{\left(\sum_{i=1}^{n} W_i x_i\right) + P_P \dfrac{h}{3} + P_A \sin\beta\, b}{P_A \cos\beta \dfrac{H}{3}} = \frac{405.9 + 0 + 98.1 \sin 10 \times 2.5}{98.1 \cos 10 \times 2.029} = 2.29 > 2$$

Bearing pressures:

Applying Equation 14.6:

$$R = \left(\sum_{i=1}^{n} W_i\right) + P_A \sin\beta = 279.4 + 98.1 \sin 10 = 296.4 \text{ kN per m}$$

$$S = 98.1 \times \cos 10 = 96.6 \text{ kN per m}$$

Inclination of the load to vertical:

$$\alpha = \tan^{-1} \frac{S}{R} = \tan^{-1} \frac{96.6}{296.4} = 18.1°$$

Substituting in Equation 14.9:

$$\overline{x} = \frac{\left(\sum_{i=1}^{n} W_i x_i\right) + P_P \dfrac{h}{3} + P_A \sin\beta\, b - P_A \cos\beta \dfrac{H}{3}}{R}$$

$$= \frac{405.9 + 0 + 98.1 \sin 10 \times 2.5 - 98.1 \cos 10 \times 2.029}{296.4} = 0.852 \text{ m}$$

∴ eccentricity $e = 1.25 - 0.852 = 0.398$ m $< B/6$ ... lies within the middle third.

$$q_{max} = \frac{Q}{B}\left(1 + \frac{6e}{B}\right) = \frac{296.4}{2.5}\left(1 + \frac{6 \times 0.398}{2.5}\right) = 231.8 \text{ kPa}$$

Bearing capacity calculations: $B' = B - 2e = 2.5 - 2 \times 0.398 = 1.70$ m

Plane-strain correction: $\phi' = 1.1 \times 34 = 37.4° \rightarrow N_q = 45, N_{\gamma, \text{Meyerhof}} = 57$

Shape factors: $s_q = s_\gamma = 1$

Depth factors: Since $D_f = 0, d_q = d_\gamma = 1$

Inclination factors:

$$i_\gamma = \left(1 - \frac{\alpha}{\phi}\right)^2 = \left(1 - \frac{18.1}{37.4}\right)^2 = 0.27$$

Applying Equation 12.7:

$$q_{\text{ult,gross}} = s_\gamma d_\gamma i_\gamma \, 0.5 \, B \, \gamma_2 N_\gamma = 0.27 \times 0.5 \times 1.70 \times 18 \times 57 = 235.5 \text{ kPa}$$

Since $D_f = 0, q_{\text{ult,net}} = 235.5$ kPa

$$\therefore F_{\text{bearing capacity}} = \frac{q_{\text{ult}}}{q_{max}} = \frac{235.5}{231.8} = 1.02 < 3.0$$

The wall is not safe with respect to sliding and bearing capacity; it is safe against overturning failure.

3. It is proposed to drive a cantilever sheet pile to a depth of $h + d$ into dry granular soil to support excavation to a depth h. It is proposed to incorporate the safety factor by (a) increasing the computed depth of penetration d_0 by 30% and (b) providing a safety factor of 1.75 on the passive resistance. Plot d/h against the friction angle for both cases (a) and (b) where d is the final depth of penetration below the excavation line.

Solution: Let's use Equation 14.12 and the expression developed in Example 14.2. The values of K_A and K_P along with those of d_0/h and d/h computed by the two methods are given in the table and figure on page 410.

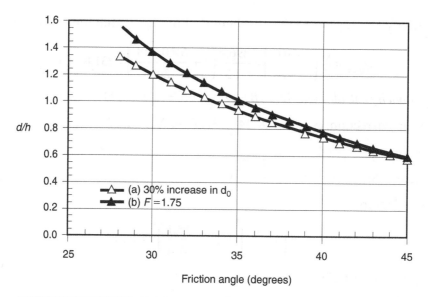

Friction angle (degrees)

φ (deg)	K_A	K_P	(a)		(b)	
			d_0/h	d/h	d_0/h	d/h
28	0.361	2.770	1.029	1.337	1.571	1.571
29	0.347	2.882	0.975	1.268	1.469	1.469
30	0.333	3.000	0.926	1.204	1.377	1.377
31	0.320	3.124	0.879	1.143	1.293	1.293
32	0.307	3.255	0.836	1.087	1.216	1.216
33	0.295	3.392	0.795	1.034	1.145	1.145
34	0.283	3.537	0.757	0.984	1.079	1.079
35	0.271	3.690	0.720	0.937	1.019	1.019
36	0.260	3.852	0.686	0.892	0.962	0.962
37	0.249	4.023	0.654	0.850	0.910	0.910
38	0.238	4.204	0.623	0.810	0.861	0.861
39	0.228	4.395	0.594	0.772	0.815	0.815
40	0.217	4.599	0.566	0.736	0.772	0.772
41	0.208	4.815	0.540	0.702	0.732	0.732
42	0.198	5.045	0.515	0.670	0.694	0.694
43	0.189	5.289	0.491	0.639	0.658	0.658
44	0.180	5.550	0.468	0.609	0.625	0.625
45	0.172	5.828	0.447	0.581	0.593	0.593

4. The water table in a granular soil is 3.5 m below the ground level. It is required to excavate the top 2.5 m. How deep would you drive the sheet pile providing a safety factor of 2
against passive resistance? The unit weights of the granular soil above and below the water
table are 17 kN/m³ and 20 kN/m³ respectively, and the friction angle is 35°.

Solution: $\phi' = 35° \rightarrow K_A = 0.271$ and $K_P = 3.690$

Let's assume that the sheet pile has to be driven to a depth of x below the water table as shown in the figure. At any depth, the active earth pressure and passive earth pressure in a granular soil are given by:

$$\sigma'_{ha} = K_A \sigma'_v \text{ and } \sigma'_{hp} = K_P \sigma'_v$$

where K_P will be replaced by K_P/F, with $F = 2$.

The values of σ'_{ha} and σ'_{hp} thus are computed as follows.

On the right (active) side, at the water table, $\sigma'_v = 3.5 \times 17 = 59.5$ kPa

$$\therefore \sigma'_{ha} = K_A \sigma'_v = 0.271 \times 59.5 = 16.12 \text{ kPa}$$

On the right (active) side, at the bottom of the sheet pile,

$$\sigma'_v = 3.5 \times 17 + x \times (20 - 9.81) = 59.5 + 10.19 \, x \text{ kPa}$$
$$\therefore \sigma'_{ha} = K_A \sigma'_v = 16.12 + 2.76 \, x \text{ kPa}$$

On the left (passive) side, at the water table, $\sigma'_v = 1.0 \times 17 = 17.0$ kPa

$$\therefore \sigma'_{hp} = K_P \sigma'_v/F = 3.690 \times 17.0/2 = 31.37 \text{ kPa}$$

On the left (passive) side, at the bottom of the sheet pile,

$$\sigma'_v = 1.0 \times 17 + x \times (20 - 9.81) = 17.0 + 10.19 \, x \text{ kPa}$$
$$\therefore \sigma'_{hp} = K_P \sigma'_v/F = 31.37 + 18.80 \, x \text{ kPa}$$

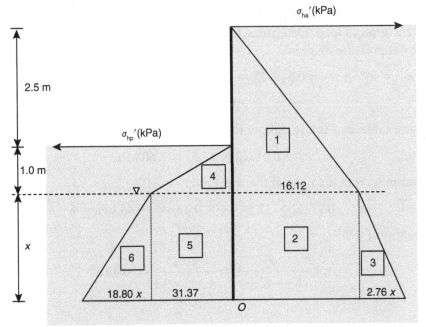

The lateral pressure diagram is divided into six triangular and rectangular blocks as shown in the figure on page 411. The horizontal load contribution from each block, the height of its location above the base, and the moment about O are summarized below. The pore water pressure acts equally on both sides, and hence is not considered in the analysis.

Block	Force (kN) per m	Height (m) above O	Moment (kN-m) per m
1	$0.5 \times 16.12 \times 3.5 = 28.21$	$x + 1.17$	$28.21x + 32.92$
2	$16.12 \times x = 16.12x$	$0.5x$	$8.06x^2$
3	$0.5 \times 2.76x \times x = 1.38x^2$	$0.33x$	$0.46x^3$
4	$0.5 \times 31.37 \times 1 = 15.69$	$x + 0.333$	$15.69x + 5.23$
5	$31.37 \times x = 31.37x$	$0.5x$	$15.69x^2$
6	$0.5 \times 18.80x \times x = 9.4x^2$	$0.33x$	$3.13x^3$

Taking moments about O:

$$28.21x + 32.92 + 8.06x^2 + 0.46x^3 - 15.69x - 5.23 - 15.69x^2 - 3.13x^3 = 0$$

$$2.67x^3 + 7.63x^2 - 12.52x - 27.69 = 0$$

Solving the above equation by trial and error, $x = 2.02$ m.

∴ The sheet pile has to be driven to 5.5 m to provide a safety factor of 2 on passive resistance.

5. The figure on the top of page 413 shows a cantilever sheet pile driven into a granular soil where the water table is 2 m below the top of the sand. The properties of the sand are: $\phi' = 40°$, $\gamma_m = 17.5$ kN/m³, and $\gamma_{sat} = 19$ kN/m³. It is proposed to excavate to a depth of 6 m below the ground level. Determine the depth to which the sheet pile must be driven, using the net pressure diagram.

Solution: $\phi' = 40° \rightarrow K_A = 0.217$, $K_P = 4.599$, and $K_P - K_A = 4.382$

$$\gamma' = 19 - 9.81 = 9.19 \text{ kN/m}^3$$

At the water table level to the right of the sheet pile:

$$\sigma'_h = 0.217(2 \times 17.5) = 7.60 \text{ kPa.}$$

At excavation level to the right of the sheet pile:

$$\sigma'_h = 0.271(2 \times 17.5 + 4 \times 9.19) = 19.45 \text{ kPa} = \sigma'_1.$$

From Equation 14.25:

$$z_0 = \frac{\sigma'_1}{(K_P - K_A)\gamma'} = \frac{19.45}{4.382 \times 9.19} = 0.48 \text{ m}$$

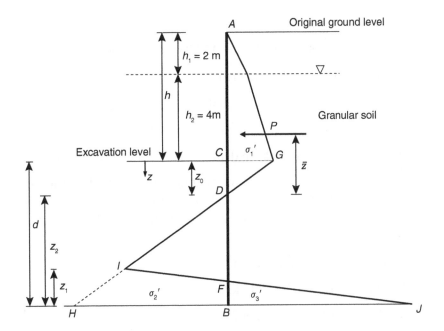

Let's compute P and $\bar{z}$:

$$P = (0.5 \times 7.60 \times 2) + (7.60 \times 4) + (0.5 \times 11.85 \times 4) + (0.5 \times 19.45 \times 0.48)$$
$$= 7.60 + 30.40 + 23.70 + 4.67 = 66.37 \text{ kN/m}$$

$$P\bar{z} = 7.60(0.67 + 4 + 0.48) + 30.40\,(2 + 0.48) + 23.70(1.33 + 0.48) + 4.67 \times 0.32$$
$$= 158.92 \text{ kNm/m}$$

$$\therefore \bar{z} = \frac{158.92}{66.37} = 2.39 \text{ m}$$

From Equation 14.28:

$$\sigma'_4 = K_P\,(\gamma_m h_1 + \gamma' h_2) + (K_P - K_A)\gamma' z_0 = 4.599(17.5 \times 2 + 9.19 \times 4)$$
$$+ 4.382 \times 9.19 \times 0.48 = 349.35 \text{ kPa}$$

$$A_1 = \frac{\sigma'_4}{\gamma'(K_P - K_A)} = \frac{349.35}{9.19 \times 4.382} = 8.68 \text{ m}$$

$$A_2 = \frac{8\,P}{\gamma'(K_P - K_A)} = \frac{8 \times 66.37}{9.19 \times 4.382} = 13.18 \text{ m}^2$$

$$A_3 = \frac{6P[2\bar{z}\gamma'(K_P - K_A) + \sigma'_4]}{\gamma'^2(K_P - K_A)^2} = \frac{6 \times 66.37[2 \times 2.39 \times 9.19 \times 4.382 + 349.35]}{9.19^2 \times 4.382^2} = 133.05 \text{ m}^3$$

$$A_4 = \frac{P(6\bar{z}\sigma'_4 + 4P)}{\gamma'^2(K_P - K_A)^2} = \frac{66.37(6 \times 2.39 \times 349.35 + 4 \times 66.37)}{9.19^2 \times 4.382^2} = 215.89 \text{ m}^4$$

Substituting in Equation 14.24:

$$z_2^4 + 8.68z_2^3 - 13.18z_2^2 - 133.05z_2 - 215.89 = 0$$

Solving the above equation by trial and error:

$$z_2 = 4.30 \text{ m}$$

$$d = z_2 + z_0 = 4.30 + 0.48 = 4.78 \text{ m}$$

Increasing d by 30%, let's provide a total depth of 12.25 m below the original ground level.

6. The top 4 m at a site consists of sand, which is underlain by clay. The water table is 1 m below the ground level. A sheet pile is to be driven into the ground to support an excavation to the top of the clay layer. How deep would you drive the sheet pile into the ground? The soil properties are as follows.

Sand: $\gamma_m = 16.0 \text{ kN/m}^3$, $\gamma_{sat} = 19.5 \text{ kN/m}^3$, $\phi' = 32°$

Clay: $\gamma_{sat} = 19.5 \text{ kN/m}^3$, $\phi_u = 0$, $c_u = 45 \text{ kPa}$

Solution: In sand, $\phi' = 32° \rightarrow K_A = 0.307$

Let's refer to Figure 14.9 and follow the procedure discussed in Section 14.3.2, with $h_1 = 1$ m and $h_2 = 3$ m.

At 1 m depth, $\sigma'_{ha} = 0.307 \times 16 \times 1 = 4.91 \text{ kPa}$.

At the bottom of the sand,

$$\sigma'_{ha} = p_1 = 0.307[1 \times 16 + 3(19.5 - 9.81) = 13.84 \text{ kPa}.$$

From the net pressure diagram (see Figure 14.9):

$$P = 0.5 \times 4.91 \times 1 + 4.91 \times 3 + 0.5 \times 8.93 \times 3 = 2.46 + 14.73 + 13.40 = 30.59 \text{ kN per m}$$

$$\bar{z} = \frac{2.46 \times 3.333 + 14.73 \times 1.5 + 13.40 \times 2.0}{30.59} = 1.87 \text{ m}$$

From Equation 14.33:

$$[4c_u - (\gamma_{m,g}h_1 + \gamma'_g h_2)]d^2 - 2Pd - \frac{P(P + 12c_u\bar{z})}{(\gamma_{m,g}h_1 + \gamma'_g h_2) + 2c_u} = 0$$

$$[4 \times 45 - (16 \times 1 + 9.69 \times 3)]d^2 - 2 \times 30.59\, d - \frac{30.59(30.59 + 12 \times 45 \times 1.87)}{(16 \times 1 + 9.69 \times 3) + 2 \times 45} = 0$$

$$134.93\, d^2 - 61.18\, d - 235.62 = 0$$

By trial and error, $d = 1.57$ m. Substituting d = 1.57 m in Equation 14.31:

$$h_4 = \frac{[4 \times 45 - (16 \times 1 + 9.69 \times 3)]1.57 - 30.59}{4 \times 45} = 1.01 \text{ m}$$

From Equation 14.29:

$$p_2 = 4c_u - (\gamma_{m,g}h_1 + \gamma'_g h_2) = 4 \times 45 - (16 \times 1 + 9.69 \times 3) = 134.93 \text{ kPa}$$

From Equation 14.30:

$$p_3 = 4c_u + (\gamma_{m,g}h_1 + \gamma'_g h_2) = 4 \times 45 + (16 \times 1 + 9.69 \times 3) = 225.07 \text{ kPa}$$

The net pressure diagram can be drawn from these.

Let's increase d by 50% to 2.35 m. The sheet pile has to be driven to 6.2 m below the ground level.

REVIEW EXERCISES

1. Write a 500-word essay on crib walls. Include pictures as appropriate.

2. Write a 500-word essay on diaphragm walls. Include pictures as appropriate.

3. The retaining wall shown in the figure at the top of page 416 is designed to retain a 4.5 m-high sandy backfill that has a friction angle of 34° and a unit weight of 17 kN/m³. The base of the wall rests on the existing ground that consists of clayey sand having an effective cohesion and a friction angle of 10 kPa and 35° respectively. The unit weights of the clayey sand and concrete are 18 kN/m³ and 23 kN/m³ respectively. Find the safety factor of the retaining wall with respect to sliding and overturning.

   Does the eccentricity at the base exceed B/6?

   What is the contact pressure beneath the toe of the wall?

   Answer: 2.34, 2.20; No; 62 kPa

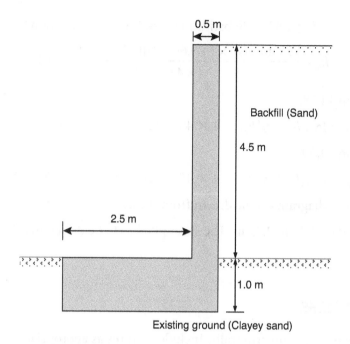

4. To retain the sandy backfill shown above (Review Exercise 3) and with the same ground conditions, an alternate design is proposed where the same retaining wall is placed as the mirror image as shown in figure. Find the safety factor of the wall with respect to sliding and overturning.

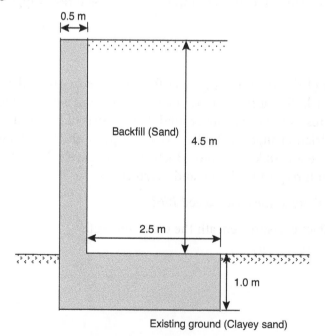

Check the stability with respect to bearing capacity, making the necessary assumptions regarding the mobilization of passive resistance and/or sliding resistance, which may not be fully mobilized.

Answer: 3.68, 3.78

5. Compare the safety factors in Review Exercises 3 and 4. Comment on how the backfill contributes to the stability. What improvements would you suggest to these design alternatives?

6. The cantilever retaining wall shown in the figure retains a sandy backfill with a unit weight of 17.2 kN/m^3 and a friction angle of 35°. The unit weight of concrete is 23.0 kN/m^3. Find the safety factor of the wall with respect to sliding and overturning.

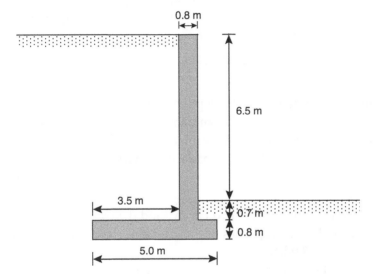

Answer: 2.4, 4.6

7. A cantilever retaining wall is proposed for retaining a loose granular backfill with $\phi' = 29°$ as shown in the figure at the top of page 418. The existing ground consists of silty sands where $\phi' = 32°$. Assuming an average unit weight of 18 kN/m^3 for both soils and 23 kN/m^3 for concrete:
   a. Find the magnitudes and locations of the active and passive thrusts on both sides of the wall.
   b. Find the safety factors with respect to sliding and overturning.
   c. Suggest any improvements to the proposed design.

Answer: 75 kN per m at 1.72 m above the base of the wall, 29 kN per m at 0.33 m above the base of the wall; 1.2, 2.7.

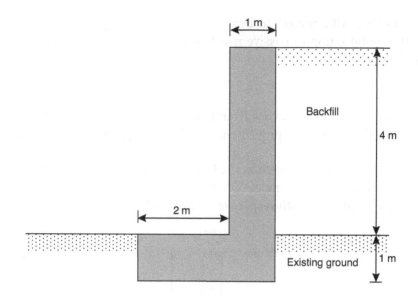

8. In a sandy soil, the water table lies at a depth of 5 m below the ground level. The properties of the sand are: $\phi' = 33°$, $\gamma_m = 17$ kN/m³, and $\gamma_{sat} = 19$ kN/m³. It is required to excavate to a depth of 4.0 m. Using method 1, estimate the depth to which the sheet pile must be driven; assume a safety factor of 1.5 on passive resistance.

Answer: 8.8 m

9. Five meters of sand overlies a saturated clayey sand deposit, and the water table lies at the top of the clayey sand. The properties of the sand and clayey sand are given below.
Sand: $\phi' = 33°$, $\gamma_m = 18.0$ kN/m³
Clayey sand: $c' = 10$ kPa, $\phi' = 31°$, $\gamma_{sat} = 19.5$ kN/m³
How deep would you drive the sheet pile (use method 1)?

Answer: 10 m

10. In a medium-dense sand deposit where the water table is at a depth of 5 m, sheet piles have to be driven to facilitate some excavation work. The properties of the sand are: $\phi' = 34°$, $\gamma_m = 17.5$ kN/m³, and $\gamma_{sat} = 19.0$ kN/m³. How deep should the sheet pile be driven into the sand to excavate to the water table level with a safety factor of 1.5 on the passive resistance? Use both methods (Method 1–Simplified analysis, and Method 2–Net lateral pressure diagram) to solve the problem.

Answer: Method 1: 12.3 m, Method 2: 11.4 m

11. An 8.0 m-deep excavation is made into a sandy soil using anchored sheet piles to support the walls of the excavation. The water table is at a depth of 4 m. The sand has a friction

angle of 37°, bulk unit weight of 17.0 kN/m³, and a saturated unit weight of 20.0 kN/m³. The tie rods are placed at a depth of 1.5 m and horizontal intervals of 2.0 m, tied to a continuous deadman anchor. Assuming a safety factor of 1.5 on passive resistance, estimate the depth to which the sheet pile must be driven. What is the force on the tie rod? Design a continuous anchor, giving a sketch.

Answer: 11.3 m, 152 kN

12. A 6.0 m-wide braced excavation shown in the figure is carried out in clay having the following properties: $c_u = 20.0$ kPa, $\phi_u = 0$, and $\gamma_{sat} = 18.5$ kN/m³. The struts are spaced 5.0 m center-to-center in plan. Determine the strut forces and the factor of safety against bottom heave.

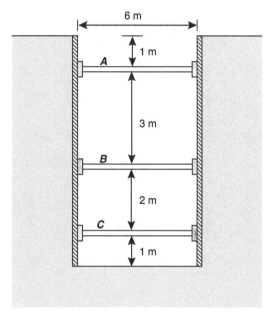

Answer: 413 kN, 546 kN, 557 kN; F = 1.07

13. A 3.0 m-wide braced excavation (see the figure at the top of page 420) is to be made to a depth of 9.0 m in a saturated clay deposit having a saturated unit weight of 17.8 kN/m³ and undrained shear strength of 30 kPa. The struts are spaced at 3.0 m horizontal intervals. Find the strut forces and the safety factor with respect to bottom heave.

Answer: 83 kN, 257 kN, 286 kN, 277 kN, 224 kN; F = 4.7

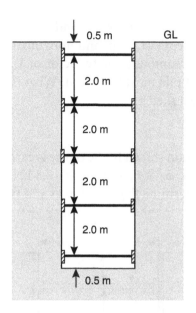

Slope Stability

15

15.1 INTRODUCTION

Slopes can be natural or artificial. Natural slopes occur in hilly terrains, or can be created by earthquakes, landslides, erosion, ground subsidence, etc. Artificial slopes are created in the process of building embankments or carrying out excavations. When the ground is not horizontal, it is possible that part of the soil mass from the higher ground will slide downward, potentially rendering the slope unstable. Figure 15.1 shows a slope failure along the banks of a river.

Figure 15.1 A slope failure (Courtesy of Dr. Kirralee Rankine, Golder Associates, Australia)

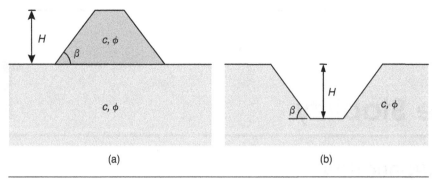

Figure 15.2 Slopes: (a) embankment (b) excavation

The stability of the slopes of the embankment (Figure 15.2a) or excavation (Figure 15.2b) depends on three major factors, height H, slope angle β, and shear strength parameters (c, ϕ). Increasing the height or slope angle reduces the stability. Larger shear strength parameters, c and ϕ, result in increased shear strength and improved stability. The slopes analyzed in this chapter are two-dimensional with the third dimension assumed infinitely long. This can be assumed a plane-strain loading situation.

15.2 SLOPE FAILURE AND SAFETY FACTOR

Let's consider the slope shown in Figure 15.3a where failure can take place along an unknown failure surface. It is a two-dimensional plane-strain problem where the dimension perpendicular to the paper is very long. Observations of previous slope failures suggest that the two-dimensional failure surface can be approximated by a circular arc. There can be thousands of potential failure circles as shown in Figure 15.3b, and the failure will take place along the most *critical slip circle* with the lowest *safety factor*. How do we define the safety factor?

Let's assume there is a possibility of failure along the arc AB in Figure 15.3a. The self-weight W of the sliding mass ABC induces instability, which is resisted by the shear strength mobilized along the failure surface AB. The mobilized shear strength τ_{mob} is the shear stress acting along the arc AB, maintaining equilibrium. If the shear stress (i.e., mobilized shear strength) acting along the arc AB is less than the shear strength τ_f, the slope is stable. The safety factor for this potential failure circle can be defined as:

$$F = \frac{\tau_f}{\tau_{mob}} \qquad (15.1)$$

The type of failure shown in Figures 15.3 and 15.4a is *rotational*, where the failure mass rotates about a center and the failure surface takes the shape of a circular arc. This is quite common in homogeneous soils. When a relatively thin layer of weak soil overlies a stiff stratum over a

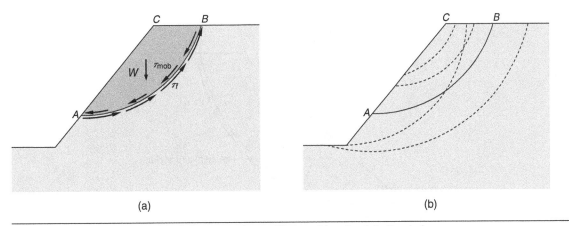

Figure 15.3 Failure circles: (a) sliding mass in equilibrium (b) potential slip circles

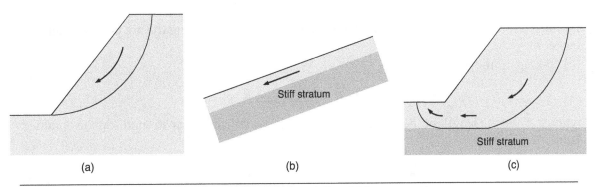

Figure 15.4 Types of slope failure: (a) rotational (b) translational (c) compound

long stretch as shown in Figure 15.4b, the failure mode is *translational*, where the failure mass slides downward along the slope. In Figure 15.4c, the failure surface cannot get through the stiff stratum due to its high shear strength, and a *compound* failure occurs. This is a combination of rotational and translational modes.

15.3 STABILITY OF HOMOGENEOUS UNDRAINED SLOPES

A homogeneous clay slope under undrained condition ($\phi_u = 0$; $\tau_f = c_u$) is shown in Figure 15.5. Failure can take place along an unknown slip surface in the form of a circular arc. Let's consider a potential slip surface AB, consisting of a circular arc with its center at O. The weight of the soil enclosed within the arc is W, acting at the centroid of the hatched area, horizontal distance of d from the center O. The shear stress acting along the arc AB is $c_{u,\text{mob}}$, which is the shear strength

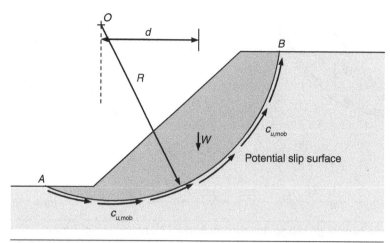

Figure 15.5 Slip circle in a homogeneous undrained slope

mobilized to maintain equilibrium. The normal stresses acting along the arc pass through the center O.

Taking moment about O:

$$W\,d = c_{u,\mathrm{mob}}\, l_{\mathrm{arc},AB}\, R$$

where $l_{\mathrm{arc},AB}$ is the length of the arc AB. Therefore, the mobilized shear strength can be obtained from:

$$c_{u,\mathrm{mob}} = \frac{W\,d}{l_{\mathrm{arc},AB}\,R} \tag{15.2}$$

The safety factor for the above slip circle can be determined from Equation 15.1, assuming $\tau_f = c_u$ and $\tau_{\mathrm{mob}} = c_{u,\mathrm{mob}}$. This can be repeated for several potential slip circles until the one with the minimum safety factor is found. Taylor (1937) proposed a shortcut to locate this *critical circle* where the safety factor is the minimum. This method is discussed in the following section.

Example 15.1: A 5.0 m-high embankment with a 2(H):1(V) slope is constructed on a clay sub-soil with $c_u = 30$ kPa. The embankment is made of a clay where $c_u = 45$ kPa. The unit weights of both clays can be assumed as 18 kN/m³. Estimate the safety factor for the slip circle shown in the figure at the top of page 425. The area $ABCDE = 45.4$ m².

Solution: Let's consider a unit thickness:

$$W = 18 \times 45.4 = 817.2 \text{ kN}$$

Continues

Example 15.1: *Continued*

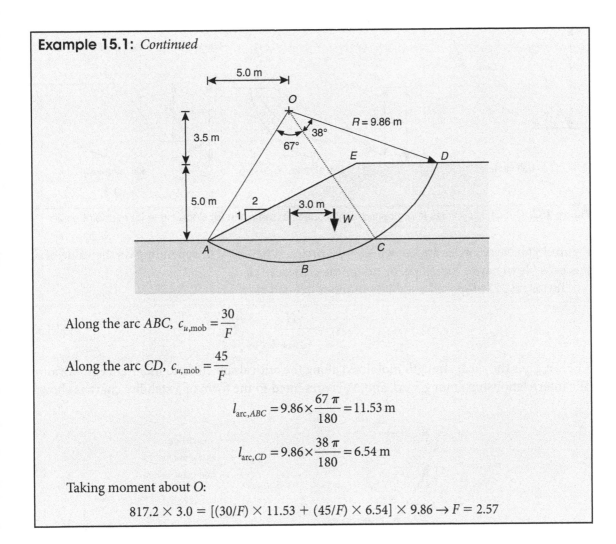

Along the arc ABC, $c_{u,\mathrm{mob}} = \dfrac{30}{F}$

Along the arc CD, $c_{u,\mathrm{mob}} = \dfrac{45}{F}$

$$l_{\mathrm{arc},ABC} = 9.86 \times \frac{67\,\pi}{180} = 11.53 \text{ m}$$

$$l_{\mathrm{arc},CD} = 9.86 \times \frac{38\,\pi}{180} = 6.54 \text{ m}$$

Taking moment about O:

$$817.2 \times 3.0 = [(30/F) \times 11.53 + (45/F) \times 6.54] \times 9.86 \rightarrow F = 2.57$$

15.3.1 Taylor's Stability Chart for Undrained Clays ($\phi_u = 0$)

Immediately following an excavation or the building of an embankment, one can assume that the clays are loaded under undrained conditions. Taylor (1937) proposed some design charts to locate the *critical circle* in undrained clays ($\phi_u = 0$). He identified three groups of failure circles: *toe circles*, *slope circles*, and *midpoint (or base) circles* as shown in Figure 15.6. When the slope angle β is greater than 53°, the failure occurs along a circular arc passing through the toe; such a circle is known as a *toe circle* (Figure 15.6a). When $n_d > 4$, the critical circle reaches the region beneath the toe as shown in Figure 15.6c, with the center directly above the middle of the slope. The failure mode is known as base failure, and the critical circle is known as a *midpoint or base circle*. When $n_d < 4$, it is possible that the critical circle exits on the face of the slope as shown in

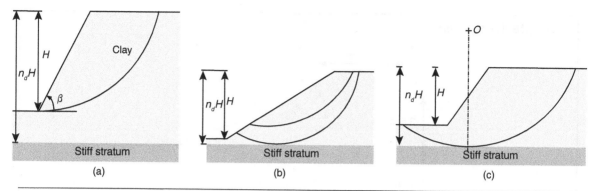

Figure 15.6 Critical slip circles in undrained clay slopes: (a) toe circle (b) slope circle (c) midpoint circle

Figure 15.6b. Such circles are known as *slope circles*. When $n_d < 4$, depending on the value of β, it is possible to have a toe, slope, or midpoint-critical circle.

Taylor (1937) proposed a stability number defined as:

$$N_S = \frac{\gamma H}{c_{u,\mathrm{mob}}}$$

(15.3)

where $c_{u,\mathrm{mob}}$ is the shear strength mobilized along the critical slip circle to maintain equilibrium. The interrelationship among n_d, β, and N_S is presented in the form of a stability chart as shown

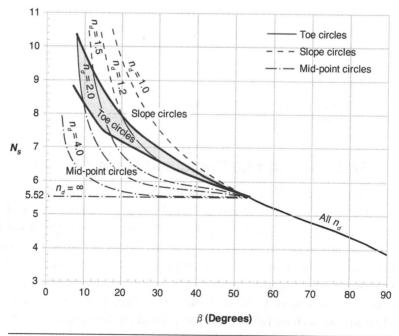

Figure 15.7 Taylor's stability chart for undrained clay slopes

in Figure 15.7. The safety factor can be computed using Equation 15.1. Such analysis is known as *short-term* or *total stress analysis*.

Example 15.2: A six meter-deep excavation is made at a 35° slope in a 9 m-thick clay deposit as shown in the figure. The clay is underlain by bedrock. The unit weight of the clay is 20 kN/m³. Find the safety factor for slope failure along the critical slip circle. What type of slip circle is it?

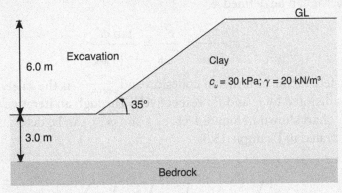

Solution: $n_d = \dfrac{9}{6} = 1.5; \beta = 35°$

From Taylor's chart (Figure 15.7):

$$N_S = 5.9 = \frac{\gamma H}{c_{u,mob}}$$

$$\therefore c_{u,mob} = \frac{20 \times 6}{5.9} = 20.3 \text{ kPa}$$

$$\therefore F = \frac{c_u}{c_{u,mob}} = \frac{30}{20.3} = 1.48 \quad \text{(a midpoint circle)}$$

15.4 TAYLOR'S STABILITY CHARTS FOR c'-ϕ' SOILS

For soils possessing cohesion and friction (e.g., clayey sands or clays in drained conditions), the procedure is slightly complex. The shear strength of a soil in terms of effective stresses can be written as:

$$\tau_f = c' + \sigma' \tan \phi' \tag{15.4}$$

As seen in Equation 15.4, shear strength derives its contribution from cohesive and frictional resistances along the slip surface. In a stable slope (i.e., $F > 1$), only a fraction of the shear strength is mobilized along the potential slip circle. This means that only fractions of the cohesive and frictional resistances are mobilized. The mobilized shear strength along a slip circle can be written as:

$$\tau_{mob} = c'_{mob} + \sigma' \tan \phi'_{mob} \qquad (15.5)$$

Assuming that the degree of mobilization is the same in cohesive as well as frictional resistances, the safety factor can be defined as:

$$F = \frac{\tau_f}{\tau_{mob}} = \frac{c'}{c'_{mob}} = \frac{\tan \phi'}{\tan \phi'_{mob}} \qquad (15.6)$$

where $\frac{c'}{c'_{mob}}$ is the safety factor in terms of cohesion and $\frac{\tan \phi'}{\tan \phi'_{mob}}$ is the safety factor in terms of friction, sometimes denoted by F_c and F_ϕ, respectively. Through an iterative process using Taylor's (1937) stability chart shown in Figure 15.8, c'_{mob} and ϕ'_{mob} can be determined such that $F = F_c = F_\phi$. This is illustrated in Example 15.3.

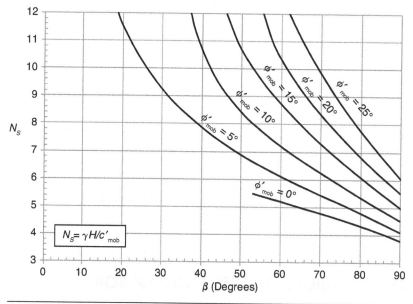

Figure 15.8 Taylor's stability chart for c'-ϕ' soils

Example 15.3: A 9.0 m-high embankment is made of the following soil parameters: $c' = 30$ kPa, $\phi' = 10°$, and $\gamma = 19$ kN/m^3. The slope is at an angle of 45° to horizontal. Find the safety factor of the critical slip circle.

Continues

Example 15.3: *Continued*

Solution:

Trial 1: Let's try $F_\phi = 2.0$:

$$F_\phi = \frac{\tan\phi'}{\tan\phi'_{mob}} \rightarrow 2.0 = \frac{\tan 10}{\tan\phi'_{mob}} \rightarrow \phi'_{mob} = 5.04°$$

For $\beta = 45°$ and $\phi'_{mob} = 5.04°$, from Figure 15.8, $N_S = 7.5$:

$$N_S = \frac{\gamma H}{c'_{mob}} \rightarrow c'_{mob} = \frac{19 \times 9}{7.5} = 22.8 \text{ kPa}$$

$$\therefore F_c = \frac{c'}{c'_{mob}} = \frac{30}{22.8} = 1.32, \text{ which is less than } F_\phi \text{ (assumed as 2.0).}$$

Trial 2: Let's try $F_\phi = 1.45$, a value between the two F_ϕ values above:

$$F_\phi = \frac{\tan\phi'}{\tan\phi'_{mob}} \rightarrow 1.45 = \frac{\tan 10}{\tan\phi'_{mob}} \rightarrow \phi'_{mob} = 6.93°$$

For $\beta = 45°$ and $\phi'_{mob} = 6.93°$, from Figure 15.8, $N_S = 8.0$:

$$N_S = \frac{\gamma H}{c'_{mob}} \rightarrow c'_{mob} = \frac{19 \times 9}{8.0} = 21.4 \text{ kPa}$$

$$\therefore F_c = \frac{c'}{c'_{mob}} = \frac{30}{21.4} = 1.40, \text{ which is very close to the assumed } F_\phi \text{ of 1.45.}$$

A few more trials would converge to $F = F_c = F_\phi = 1.43$, which is the true safety factor for the critical slip circle.

15.5 INFINITE SLOPES

Figure 15.9 shows an infinitely long slope where a soil layer of thickness z overlies a stiff stratum, along which failure occurs. The slip is purely translational along the sliding plane, which is the interface between the soil and the stiff stratum. The water table is at a height of mz above the failure plane, where $m < 1$. It is assumed that the ground surface, water table, and the failure plane are parallel, inclined at an angle of β to the horizontal. Let's consider the vertical slice shown in the figure, which is equilibrium under the following forces: W, T, N', U, and R where:

W = weight of the slice
T = tangential shear force along the failure plane, resisting the slide
N' = normal force at the failure plane due to the effective stresses
U = normal force at the failure plane due to pore water pressure
R = end force acting on both vertical sides, in opposite directions parallel to the slope

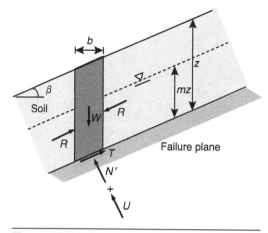

Figure 15.9 Infinite slope

Considering a unit thickness perpendicular to the plane:

$$W = b\, mz\, \gamma_{sat} + b\,(1-m)z\, \gamma_m \tag{15.7}$$

$$T = W \sin \beta \tag{15.8}$$

$$N = N' + U = W \cos \beta \tag{15.9}$$

N is the total normal load that includes the contributions from the effective stresses and the pore water pressure. From Equations 15.7 and 15.9:

$$N = [b\, mz\, \gamma_{sat} + b\,(1-m)z\, \gamma_m]\cos\beta$$

$$N = b\, mz\, \gamma' \cos\beta + b\, mz\, \gamma_w \cos\beta + b\,(1-m)z\, \gamma_m \cos\beta$$

$$\therefore N' = b\, mz\, \gamma' \cos\beta + b\,(1-m)z\, \gamma_m \cos\beta$$

and

$$U = b\, mz\, \gamma_w \cos\beta$$

The effective normal stress and the pore water pressure on the failure plane are given by:

$$\sigma' = \frac{N'}{b/\cos\beta} = mz\gamma'\cos^2\beta + (1-m)z\gamma_m \cos^2\beta \tag{15.10}$$

$$u = \frac{U}{b/\cos\beta} = mz\,\gamma_w \cos^2\beta \tag{15.11}$$

The shear strength along the failure plane is given by:

$$\tau_f = c' + \sigma' \tan \phi' = c' + [m\gamma' + (1-m)\gamma_m]z \cos^2\beta \tan \phi'$$

The shear strength mobilized along the failure plane is given by:

$$\tau_{mob} = \frac{T}{b/\cos\beta} = \frac{W\sin\beta}{b/\cos\beta} = [m\gamma_{sat} + (1-m)\gamma_m]z\sin\beta\cos\beta$$

$$\therefore F = \frac{\tau_f}{\tau_{mob}} = \frac{c' + [m\gamma' + (1-m)\gamma_m]z\cos^2\beta\tan\phi'}{[m\gamma_{sat} + (1-m)\gamma_m]z\sin\beta\cos\beta} \qquad (15.12)$$

The general expression for the safety factor in Equation 15.12 above can be used to investigate some special cases.

Special case 1: Dry granular soil

In dry granular soil, $c' = 0$ and $m = 0$. Substituting these in Equation 15.12 gives:

$$F = \frac{\tan\phi'}{\tan\beta} \qquad (15.13)$$

Special case 2: Fully submerged granular soil with steady seepage down the slope

In a critical situation where the soil is fully submerged and seepage occurs down the slope with $c' = 0$ and $m = 1$, Equation 15.12 becomes:

$$F = \frac{\gamma'}{\gamma_{sat}} \frac{\tan\phi'}{\tan\beta} \qquad (15.14)$$

Special case 3: Cohesive soil with no water table present

In a cohesive soil where there is no water table present (i.e., where $m = 0$), Equation 15.12 becomes:

$$F = \frac{c'}{\gamma_m z \sin\beta\cos\beta} + \frac{\tan\phi'}{\tan\beta} \qquad (15.15)$$

Special case 4: Fully submerged cohesive soil with steady seepage down the slope

In a fully submerged cohesive slope ($m = 1$), Equation 15.12 becomes:

$$F = \frac{c'}{\gamma_{sat} z \sin\beta\cos\beta} + \frac{\gamma'}{\gamma_{sat}} \frac{\tan\phi'}{\tan\beta} \qquad (15.16)$$

Example 15.4: Steady seepage occurs down an infinite granular slope with a water table at the ground level. The saturated unit weight of the sand is 19 kN/m³ and the effective friction angle is 32°. What is the maximum possible slope such that there will be no failure?

Solution: The situation is the same as in Special case 2, where the safety factor is given by Equation 15.14. Substituting $F = 1$:

$$1 = \frac{(19 - 9.81)}{19} \frac{\tan 32}{\tan \beta} \rightarrow \beta = 16.8°$$

15.6 METHOD OF SLICES

The soil enclosed within the slip circle is not always homogeneous. Part of the soil here may be submerged. The simple methods discussed in the previous sections cannot be applied in these circumstances. Such complex situations can be analyzed by the *method of slices*.

Figure 15.10a shows a slip circle of radius R where the enclosed soil is subdivided into vertical sections or *slices*. The circular arc at the base of the ith slice is inclined at α_i to the horizontal. The free body diagram of the ith slice is shown in Figure 15.10b. This slice is in equilibrium under the following forces:

- W_i = self-weight
- T_i = tangential force resisting the slide
- N'_i = normal load at the base due to effective stress σ'_i
- U_i = normal load at the base due to pore water pressure u_i
- E_i, E_{i+1} = the horizontal end forces
- X_i, X_{i+1} = vertical shear forces along the sides of the slice

To define the safety factor for this potential slip circle, Equation 15.1 can still be applied. The inherent assumption is that the safety factor is the same along the entire slip circle, and hence at each slice. Let's consider a *unit thickness* and take moment about the center O, and add them up for all the slices:

$$\sum_{i=1}^{n} T_i R = \sum_{i=1}^{n} W_i R \sin \alpha_i$$

$$\sum_{i=1}^{n} \tau_{\text{mob},i} l_i = \sum_{i=1}^{n} W_i \sin \alpha_i$$

where l_i is the arc length measured along the bottom of the slice. Substituting $\tau_{\text{mob}} = \tau_f / F$:

$$\sum_{i=1}^{n} \frac{\tau_{f,i}}{F} l_i = \sum_{i=1}^{n} W_i \sin \alpha_i$$

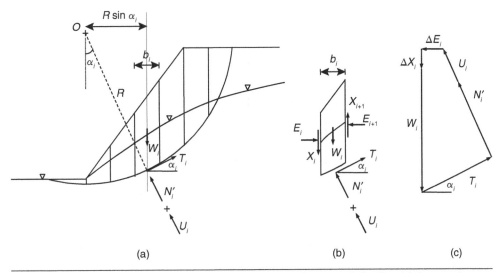

Figure 15.10 Method of slices: (a) slices (b) free body diagram of ith slice (c) force polygon for ith slice

$$\therefore F = \frac{\sum\limits_{i=1}^{n} \tau_{f,i}\, l_i}{\sum\limits_{i=1}^{n} W_i \sin\alpha_i} = \frac{\sum\limits_{i=1}^{n} c_i'\, l_i + \sum\limits_{i=1}^{n} \sigma_i'\, l_i \tan\phi_i'}{\sum\limits_{i=1}^{n} W_i \sin\alpha_i} = \frac{\sum\limits_{i=1}^{n} c_i'\, l_i + \sum\limits_{i=1}^{n} N_i' \tan\phi_i'}{\sum\limits_{i=1}^{n} W_i \sin\alpha_i} \qquad (15.17)$$

The safety factor can be determined from Equation 15.17, provided all the parameters are known for each slice. The parameters c_i', l_i, ϕ_i', W_i, and α_i can be easily determined, but not N_i'. This makes it an indeterminate problem where N_i' depends on the end forces E_i and X_i. Figure 15.10c shows the force polygon for the ith slice where $\Delta E_i = E_{i+1} - E_i$ and $\Delta X_i = X_i - X_{i+1}$.

15.6.1 Ordinary Method of Slices

The *ordinary method of slices*, also known as the *Swedish* or *Fellenius method of slices*, is the simplest and earliest of the different methods of slices reported in the literature. It is assumed that at each slice, $\Delta X_i = 0$ and $\Delta E_i = 0$ (Fellenius 1936). Therefore (see Figure 15.11a):

$$W_i \cos\alpha_i = N_i' + U_i$$

$$N_i' = W_i \cos\alpha_i - u_i\, l_i$$

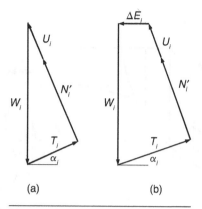

Figure 15.11 Force polygons:
(a) ordinary method (b) Bishop's
simplified method

Substituting this expression for N_i' in Equation 15.17:

$$F = \frac{\displaystyle\sum_{i=1}^{n} c_i' l_i + \sum_{i=1}^{n} (W_i \cos\alpha_i - u_i l_i) \tan\phi_i'}{\displaystyle\sum_{i=1}^{n} W_i \sin\alpha_i} \tag{15.18}$$

Hand calculations can be made to compute the safety factor based on Equation 15.18. This can also be easily implemented in a spreadsheet. By neglecting the interslice forces, the ordinary method of slices violates force equilibrium, but satisfies moment equilibrium.

15.6.2 Bishop's Simplified Method of Slices

Bishop (1955) proposed a method where he assumed $\Delta X_i = 0$ at each slice. The resulting force polygon is shown in Figure 15.11b. Here, T_i can be written as:

$$T_i = \frac{\tau_{f,i}}{F} l_i = \frac{1}{F}\{c_i' l_i + N' \tan\phi_i'\}$$

From the force polygon (Figure 15.10b):

$$W_i = T_i \sin\alpha_i + N_i' \cos\alpha_i + U_i \cos\alpha_i$$

Substituting $U_i = u_i l_i$, and for T_i in the above equation:

$$W_i = \frac{1}{F}\{c_i' l_i + N' \tan\phi_i'\}\sin\alpha_i + N_i' \cos\alpha_i + u_i l_i \cos\alpha_i$$

$$N_i' = \frac{W_i - \dfrac{c_i' l_i \sin\alpha_i}{F} - u_i l_i \cos\alpha_i}{\cos\alpha_i + \dfrac{\sin\alpha_i \tan\phi_i'}{F}} \tag{15.19}$$

Substituting $l_i = b_i / \cos\alpha_i$ and the above expression for N_i' in Equation 15.17:

$$F = \frac{\displaystyle\sum_{i=1}^{n} c_i' l_i + \sum_{i=1}^{n} N_i' \tan\phi_i'}{\displaystyle\sum_{i=1}^{n} W_i \sin\alpha_i}$$

$$F \times \sum_{i=1}^{n} W_i \sin\alpha_i = \sum_{i=1}^{n} \left[\frac{c_i' b_i}{\cos\alpha_i} + \left(\frac{W_i - \dfrac{c_i' l_i \sin\alpha_i}{F} - u_i l_i \cos\alpha_i}{\cos\alpha_i + \dfrac{\sin\alpha_i \tan\phi_i'}{F}} \right) \tan\phi_i' \right]$$

$$F \times \sum_{i=1}^{n} W_i \sin\alpha_i = \sum_{i=1}^{n} \left[\frac{c_i' b_i}{\cos\alpha_i} + \left(\frac{W_i - \dfrac{c_i' b_i \tan\alpha_i}{F} - u_i b_i}{\cos\alpha_i + \dfrac{\sin\alpha_i \tan\phi_i'}{F}} \right) \tan\phi_i' \right]$$

$$F = \frac{\displaystyle\sum_{i=1}^{n} \{c_i' b_i + (W_i - u_i b_i)\tan\phi_i'\} \left[\dfrac{\sec\alpha_i}{1 + \dfrac{\tan\alpha_i \tan\phi_i'}{F}} \right]}{\displaystyle\sum_{i=1}^{n} W_i \sin\alpha_i} \tag{15.20}$$

The problem with Equation 15.20 is that since the safety factor F is appearing on both sides of the equation, it can only be solved by trial and error.

The stability analysis methods discussed herein are known as *limit equilibrium methods*. They are based on equilibrium considerations only and do not give any idea regarding the magnitudes of displacements. Further extensions to the method of slices were proposed by Morgenstern and Price (1965), Spencer (1967), and several others. Some of these methods allow for noncircular slip surfaces. Today, computer programs incorporating the above methods are available for analyzing slope stability problems.

15.7 STABILITY ANALYSIS USING *SLOPE/W*

For analyzing slope stability and for determining the safety factor of a soil or rock slope, there are several limit equilibrium methods available. The *Student Edition* of *SLOPE/W 2007* accommodates a few of them (e.g., Fellenius, Bishop, Morgenstern-Price, etc). *SLOPE/W* is a slope stability software that is used in more than 100 countries. It works on the basis of limit equilibrium principles, and incorporates several different methods of analysis. Its user-friendly interface and versatility make it one of the most popular software packages worldwide when it comes to slope stability analysis. It is part of the *GeoStudio 2007* suite of software. A DVD containing the *Student Edition* of *GeoStudio 2007* is included with this book. This section describes how to use the *Student Edition* of *SLOPE/W* in solving slope stability problems. The student version has a few limitations that make it suitable mainly for learning and evaluation. The full version is available from *GEO-SLOPE* International, Canada (http://www.geo-slope.com).

The full version has several advanced features (e.g., external loads, tension cracks, noncircular slip surfaces, more constitutive models, soil reinforcements, auto-search) that are not available in the *Student Edition*. Nevertheless, the *Student Edition* is adequate to try out a wide range of simpler problems and to get a feel for a versatile slope stability analysis software. There is a very good chance that some of you will use it in professional practice, sooner rather than later.

15.7.1 Getting Started with *SLOPE/W*

When running *GeoStudio*, select *Student License* from the start page. All *GeoStudio* project files are saved with extension *.gsz* so that they can be called by any of the applications (e.g., *SIGMA/W*, *SEEP/W*) within the suite.

Familiarize yourself with the different toolbars that can be made visible through the $\boxed{View/Toolbar...}$ menu. Moving the cursor over an icon displays its function. In the $\boxed{Analysis}$ toolbar, you will see three icons, $\boxed{DEFINE}$, $\boxed{SOLVE}$, and $\boxed{CONTOUR}$, next to each other. $\boxed{DEFINE}$ and $\boxed{CONTOUR}$ are two separate windows and you can switch between them. The problem is fully defined in the $\boxed{DEFINE}$ window and saved. Clicking the $\boxed{SOLVE}$ icon solves the problem as specified. Clicking the $\boxed{CONTOUR}$ icon displays the results in the $\boxed{CONTOUR}$ window. The input data can be changed by switching to the $\boxed{DEFINE}$ window and $\boxed{SOLVE}$d again for different output.

The major components in solving a slope stability problem are:

1. Defining the geometry
2. Defining the soil properties and assigning them to the regions
3. Defining the piezometric line (water table)
4. Defining the method of analysis (e.g., Morgenstern-Price) and the slip circles
5. Solving the problem
6. Displaying the results (e.g., critical slip circle, plots, etc.)

1. *Defining the geometry*:

   Always have a rough sketch of your problem geometry with the right dimensions before you start *SLOPE/W*. When *SLOPE/W* is started, it is in the $\boxed{DEFINE}$ window. The $\boxed{Set}$ menu has two different but related entries, $\boxed{Page...}$ and $\boxed{Units\ and\ Scales...}$, which can be used to define your working area and units. A good start is to use a 260 mm (width) $\times$ 200 mm (height) area that fits nicely on an A4 sheet. Here, a scale of 1:200 would represent 52 m (width) $\times$ 40 m (height) of a problem geometry. Try to use the same scale in *x* and *y* directions so that the geometry is not distorted. The $\boxed{Grid...}$ feature will allow you to select the grid spacing, make it visible, and snap it to the grid points. The $\boxed{Axes...}$ feature will allow you to draw the axes and label them. $\boxed{Sketch/Axes...}$ may be a better way to draw the axes and label them. Use $\boxed{View/Preferences...}$ to change the way the geometry and fonts are displayed and to change the way the slip circles are graphically presented.

   Use $\boxed{Sketch/Lines}$ to sketch the geometry using free lines. Use $\boxed{Modify/Objects...}$ to delete or move them. $\boxed{Sketch}$ is different from $\boxed{Draw}$. Use the $\boxed{Draw/Regions...}$ feature on the sketched outlines to create the real geometry and define the different material zones. One may also omit $\boxed{Sketch}$ and start from $\boxed{Draw}$ feature instead. While $\boxed{Sketch}$ing, $\boxed{Draw}$ing, or $\boxed{Modify}$ing, right-clicking the mouse ends the action. The $\boxed{Sketch}$ menu has commands to draw dimension lines with arrowheads and to label the dimensions and objects. Sketch objects are not used in any computations.

2. *Defining soil properties and assigning to regions*:

   Use $\boxed{Draw/Materials...}$ to assign the material properties (i.e., *c*, ϕ, γ) and apply them to the regions by dragging. The *Student Edition* can accommodate up to three different materials. They can be either Mohr-Coulomb materials or impenetrable bedrock.

3. *Defining the piezometric line*:

   From $\boxed{KeyIn/Analyses...}$, select piezometric line for PWP conditions in the settings. Use $\boxed{Draw/PoreWater\ Pressure....}$ to draw the piezometric line. It does not have to be horizontal.

4. *Defining the method of analysis and the slip circles*:

   In $\boxed{KeyIn/Analyses...}$ select the method of analysis (e.g., Spencer) and give a name and description to the problem. Under the $\boxed{Settings...}$ tab, select how the pore water pressure is specified (e.g., piezometric line). A series of circular trial slip surfaces can be defined in two ways: (a) *Entry and Exit* (b) *Grid and Radius*, through $\boxed{KeyIn/Analyses.../Slip\ Surface}$. With the *Entry and Exit* method, it is required to specify where the circular arc enters and exits the slope. The number of slip circles can be varied by adjusting the increments. In the *Grid and Radius* method, a grid has to be specified (the four corners defined counterclockwise, starting from the top left) where each of the grid points will be a center. The radius is defined by the lines confined within a box (the four corners defined counterclockwise, starting from top left) that are tangent to the circles. In both methods, the slip circles are defined through the $\boxed{Draw/Slip\ Surface}$ feature. The number of slices (default = 30) can be varied through the $\boxed{Advanced}$ tab.

A single slip circle can be defined with the *Grid and Radius* method by collapsing the center-grid into a single point and by collapsing the tangential lines into a single line.

5. *Solving the problem:*
Once the problem is fully defined through the above steps, it can be $\boxed{SOLVE}$d, and the results can be viewed in a $\boxed{CONTOUR}$ window. If $\boxed{SOLVE}$ does not really solve and suggests an error, you can view the errors in the $\boxed{Draw/Slip\ Surface}$ dialog box. You can switch between the $\boxed{DEFINE}$ and $\boxed{CONTOUR}$ windows while experimenting with the output. This can be very effective for a parametric study. The $\boxed{Tools/Verify}$ feature can be used for checking the problem definition before solving.

6. *Displaying the results:*
Once the problem is solved, the critical slip surface appears in the $\boxed{CONTOUR}$ window by default. By selecting the number of slip circles from $\boxed{View/Preferences...}$, multiple slip surfaces with the lowest safety factors can be viewed. Selecting the $\boxed{Draw/Slip\ Surfaces...}$ menu, it is possible to access all the trial circles and see what they look like. The critical one appears at the top of the list, along with the safety factors, center coordinates and the radii for all in the list. The slice details are available only for the critical slip circle, for which various plots can be generated using the $\boxed{Draw/Graph...}$ feature in the $\boxed{CONTOUR}$ window. The critical slip circle, force polygons, graphs, or data can be copied to the clipboard. $\boxed{Draw/Contours...}$ can be used to draw safety factor contours when the slip circles are specified using the *Grid and Radius* method. The contour intervals and the number of contours can be specified. To show the contour labels, click $\boxed{Draw/Contour\ labels}$, which will change the cursor from an arrow to a crosshair. Place the cursor on a contour line and left-click the mouse. This will display the contour value.

Not all the defined circles will be geometrically sensible. When error messages are displayed, go to the help menu and find the appropriate error message number that corresponds to the error message number that was displayed in the problem; this should help you understand the reason. It is a good practice to do a coarse run to identify the approximate slip circle and then do some fine-tuning.

Example 15.5: Solve Example 15.1 using *SLOPE/W*, dividing the soil into 6 slices. Show the force polygon for the 3rd slice using the ordinary (Fellenius) method. What is the area of soil enclosed within the critical slip circle? Summarize the forces in the 4th slice for the Fellenius, simplified Bishop, Janbu, and Morgenstern-Price methods.

Solution:

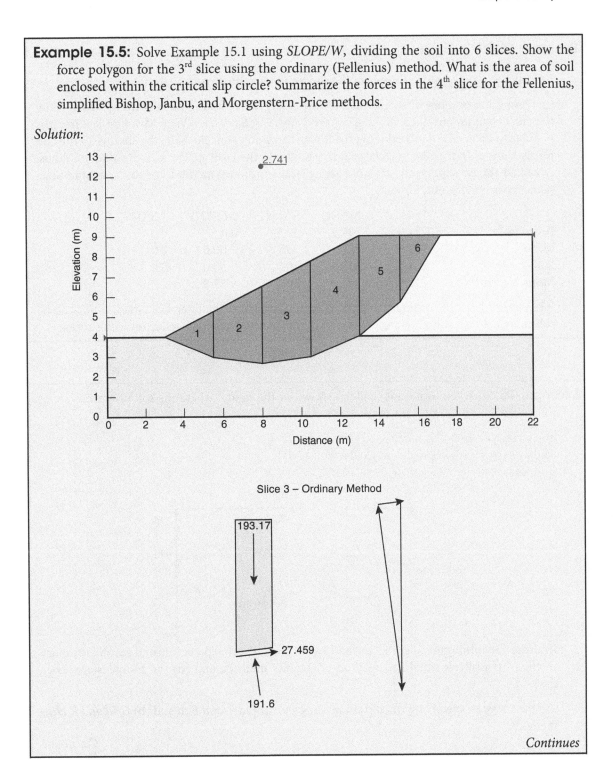

Continues

Example 15.5: *Continued*

From $\boxed{\text{View.../Slide Mass}}$, the area of the sliding mass = 46.27 m².

Increasing the number of slices will alter the safety factor and make it converge. By placing the cursor within any slice while using the *Draw/Slip Surfaces.../View Slice Info* feature, it is possible to access the data including the forces, force polygon, etc. Note the absence of the end forces X_i and E_i (remember it is the ordinary method); they will appear with the other methods. Note that the force polygons are not closing in the ordinary method due to the assumptions made regarding the end forces.

	W (kN)	N (kN)	T (kN)	ΔE (kN)	ΔX (kN)
Fellenius	221.9	204.9	30.1	0.0	0.0
Bishop	221.9	227.8	30.1	−60.0	0.0
Janbu	221.9	227.5	30.5	−59.0	0.0
Morgenstern-Price	221.9	223.3	30.1	−57.8	−4.2

It can be seen that the ordinary method (Fellenius) neglects the interslice forces, which can be substantial. As a result, the safety factor from this method has to be relied on with caution.

Example 15.6: An excavation is made as shown in the figure, where the soil properties are as follows:

Top layer: $\gamma = 18.0$ kN/m³, $c' = 20$ kPa, $\phi' = 24°$
Midlayer: $\gamma = 19.0$ kN/m³, $c' = 15$ kPa, $\phi' = 26°$
Bottom layer: $\gamma = 19.5$ kN/m³, $c' = 10$ kPa, $\phi' = 22°$

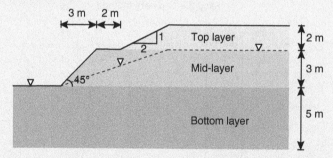

Evaluate the minimum safety factor and locate the critical slip circle using (a) the ordinary method (b) Bishop's simplified method (c) Janbu's method, and (d) the Morgenstern-Price method.

Try the above by specifying the slip circle using the (a) *Entry and Exit* and (b) *Grid and Radius* methods.

Continues

Example 15.6: *Continued*

Show the critical slip circle obtained from Bishop's method with the Grid and Radius method, showing the safety factor contour.

Solution:

Method	F_{min}	Center X,Y (m)	Radius (m)
Ordinary	1.56	11.21,10.48	9.72
Bishop	1.84	11.13,13.46	10.38
Janbu	1.71	11.22,12.36	9.98
M-P	1.84	11.13,13.46	10.38

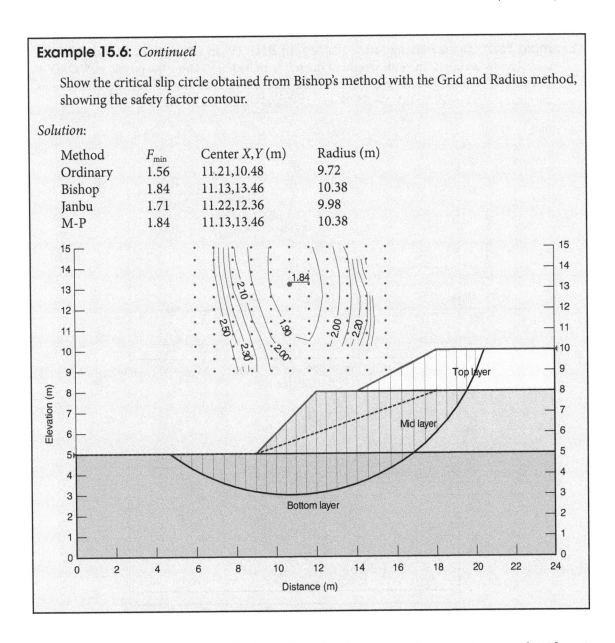

The position of the critical slip circle depends on the shear strength parameters, c and ϕ. If $c = 0$ (i.e., shear resistance is purely frictional), the slip circle tends to be shallow and the failure zone is parallel to the slope. If $\phi = 0$ (i.e., shear resistance is purely cohesive), the critical slip surface can be deeper. This is illustrated in Example 15.7.

Example 15.7: An excavation is made at a slope of 2(H):1(V) in a 20 m-thick soil bed, which is underlain by bedrock. The unit weight of the soil is 18.5 kN/m³. Draw the profile in *SLOPE/W* and vary the values of c and ϕ to see the effects on the location of the critical slip circle. Show the critical slip circles for (a) $c = 5$ kPa, $\phi = 35°$; and (b) $c = 35$ kPa, $\phi = 0°$.

Solution:

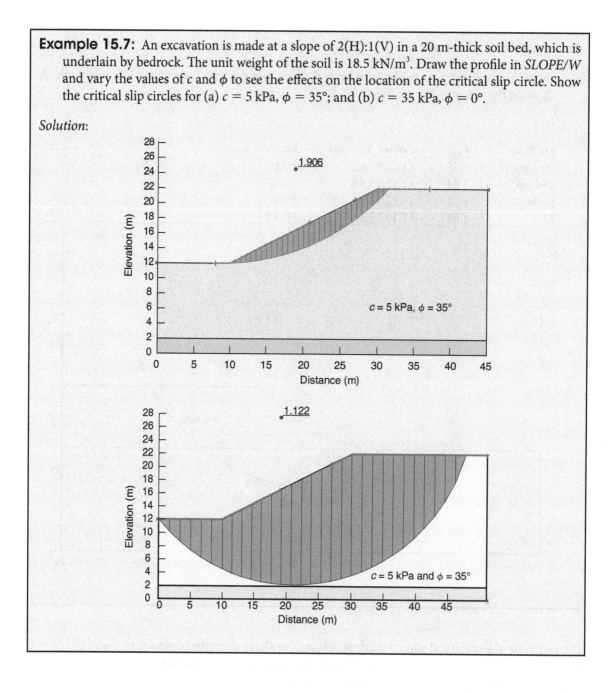

Reminder

❖ It is an assumption that the failure surface of a slope follows a circular arc. The limit equilibrium methods discussed herein are based on this assumption.

❖ It is possible to have noncircular slip surfaces, and there are methods to analyze them.

❖ Taylor's chart can be used for computing the safety factor along the critical slip circle.

❖ By neglecting the interslice forces ΔE and ΔX, the ordinary method of slices violates force equilibrium, but satisfies moment equilibrium. Therefore, the force triangles do not close. Bishop's method includes ΔE (ΔX is neglected), and hence the force polygons close better than in the ordinary method, but it still violates force equilibrium. Note: The Morgenstern-Price and Spencer methods satisfy both force and moment equilibrium.

❖ *SLOPE/W 2007 Student Edition* is a versatile tool that can be used for solving simple slope stability problems.

❖ *GeoStudio 2007 Student Edition* includes *SEEP/W, SIGMA/W*, etc. that work well with *SLOPE/W*.

WORKED EXAMPLES

1. An 8 m-deep excavation is made into a clay deposit with $c_u = 30$ kPa and $\gamma = 19$ kN/m³. A hard stratum consisting of very stiff clays lies at a depth of 10 m below the ground level.

   a. What would be the steepest slope at which a cut could be made before any failure occurs?

   b. Find the slope that would give a short-term safety factor of 1.2.

   c. What would be the safety factor against any short-term failure if the excavation was made at 30° to horizontal?

   Solution:

   a.

$$n_d = \frac{10}{8} = 1.25$$

For the steepest slope without failure:

$$F = 1 \rightarrow c_{u,mob} = 30 \text{ kPa}$$

$$\therefore N_S = \frac{\gamma H}{c_{u,mob}} = \frac{19 \times 8}{30} = 5.07$$

From Taylor's chart (Figure 15.7), $\beta = 63°$ ... toe circle

b.

$$F = 1.2 \rightarrow c_{u,mob} = 30/1.2 = 25.0 \text{ kPa} \rightarrow N_S = \frac{19 \times 8}{25} = 6.08$$

From Taylor's chart, $\beta = 39°$ ... midpoint circle

c. For $\beta = 30°$ and $n_d = 1.25$, from Taylor's chart:

$$N_S = 6.4 = \frac{19 \times 8}{c_{u,mob}}$$

$$\therefore c_{u,mob} = 23.8 \text{ kPa} \rightarrow F = 30/17.8 = 1.26$$

2. A cohesive infinite soil slope has 2.5 m soil overburden above the underlying stiff stratum. The slope is inclined at 20° to the horizontal and there is no water table within the overburden soil. The soil properties are $\gamma_m = 19.0 \text{ kN/m}^3$, $\phi' = 15°$, and $c' = 35 \text{ kPa}$.

Determine the following:

a. Safety factor against sliding
b. Maximum shear stress developed within the overburden soil
c. Shear strength along the potential failure plane above
d. Critical height of overburden that would have caused sliding

Solution:

a. Substituting $m = 0$ in Equation 15.15:

$$F = \frac{c'}{\gamma_m z \sin\beta \cos\beta} + \frac{\tan\phi'}{\tan\beta} = \frac{35}{19 \times 2.5 \times \sin 20 \times \cos 20} + \frac{\tan 15}{\tan 20} = 3.03$$

b. Substituting $m = 0$ in the expression for mobilized shear stress at depth z:

$$\tau_{mob} = \gamma_m z \sin\beta \cos\beta = 19 \times 2.5 \times \sin 20 \cos 20 = 15.3 \text{ kPa}$$

c. Substituting $m = 1$ in the expression for shear strength at depth z:

$$\tau_f = c' + \gamma_m z \cos^2\beta \tan\phi' = 35 + 19 \times 2.5 \times \cos^2 20 \times \tan 15 = 46.2 \text{ kPa}$$

d. Substituting $F = 1$ in Equation 15.15:

$$1 = \frac{35}{19 \times H \times \sin 20 \cos 20} + \frac{\tan 15}{\tan 20} \rightarrow H = 21.7 \text{ m}$$

3. Use *SLOPE/W* to identify the type of failure, the location, and the short-term safety factor of the critical slip circle for the following three undrained clay slopes where $c_u = 35$ kPa, $\phi_u = 0$, and $\gamma = 19$ kN/m³:

    a. Bedrock 11 m below ground; height = 9 m; slope = 1(H):1.5(V)
    b. Bedrock 8 m below ground; height = 6 m; slope = 1.5(H):1(V)
    c. Bedrock 8 m below ground; height = 8 m; slope = 2(H):1(V)

Use simplified Bishop's method.

Solution:

    a. Toe circle with $F = 1.113$ (see figure a)
       Also shown in the figure are the five slip circles with the lowest safety factors.

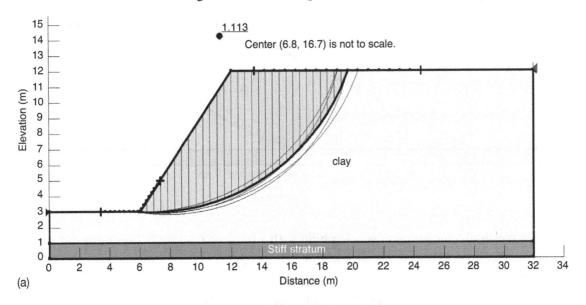

(a)

    b. Midpoint circle with $F = 1.886$ (see figure b on page 446)
    c. Compound (not in Taylor's chart); a slope circle consisting of circular arcs and a straight line, with $F = 1.833$ (see figure c on page 446)

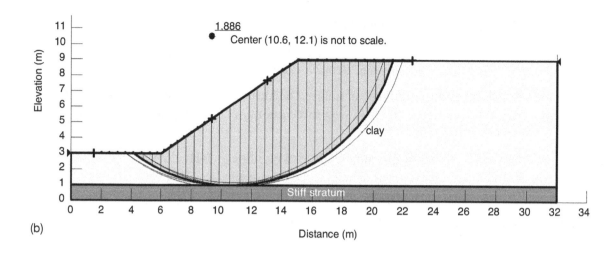

(b)

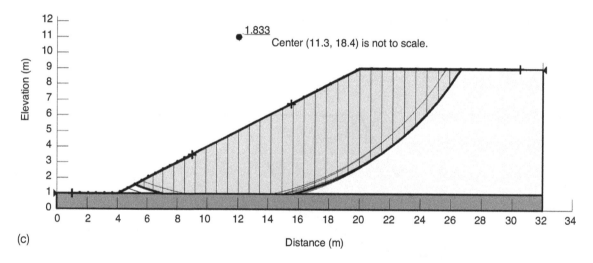

(c)

4. An 8.0 m-high embankment is being built with a slope of 1.5(H):1(V) on a ground where $\phi' = 34°$, $c' = 5$ kPa, and $\gamma = 20.0$ kN/m³. The properties of the embankment soil are as follows:

$$\phi' = 24°, c' = 10 \text{ kPa, and } \gamma = 19.0 \text{ kN/m}^3$$

Using *SLOPE/W*, find the safety factor of the embankment against slope instability based on the Morgenstern-Price method when there is 5.0 m of water in the reservoir with the water level at *aeb* (see the top figure on page 447).

If the water level in the reservoir is drawn down rapidly to the ground level, estimate the new safety factor of the slope, assuming that the phreatic line is *cdeb*.

In both cases, use the Grid and Radius method to specify the slip circles and show five safety factor contours with intervals of 0.05.

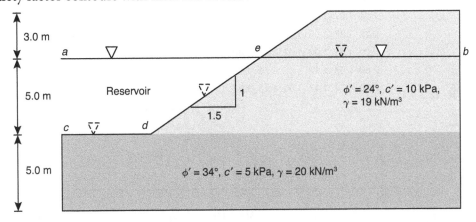

Solution: With 5.0 m of water, $F = 1.53$ (safety factor contours from 1.55 to 1.75).

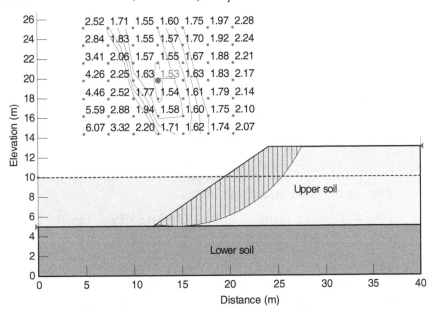

When the water level drops suddenly to *cdeb*, $F = 1.03$; safety factor contours from 1.05 to 1.25.

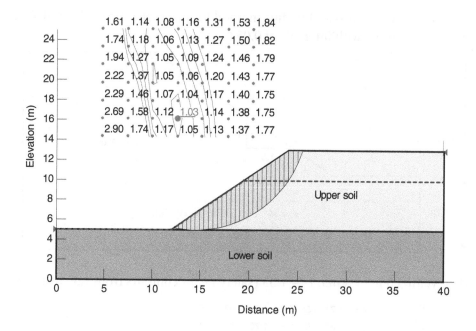

5. A 5 m-thick sand ($\phi' = 34°$; $\gamma_m = \gamma_{sat} = 19$ kN/m³) overlies a stiff stratum on an infinite slope at an angle of 12(H):5(V). The water table lies at a depth of 2.5 m within the sand and is parallel to the slope. Using the theory of infinite slope, find the safety factor.

Carry out this analysis using *SLOPE/W*. (Hint: Set to a very small scale to be able to draw a slip circle of a very large radius, tangent to the stiff stratum.)

What would be the safety factor if the sand were dry?

Solution: Substituting $c' = 0$ in Equation 15.12:

$$F = \frac{(m\gamma' + (1-m)\gamma_m)\tan\phi'}{(m\gamma_{sat} + (1-m)\gamma_m)\tan\beta} = \frac{(0.5 \times 9.19 + 0.5 \times 19)}{(0.5 \times 19 + 0.5 \times 19)} \times \frac{\tan 34}{5/12} = 1.201$$

The critical slip circle shown in the figure on the top of page 449, as obtained from *SLOPE/W*, gives a safety factor of 1.238 (the center is not marked to scale). Reducing the scale, thus enabling larger slip circles, will make the safety factor converge to the above value of 1.201 (see figure on page 449).

If the sand were dry, from Equation 15.13:

$$F = \frac{\tan\phi'}{\tan\beta} = \frac{\tan 34}{5/12} = 1.618$$

Using the above slip surface with *SLOPE/W*, $F = 1.628$.

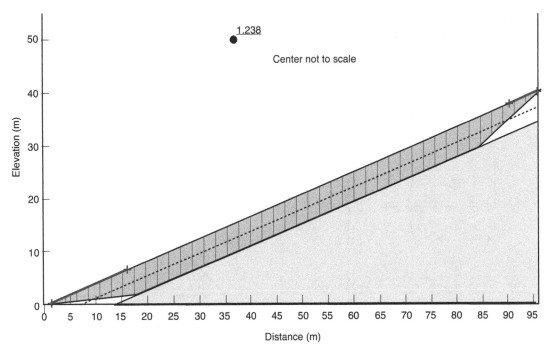

REVIEW EXERCISES

1. A cut is made into a layered clayey soil as shown in the figure below. Assuming the clays to be under undrained conditions with unit weights of 18.0 kN/m³, find the safety factor for the 13.0 m-radius slip circle shown. The area of *ABCDEH* is 65.7 m².

Answer 2.22

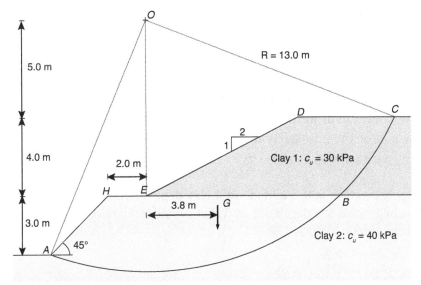

2. Show that the centroid G of the circular sector area shown in part (a) of the figure below is located at a distance $\bar{x}$ from the center, given by:

$$\bar{x} = \frac{2}{3} \frac{R\sin\theta}{\theta}$$

where θ is in radians.

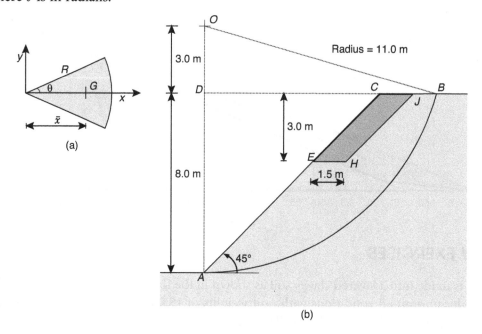

(a)

(b)

An 8.0 m-high and 45° clay slope is shown in part (b) of the figure, with a potential slip circle of 11.0 m radius. The undrained shear strength of the clay is 30 kPa and the unit weight is 18.0 kN/m³.

a. Find the angle AOB.
b. Find the weight of the soil mass enclosed within the arc (i.e., ABC) and the horizontal distance of its centroid from O.
c. Find the safety factor for possible failures along this slip circle.
d. What would be the new safety factor if the dark shaded section $EHJC$ is removed?

Answer 74.2°; 548.1 kN, 5.95 m; 1.44; 1.76

3. A 6 m-deep excavation is to be made in a clay deposit where $c_u = 25$ kPa and $\gamma = 19.0$ kN/m³. The bedrock is at a depth of 9 m. What should be the slope of the excavation so that the short-term safety factor is 1.5?

Answer: 20°, midpoint circle

4. Summarize the key features of the Morgenstern-Price, Janbu, and Spencer methods, along with the simpler Fellenius and Bishop methods.

5. Solve Worked Example 15.3 using Taylor's chart, identifying the type of critical slip circle and the safety factor. Compare the results.

   Answer: Toe circle, 1.08; Midpoint circle, 1.90; Slope circle, 1.88

6. *Rapid drawdown* is a critical situation associated with failure of slopes when there is a sudden drop in water levels. Discuss this.

7. A 6.0 m-high clay slope is constructed at an inclination of 2.5(H):1(V) in a clay with $c_u = 45$ kPa and $\gamma = 19$ kN/m^3. The bedrock lies 6 m below the bottom of the slope. Using Taylor's chart, find the safety factor of the slope. What is the type of critical slip circle? Repeat this exercise using *SLOPE/W* and locate the critical slip circle.

   Answer: F = 2.41, Midpoint circle; F = 2.40, Midpoint circle, Center = 22.6 m, 20.5 m, and radius = 18.5 m.

8. Carry out Example 15.3 using *SLOPE/W* and find the safety factor of the slope.

   Answer: 1.424

9. An infinite slope of sands is at inclination of 2(H):1(V). The friction angle of the sands is 34°. Find the safety factor of the slope using Equation 15.13 and *SLOPE/W*.

   Answer: 1.349, 1.353

10. A 10.0 m-high, undrained clay slope with a unit weight of 19 kN/m^3 stands vertically. Using Bishop's simplified method in *SLOPE/W*, estimate the minimum undrained shear strength required for the slope to remain stable.

    Answer: 40 kPa

11. In a 10 m-high slope where $c' = 20$ kPa and $\gamma = 19$ kN/m^3, use Bishop's simplified method in *SLOPE/W* and complete the flowing table.

    Extend the table on the top of page 452 and develop a design chart similar to Taylor's (or make it even better!).

ϕ' (°)	β (°)	$F = F_c = F_\phi$	ϕ'_{mob} (°)	c'_{mob} (kPa)	$N_s = \gamma H/c'_{mob}$
5	45	0.814	6.13	24.57	7.73
10	45	0.982	10.18	20.37	9.33
20	45	1.305	15.58	15.33	12.40
25	45	1.467	17.63	13.63	13.94
5	26.6*				
10	26.6*				
20	26.6*				
25	26.6*				

*2(H):1(V) slope

12. For an undrained clay slope with $H = 8$ m, $\gamma = 19$ kN/m³, and $c_u = 30$ kPa, use Bishop's simplified method in *SLOPE/W* to complete the following table.

Expand this table and develop Taylor's chart for undrained clay slopes using *SLOPE/W*.

n_d	Slope, H:V (°)	F_{min}	$c_{u,mob}$ (kPa)	N_s (Eq. 15.3)	Type of circle
1.0	8:1 (7.13)				
1.0	5:1 (11.31)				
1.0	4:1 (14.04)				
1.0	3:1 (18.43)	1.925	15.59	9.75	S2
1.0	2:1 (26.57)				
1.0	1.5:1 (33.69)				
1.0	1:1 (45.00)				
1.0	1:1.5 (56.31)				
1.0	1:2 (63.43)				
1.0	1:3 (71.57)	0.974	30.80	4.93	T
1.0	1:4 (75.96)				
1.0	1:5 (78.69)				
1.0	1:8 (82.88)				
1.2	8:1 (7.13)				
1.2	5:1 (11.31)				
1.2	4:1 (14.04)				
1.2	3:1 (18.43)				
1.2	2:1 (26.57)				
1.2	1.5:1 (33.69)	1.244	24.12	6.30	T
1.5	2:1 (26.57)	1.241	24.17	6.29	B1
1.5	1.5:1 (33.69)				
4.0	2:1 (26.57)	1.109	27.05	5.62	B1

S1 = Slope circle; S2 = Compound slope circle (part of the critical surface is a straight line along the bedrock); T = Toe circle; B1 = midpoint or base circle; B2 = Compound base circle (part of the critical surface is a straight line along the bedrock).
$c_{u,mob}$ is the shear strength required to maintain equilibrium and therefore it can exceed c_u, implying $F < 1$.

Vibrations of Foundations

16

16.1 INTRODUCTION

Foundations that support vibrating equipment experience rigid body displacements. The cyclic displacement of a foundation can have the six possible modes that follow (see Figure 16.1):

- translation in the vertical direction
- translation in the longitudinal direction
- translation in the lateral direction
- rotation about the vertical axis (*yawing*)
- rotation about the longitudinal axis (*rocking*)
- rotation about the lateral axis (*pitching*)

In this chapter, we will explore the fundamentals of foundation vibration in the various modes supported on an *elastic medium*. The elastic medium that supports the foundation is considered both homogeneous and isotropic. In general, the behavior of soils departs considerably from that of an elastic material. Only at low strain levels is it considered a reasonable approximation of an elastic material.

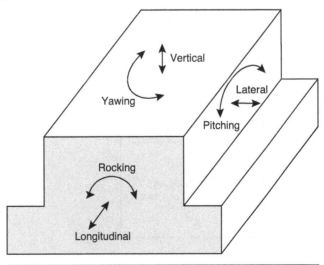

Figure 16.1 Six modes of vibration for a foundation

16.2 VIBRATION THEORY—GENERAL

In this section, we will discuss the elements of vibration theory. This knowledge is essential to foundation designs that are subjected to cyclic loading. We will discuss free vibration of a spring-mass system with and without damping, and extend the discussion to steady-state forced vibration due to a sinusoidally varying force or rotating mass.

16.2.1 Free Vibration of a Spring-Mass System

Figure 16.2 shows a foundation resting on a spring. Let the spring represent the elastic properties of the soil. The load W represents the weight of the foundation plus the weight that comes from the machinery supported by the foundation. Due to the load W, a static deflection z_s will develop. By definition:

$$k = \frac{W}{z_s} \qquad (16.1)$$

where k = spring constant for the elastic support.

  If the foundation is disturbed from its static equilibrium position, the system will oscillate. The equation of motion of the foundation when it has been disturbed through a distance z can be written from Newton's second law of motion as:

$$\left(\frac{W}{g}\right)\ddot{z} + kz = 0$$

or

$$\ddot{z} + \left(\frac{k}{m}\right)z = 0 \qquad (16.2)$$

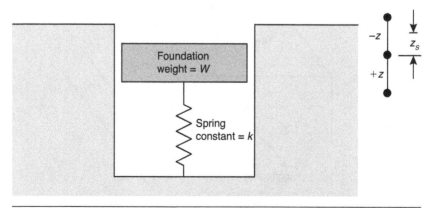

Figure 16.2 Free vibration of a spring-mass system

where

g = the acceleration due to gravity, $\ddot{z} = \dfrac{d^2 z}{dt^2}$, t is time, and m is mass = W/g.

The preceding equation can be solved to obtain the *frequency of vibration* (that is, the number of cycles per unit time) as:

$$f = f_n = \frac{\omega_n}{2\pi} = \frac{1}{2\pi}\sqrt{\frac{k}{m}} \qquad (16.3)$$

where

f = frequency of oscillation (cps)

f_n = undamped natural frequency (cps)

ω_n = undamped natural circular frequency (radians/s) = $\sqrt{\dfrac{k}{m}}$

Under idealized situations, the vibration can continue forever.

Example 16.1: A mass is supported by a spring. The static deflection of a spring z_s due to the mass is 0.4 mm. Determine the natural frequency of vibration.

Solution: From Equation 16.1:

$$k = \frac{W}{z_s}$$

However, $W = mg$; $g = 9.81$ m/s². Therefore:

$$k = \frac{mg}{z_s}$$

$$f_n = \frac{1}{2\pi}\sqrt{\frac{k}{m}} = \frac{1}{2\pi}\sqrt{\left(\frac{mg}{z_s}\right)\frac{1}{m}} = \frac{1}{2\pi}\sqrt{\frac{g}{z_x}} = \frac{1}{2\pi}\sqrt{\frac{9.81}{\dfrac{0.4}{1000}\,\text{m}}} = 24.9\,\text{cps}$$

16.2.2 Free Vibration with Viscous Damping

In the case of *undamped free vibration* discussed above, the vibration would continue indefinitely once the system had been set in motion. However, in practical cases, all vibrations undergo a gradual decrease in amplitude over time. This characteristic of vibration is referred to as *damping*. Figure 16.3 shows a foundation supported by a spring and dashpot. The dashpot represents

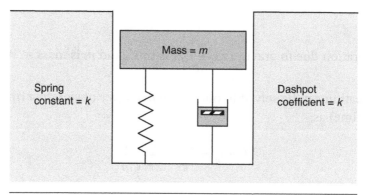

Figure 16.3 Free vibration of a spring-mass system with viscous damping

the *damping characteristic* of the soil. The dashpot coefficient is equal to *c*. For free vibration of the foundation, the differential equation of motion can be given by:

$$m\ddot{z} + c\dot{z} + kz = 0 \tag{16.4}$$

The preceding equation can be solved to show three possible cases of vibration that are functions of a quantity called the *damping ratio D*. The damping ratio is defined as:

$$D = \frac{c}{c_c} \tag{16.5}$$

where

$$c_c = \text{critical damping coefficient} = 2\sqrt{km} \tag{16.6}$$

- If $D > 1$, it will be an *overdamped* case. In this case, the system will not oscillate at all. The variation of displacement *z* with time will be as shown in Figure 16.4a.
- If $D = 1$, it will be a case of *critical* damping (see Figure 16.4b). In this case, the sign of *z* changes only once.
- If $D < 1$, it is an *underdamped* condition. Figure 16.4c shows the nature of vibration over time for this case. For this condition, the *damped natural frequency* of vibration *f* can be given as:

$$f = \frac{\omega_d}{2\pi} \tag{16.7}$$

where

$$\omega_d = \text{damped natural circular frequency (radians/s)}$$

$$\omega_d = \omega_n\sqrt{1 - D^2} \tag{16.8}$$

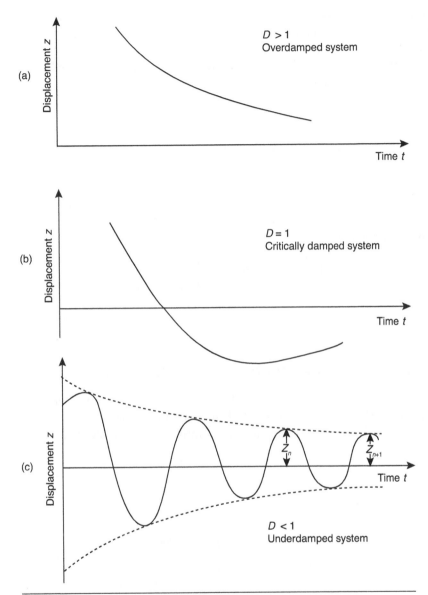

Figure 16.4 Free vibration of a mass-spring-dashpot system: (a) over-damped case; (b) critically damped case; (c) underdamped case

Combining Equations 16.7, 16.8, and 16.3:

$$f = f_m = \frac{\omega_n \sqrt{1-D^2}}{2\pi} = f_n \sqrt{1-D^2} \qquad (16.9)$$

where f_n and f_m are the undamped and damped natural frequencies.

Example 16.2: For a machine foundation, it is given that: $W = 70$ kN, $k = 12,500$ kN/m, and $c = 250$ kN-s/m. Determine:

a. whether the system is overdamped, underdamped, or critically damped
b. the damped natural frequency

Solution:

a.

$$c_c = 2\sqrt{km} = 2\sqrt{k\left(\frac{W}{g}\right)} = 2\sqrt{(12,500)\left(\frac{70}{9.81}\right)} = 597.3 \text{ kN-s/m}$$

$$D = \frac{c}{c_c} = \frac{250}{597.3} = 0.419 < 1 \leftarrow \text{The system is underdamped.}$$

b. From Equation 16.9:

$$f_m = f_n\sqrt{1-D^2} = \frac{1}{2\pi}\left(\sqrt{\frac{k}{m}}\right)\left(\sqrt{1-D^2}\right) = \frac{1}{2\pi}\left[\sqrt{\frac{12,500}{\left(\frac{70}{9.81}\right)}}\right]\left[\sqrt{1-(0.419)^2}\right] = 6.05 \text{ cps}$$

16.2.3 Steady-State Forced Vibration with Damping

Figure 16.5 shows the case of a foundation resting on a soil that can be approximated to be an equivalent spring and dashpot. This foundation is being subjected to a sinusoidally varying force $Q = Q_0 \sin \omega t$. The differential equation of motion for this system can be given by:

$$m\ddot{z} + kz + c\dot{z} = Q_0 \sin \omega t \tag{16.10}$$

where ω = circular frequency of vibration (rad/s).

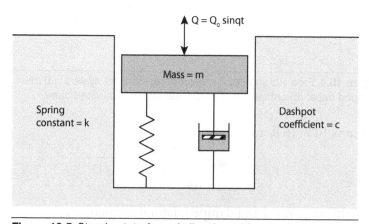

Figure 16.5 Steady-state forced vibration with damping

Equation 16.10 can be solved to obtain the amplitude (i.e., maximum displacement) of vibration Z of the foundation as:

$$Z = \frac{\left(\dfrac{Q_0}{k}\right)}{\sqrt{\left[1-\left(\dfrac{\omega^2}{\omega_n^2}\right)\right]^2 + 4D^2\left(\dfrac{\omega^2}{\omega_n^2}\right)}} \tag{16.11}$$

where $\omega_n = \sqrt{k/m}$ is the undamped natural frequency and D is the damping ratio.

Equation 16.11 is plotted in a nondimensional form as $Z/(Q_0/k)$ versus ω/ω_n in Figure 16.6. Note that the maximum value of $Z/(Q_0/k)$ (and hence Z) occurs at:

$$\omega = \omega_n\sqrt{1-2D^2} \tag{16.12}$$

or

$$f_m = f_n\sqrt{1-2D^2} \tag{16.13}$$

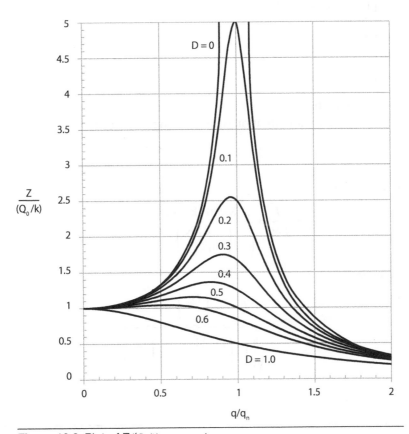

Figure 16.6 Plot of $Z/(Q_0/k)$ versus ω/ω_n

where f_m is the frequency that gives the *maximum amplitude* (the *resonant frequency for vibration with damping*) and f_n is the natural frequency $= (1/2\pi)\sqrt{k/m}$. Note the slight difference from Equation 16.9. Hence, the *amplitude of vibration at resonance* can be obtained by substituting Equation 16.12 into Equation 16.11, which gives:

$$Z_{res} = \frac{Q_0}{k} \frac{1}{\sqrt{\left[1-(1-2D^2)\right]^2 + 4D^2(1-2D^2)}} = \frac{Q_0}{k} \frac{1}{2D\sqrt{1-D^2}} \tag{16.14}$$

Example 16.3: Refer to Figure 16.5.

Given:

The weight of machine and foundation = 200 kN
The spring constant $k = 18 \times 10^4$ kN/m
The damping ratio $D = 0.3$
Q (kN) $= Q_0 \sin \omega t$
$Q_0 = 60$ kN
$\omega = 130$ rad/s

Determine:

a. the amplitude of motion Z
b. the resonant frequency for vibration with damping and the amplitude of vibration at resonance

Solution:

a. From Equation 16.3:

$$\omega_n = \sqrt{\frac{k}{m}} = \sqrt{\frac{(18 \times 10^4 \text{ kN/m})}{\left(\dfrac{200 \text{ kN}}{9.81}\right)}} = 93.96 \text{ rad/s}$$

From Equation 16.11:

$$Z = \frac{\left(\dfrac{Q_0}{k}\right)}{\sqrt{\left[1-\left(\dfrac{\omega^2}{\omega_n^2}\right)\right]^2 + 4D^2\left(\dfrac{\omega^2}{\omega_n^2}\right)}}$$

Hence:

Continues

Example 16.3: *Continued*

$$Z = \frac{\left(\dfrac{60}{18 \times 10^4}\right)}{\sqrt{\left[1 - \left(\dfrac{130}{93.96}\right)\right]^2 + (4)(0.3)^2\left(\dfrac{130}{93.96}\right)^2}} = 0.00027\ m = 0.27\ mm$$

b. From Equation 16.13:

$$f_m = f_n\sqrt{1 - 2D^2}$$

$$f_n = \frac{\omega_n}{2\pi} = \frac{93.96}{(2)(\pi)} = 14.95\ cps$$

Thus:

$$f_m = (14.95)\sqrt{1 - (2)(0.3)^2} = 13.54\ cps$$

From Equation 16.14:

$$Z_{res} = \frac{60}{18 \times 10^4} \times \frac{1}{2 \times 0.3 \times \sqrt{1 - 0.3^2}} = m = 0.58\ mm$$

16.2.4 Rotating Mass-Type Excitation

In many cases of foundation equipment, vertical foundation vibration is produced by coun-ter-rotating masses as shown in Figure 16.7a. Since horizontal forces on the foundation at any instance cancel, the net vibrating force on the foundation can be determined to be equal to $2m_e e\omega^2$ (where m_e = mass of each counter-rotating element, e = eccentricity, and ω = angular frequency of the masses). In such cases, the equation of motion with viscous damping (see Equation 16.10) can be modified to the following form:

$$m\ddot{z} + kz + c\dot{z} = Q_0 \sin \omega t \qquad (16.15)$$

$$Q_0 = 2m_e e\omega^2 = U\omega^2 \qquad (16.16)$$

$$U = 2m_e e \qquad (16.17)$$

In Equation 16.15, m is the mass of the foundation, which *includes* $2m_e$. Solving Equation 16.15, the amplitude of motion becomes:

$$Z = \frac{\left(\dfrac{U}{m}\right)\left(\dfrac{\omega^2}{\omega_n^2}\right)}{\sqrt{\left[1 - \left(\dfrac{\omega^2}{\omega_n^2}\right)\right]^2 + 4D^2\left(\dfrac{\omega^2}{\omega_n^2}\right)}} \qquad (16.18)$$

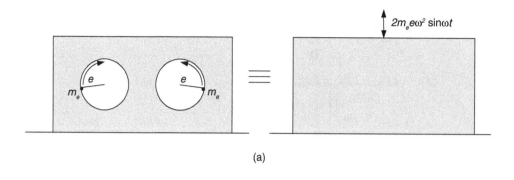

(a)

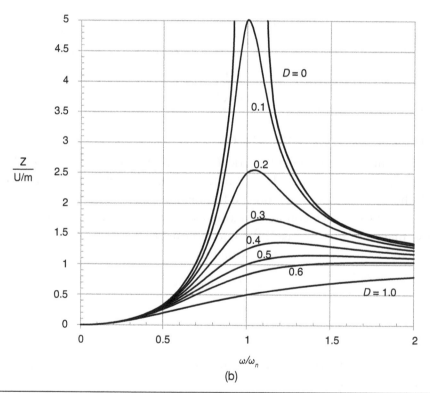

$\dfrac{Z}{U/m}$

ω/ω_n

(b)

Figure 16.7 (a) rotating mass-type excitation; (b) plot of $Z/(U/m)$ against ω/ω_n

Figure 16.7b shows a nondimensional plot of $Z/(U/m)$ versus ω/ω_n for various values of the damping ratio. For this type of excitation, the angular resonant frequency can be obtained as:

$$\omega = \frac{\omega_n}{\sqrt{1-2D^2}} \qquad (16.19)$$

or

$$f_m = \text{damped resonant frequency} = \frac{f_n}{\sqrt{1-2D^2}} \qquad (16.20)$$

The amplitude at damped resonant frequency (similar to Equation 16.14) can be given as:

$$Z_{res} = \frac{\left(\dfrac{U}{m}\right)}{2D\sqrt{1-D^2}} \qquad (16.21)$$

16.3 SHEAR MODULUS AND POISSON'S RATIO

For solving practical problems in foundation vibration, relationships for the spring constant k and dashpot coefficient c are necessary. Those relationships presently available are functions of shear modulus G and Poisson's ratio v of various soils. In this section, we will discuss some of the available relationships for shear modulus of sand and clayey soils.

16.3.1 Shear Modulus G for Sand

At *low strain amplitudes* ($\leq 10^{-4}\%$), the shear modulus of sand was correlated by Hardin and Black (1968) as:

$$G = \frac{6908(2.17-e)^2}{1+e}(\overline{\sigma}_0')^{0.5} \quad \text{for round-grained soil} \qquad (16.22)$$

and

$$G = \frac{3230(2.97-e)^2}{1+e}(\overline{\sigma}_0')^{0.5} \quad \text{for angular-grained soil} \qquad (16.23)$$

where

G = shear modulus (kN/m²)
e = void ratio
$\overline{\sigma}_0'$ = average effective confining pressure (kN/m²)

In the field:

$$\overline{\sigma}_0' \approx \frac{\sigma_v' + 2\sigma_v'(1-\sin\phi)}{3} \qquad (16.24)$$

where

σ'_v = vertical effective stress at a certain point in a soil mass, and
ϕ = drained friction angle.

Example 16.4: For a dry angular-grained sand deposit—

Given:

Dry unit weight $\gamma_d = 17.5$ kN/m³
Angle of friction $\phi = 34°$
Specific gravity of soil solids $G_s = 2.67$

Estimate the shear modulus of the soil at a depth of 7 m from the ground surface.

Solution:

$$\gamma_d = \frac{G_s \gamma_w}{1+e}$$

$$e = \frac{G_s \gamma_w}{\gamma_d} - 1 = \frac{(2.67)(9.81)}{17.5} - 1 \approx 0.497$$

At a depth of 7 m:

$$\sigma'_v = (17.5)(7) = 122.5 \text{ kN/m}^2$$

$$\overline{\sigma}'_0 \approx \frac{\sigma'_v + 2\sigma'_v(1 - \sin \phi)}{3} = \frac{122.5 + (2)(122.5)(1 - \sin 30)}{3} = 81.7 \text{ kN/m}^2$$

From Equation 16.23:

$$G = \frac{3230(2.97 - e)^2}{1+e}(\overline{\sigma}'_0)^{0.5} = \frac{3230(2.97 - 0.497)^2}{1 + 0.497}(81.7)^{0.5} \approx 199,273 \text{ kN/m}^2$$

16.3.2 Shear Modulus *G* for Clay

The shear modulus at low strain amplitudes in clay soils was proposed by Hardin and Drnevich (1972) in this form:

$$G \text{ (kN/m}^2) = \frac{3230(2.97 - e)^2}{1+e}(\text{OCR})^K [\overline{\sigma}'_0 (\text{kN/m}^2)]^{0.5} \tag{16.25}$$

where

OCR = overconsolidation ratio
K = a constant, which is a function of plasticity index PI

The average effective stress $\bar{\sigma}_0'$ was defined by Equation 16.24. The suggested variation of K with plasticity index PI is given in Table 16.1.

Table 16.1 K versus plasticity index variation

PI (%)	K
0	0
20	0.18
40	0.30
60	0.41
80	0.48
≥ 100	0.50

16.4 VERTICAL VIBRATION OF FOUNDATIONS—ANALOG SOLUTION

16.4.1 Constant Force Excitation

Lysmer and Richart (1966) provided an analog solution for the vertical vibration of a rigid circular foundation. According to this solution, it was proposed that satisfactory results could be obtained within the range of practical interest by expressing the rigid circular foundation vibration in the following form (see Figure 16.8):

$$m\ddot{z} + c_z\dot{z} + k_z z = Q_0 e^{i\omega t} \tag{16.26}$$

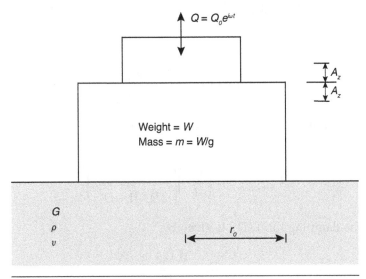

Figure 16.8 Vertical vibration of a foundation

where

$$k_z = \text{static spring constant for rigid circular foundation} = \frac{4Gr_0}{1-v} \tag{16.27}$$

$$c_z = \text{dashpot coefficient} = \frac{3.4r_0^2}{1-v}\sqrt{G\rho} \tag{16.28}$$

m = mass of the foundation and the machine the foundation is supporting

r_0 = radius of the foundation

v = Poisson's ratio of the soil

G = shear modulus of the soil

ρ = density of the soil

If a foundation is rectangular with a length L and width B, then the equivalent radius of a circular foundation can be given as:

$$r_0 \approx \sqrt{\frac{BL}{\pi}} \tag{16.29}$$

The resonant frequency f_m (frequency at maximum displacement) for *constant force excitation* can be obtained by solving Equations 16.26 to 16.28 (similar to solving Equation 16.10). It becomes:

$$f_m = \left(\frac{1}{2\pi}\right)\left(\sqrt{\frac{G}{\rho}}\right)\left(\frac{1}{r_0}\right)\sqrt{\frac{B_z - 0.36}{B_z}} \quad \text{for } B_z \geq 0.3 \tag{16.30}$$

where

$$B_z = \text{mass ratio} = \left(\frac{1-v}{4}\right)\left(\frac{m}{\rho r_0^3}\right) \tag{16.31}$$

The amplitude of vibration A_z at resonance for *constant force-type excitation* can be determined from Equation 16.14 as:

$$A_{z(\text{resonance})} = \left(\frac{Q_0}{k}\right)\left(\frac{1}{2D_z\sqrt{1-D_z^2}}\right) \tag{16.32}$$

where $k_z = \frac{4Gr_0}{1-v}$. The damping ratio (D_z) is given by:

$$D_z = \frac{0.425}{\sqrt{B_z}} \tag{16.33}$$

Substituting the above relationships for k_z (Equation 16.27) and D_z (Equation 16.33) into Equation 16.32 yields:

$$A_{z(\text{resonance})} = \frac{Q_0(1-v)}{4Gr_0}\frac{B_z}{0.85\sqrt{B_z-0.18}} \tag{16.34}$$

The amplitude of vibration at frequencies other than resonance can be obtained using Equation 16.11 as:

$$A_z = \frac{\left(\dfrac{Q_0}{k_z}\right)}{\sqrt{\left[1-\left(\dfrac{\omega^2}{\omega_n^2}\right)\right]^2 + 4D_z^2\left(\dfrac{\omega^2}{\omega_n^2}\right)}} \tag{16.35}$$

The relationships for k_z and D_z are given by Equations 16.27 and 16.33, and:

$$\omega_n = \sqrt{\frac{k_z}{m}} \tag{16.36}$$

Example 16.5: A foundation 6 m-long and 2 m-wide is subjected to a constant force-type vertical vibration. Given:

The total weight of the machinery and foundation block $W = 670$ kN

The unit weight of soil $\gamma = 18$ kN/m³

$v = 0.4$
$G = 21,000$ kN/m²
Amplitude of the vibrating force $Q_0 = 7$ kN
Operating frequency $f = 180$ cpm

Determine:

a. the resonant frequency
b. the amplitude of vibration at resonance

Solution:

a. This is a rectangular foundation, so the equivalent radius (see Equation 16.29) is:

$$r_0 = \sqrt{\frac{BL}{\pi}} = \sqrt{\frac{(2)(6)}{\pi}} = 1.95 \text{ m}$$

Continues

Example 16.5: *Continued*

The mass ratio (Equation 16.31) is:

$$B_z = \left(\frac{1-v}{4}\right)\left(\frac{m}{\rho r_0^3}\right) = \left(\frac{1-v}{4}\right)\left(\frac{W}{\gamma r_0^3}\right) = \left(\frac{1-0.4}{4}\right)\left[\frac{670}{(18)(1.95)^3}\right] = 0.753$$

From Equation 16.30, the resonant frequency is:

$$f_m = \left(\frac{1}{2\pi}\right)\left(\sqrt{\frac{G}{\rho}}\right)\left(\frac{1}{r_0}\right)\sqrt{\frac{B_z - 0.36}{B_z}}$$

$$= \left(\frac{1}{2\pi}\right)\left[\sqrt{\frac{21,000}{\left(\frac{18}{9.81}\right)}}\right]\left(\frac{1}{1.95}\right)\sqrt{\frac{0.753-0.36}{0.753}} = 6.3\,\text{cps} \approx 378\,\text{cpm}$$

b. From Equation 16.34:

$$A_{z(\text{resonance})} = \frac{Q_0(1-v)}{4Gr_0}\frac{B_z}{0.85\sqrt{B_z - 0.18}}$$

$$= \left[\frac{(7)(1-0.4)}{(4)(21,000)(1.95)}\right]\left[\frac{0.753}{0.85\sqrt{0.753-0.18}}\right] = 0.00003\,m = 0.03\,\text{mm}$$

16.4.2 Rotating Mass Excitation

If a structure is subjected to vertical vibration due to rotating mass excitation as shown in Figure 16.9 (similar to that shown in Figure 16.7a), the corresponding relationships will be as follows:

Resonant frequency:

$$f_m = \left(\frac{1}{2\pi}\right)\left(\sqrt{\frac{G}{\rho}}\right)\left(\frac{1}{r_0}\right)\sqrt{\frac{0.9}{B_z - 0.45}} \tag{16.37}$$

Amplitude of vibration at resonance A_z:

$$A_{z(\text{resonance})} = \frac{m_1 e}{m}\frac{B_z}{0.85\sqrt{B_z - 0.18}} \tag{16.38}$$

where m_1 = total rotating mass causing excitation, and m = mass of the foundation and the supporting machine.

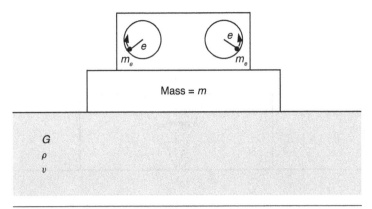

Figure 16.9 Foundation vibration (vertical) by a frequency-dependent exciting force

Amplitude of vibration at frequencies other than resonance:

$$A_z = \frac{\left(\dfrac{m_1 e}{m}\right)\left(\dfrac{\omega}{\omega_n}\right)^2}{\sqrt{\left[1-\left(\dfrac{\omega^2}{\omega_n^2}\right)\right]^2 + 4D_z^2\left(\dfrac{\omega^2}{\omega_n^2}\right)}} \tag{16.39}$$

Note that B_z, D_z, and ω_n are defined by Equations 16.31, 16.33, and 16.36 respectively.

16.5 ROCKING VIBRATION OF FOUNDATIONS

16.5.1 Constant Force Excitation

Hall (1967) developed a mass-spring-dashpot model for the rocking vibration of a rigid circular foundation (Figure 16.10). According to this model:

$$I_0\ddot{\theta} + c_\theta\dot{\theta} + k_\theta\theta = M_y e^{i\omega t} \tag{16.40}$$

where

M_y = amplitude of the exciting moment

θ = rotation of the vertical axis of the foundation at any time t

I_0 = mass moment of inertia about the y axis (i.e., axis perpendicular to the

cross section passing through O) = $\dfrac{W_0}{g}\left(\dfrac{r_0^2}{4}+\dfrac{h^2}{3}\right)$ (16.41)

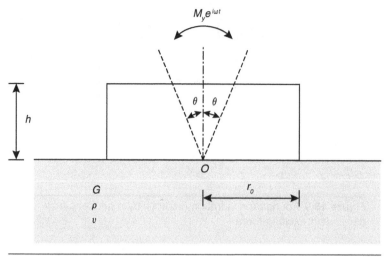

Figure 16.10 Rocking vibration of a foundation

W_0 = weight of the foundation and machine

g = acceleration due to gravity

h = height of the foundation

k_θ = static spring constant = $\dfrac{8Gr_0^3}{3(1-v)}$ (16.42)

c_θ = dashpot coefficient = $\dfrac{0.8r_0^4\sqrt{G}}{(1-v)(1+B_\theta)}$ (16.43)

B_θ = inertia ratio = $\dfrac{3(1-v)}{8}\dfrac{I_0}{\rho r_0^5}$ (16.44)

Based on the solution of Equation 16.40, the resonant frequency f_m, the amplitude of vibration at resonant frequency $\theta_{\text{resonance}}$, and the amplitude of vibration at a nonresonant frequency θ are given by the following relationships:

$$f_m = \left(\frac{1}{2\pi}\sqrt{\frac{k_\theta}{I_0}}\right)\left(\sqrt{1-2D_\theta^2}\right)$$ (16.45)

$$D_\theta = \text{damping ratio} = \frac{0.15}{\sqrt{B_\theta}(1+B_\theta)}$$ (16.46)

$$\theta_{resonance} = \frac{M_y}{k_\theta} \frac{1}{2D_\theta \sqrt{1 - D_\theta^2}} \tag{16.47}$$

$$\theta = \frac{\left(\dfrac{M_y}{k_\theta}\right)}{\sqrt{\left[1 - \left(\dfrac{\omega^2}{\omega_n^2}\right)\right]^2 + 4D_\theta^2 \left(\dfrac{\omega^2}{\omega_n^2}\right)}} \tag{16.48}$$

$$\omega_n = \sqrt{\frac{k_\theta}{I_0}} \tag{16.49}$$

In the case of rectangular foundations, the preceding relationships can be used by determining the equivalent radius as:

$$r_0 = \sqrt[4]{\frac{BL^3}{3\pi}} \tag{16.50}$$

The definitions of B and L are shown in Figure 16.11.

16.5.2 Rotating Mass Excitation

Referring to Figure 16.12, for rocking vibration with rotating mass excitation, the relationships for f_m, $\theta_{resonance}$, and θ are as follows:

$$f_m = \left(\frac{1}{2\pi} \sqrt{\frac{k_\theta}{I_0}}\right)\left(\frac{1}{\sqrt{1 - 2D_\theta^2}}\right) \tag{16.51}$$

$$\theta_{resonance} = \frac{m_1 e z'}{I_0} \frac{1}{2D_\theta \sqrt{1 - D_\theta^2}} \tag{16.52}$$

$$\theta = \frac{\left(\dfrac{m_1 e z'}{I_0}\right)\left(\dfrac{\omega^2}{\omega_n^2}\right)}{\sqrt{\left[1 - \left(\dfrac{\omega^2}{\omega_n^2}\right)\right]^2 + 4D_\theta^2 \left(\dfrac{\omega^2}{\omega_n^2}\right)}} \tag{16.53}$$

The relationships for D_θ and ω_n are given in Equations 16.46 and 16.49 respectively.

Figure 16.11 Equivalent radius of rectangular rigid foundation-rocking motion

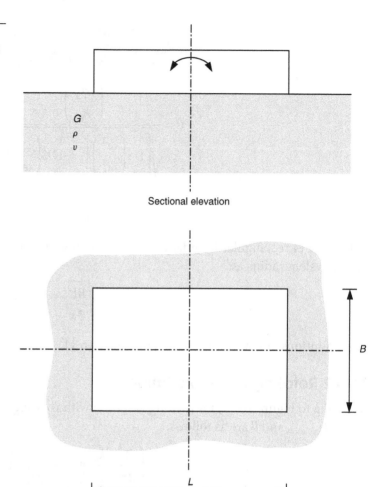

Sectional elevation

Plan

Figure 16.12 Rocking vibration due to rotating mass excitation

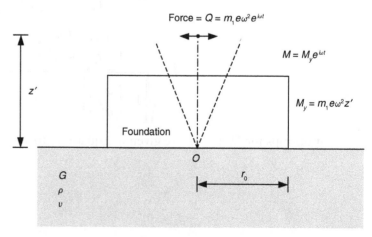

Example 16.6: A horizontal piston-type compressor is shown in part (a) of the figure on page 474. The operating frequency is 600 cpm. The amplitude of the horizontal unbalanced force of the compressor is 30 kN, and it creates a rocking motion of the foundation about point O (see part (b) on page 474). The mass moment of inertia of the compressor assembly about the axis $b'Ob'$ is 16×10^5 kg-m^2 (see part (c) on page 474).

Determine:

a. the resonant frequency
b. the amplitude of rocking at resonance

Solution: The moment of inertia of the foundation block and the compressor assembly about $b'Ob'$:

The moment of inertia of the foundation block about the axis through its centroid $= \frac{1}{12}m(L^2 + h^2)$. Therefore, the moment of inertia about the axis through:

$$O = \frac{1}{12}m(L^2 + h^2) + m\left(\frac{h}{2}\right)^2 = \frac{1}{12}mL^2 + \frac{1}{3}mh^2 = \frac{m}{3}\left[\left(\frac{L}{2}\right)^2 + h^2\right]$$

$$I_0 = \left(\frac{W_{\text{foundation block}}}{3g}\right)\left[\left(\frac{L}{2}\right)^2 + h^2\right] + 16 \times 10^5 \text{ kg-m}^2$$

Assume the unit weight of concrete is 23.58 kN/m^3:

$$W_{\text{foundation block}} = (8 \times 6 \times 3)(23.58) = 3395.52 \text{ kN} = 3395.52 \times 10^3 \text{ N}$$

$$I_0 = \frac{3395.52 \times 10^3}{(3)(9.81)}(3^2 + 3^2) + 16 \times 10^5 = 36.768 \times 10^5 \text{ kg-m}^2$$

Equivalent radius of the foundation: From Equation 16.50, the equivalent radius is:

$$r_0 = \sqrt[4]{\frac{BL^3}{3\pi}} = \sqrt[4]{\frac{8 \times 6^3}{3\pi}} = 3.67 \text{ m}$$

a. Resonant frequency:

$$k_\theta = \frac{8Gr_0^3}{3(1-v)} = \frac{(8)(18,000)(3.67)^3}{(3)(1-0.35)} = 3650279 \text{ kN-m/rad}$$

$$B_\theta = \frac{3(1-v)}{8}\frac{I_0}{\rho r_0^5} = \frac{3(1-0.35)}{8}\frac{36.768 \times 10^5}{1800(3.67)^5} = 0.748$$

$$D_\theta = \frac{0.15}{\sqrt{B_\theta}(1+B_\theta)} = \frac{0.15}{\sqrt{0.748}(1+0.748)} = 0.099$$

Continues

Example 16.6: *Continued*

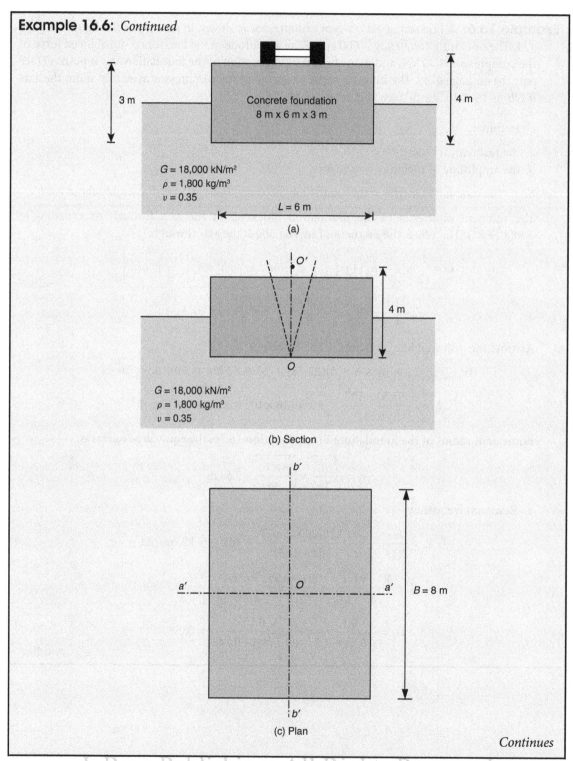

Concrete foundation
8 m × 6 m × 3 m

3 m

4 m

$G = 18,000$ kN/m²
$\rho = 1,800$ kg/m³
$\upsilon = 0.35$

$L = 6$ m

(a)

O'

4 m

O

$G = 18,000$ kN/m²
$\rho = 1,800$ kg/m³
$\upsilon = 0.35$

(b) Section

b'

a' ———— O ———— a'

$B = 8$ m

b'

(c) Plan

Continues

Example 16.6: *Continued*

From Equation 16.51:

$$f_n = \left(\frac{1}{2\pi}\sqrt{\frac{k_\theta}{I_0}}\right)\left(\frac{1}{\sqrt{1-2D_\theta^2}}\right)$$

$$= \left(\frac{1}{2\pi}\sqrt{\frac{3650279\times10^3 \text{ N-m/rad}}{36.768\times10^5}}\right)\left[\frac{1}{\sqrt{1-2(0.099)^2}}\right]$$

$$= 5.05 \text{ cps} = 303 \text{ cpm}$$

b. Amplitude of vibration at resonance:

$$M_{y(\text{operating frequency})} = \text{unbalanced force} \times 4 = 30 \times 4 = 120 \text{ kN-m}$$

$$M_{y(\text{at resonance})} = 120\left(\frac{f_m}{f_{\text{operating}}}\right)^2 = 120\left(\frac{303}{600}\right)^2 = 30.6 \text{ kN-m}$$

$$(m_1 e\omega^2)z' = M_y$$

$$\omega_{\text{resonance}} = \frac{(2\pi)(303)}{60} = 31.73 \text{ rad/s}$$

$$m_1 ez' = \frac{M_y}{\omega^2} = \frac{30.6\times10^3 \text{ N-m}}{(31.73)^2} = 0.0304\times10^3$$

From Equation 16.52:

$$\theta_{\text{resonance}} = \frac{m_1 ez'}{I_0}\frac{1}{2D_\theta\sqrt{1-D_\theta^2}}$$

$$= \left(\frac{0.0304\times10^3}{36.768\times10^5}\right)\left[\frac{1}{(2)(0.099)\sqrt{1-(0.099)^2}}\right] = 4.2\times10^{-5} \text{ rad}$$

16.6 SLIDING VIBRATION OF FOUNDATIONS

Hall (1967) developed the mass-spring-dashpot analog for the sliding vibration of a rigid circular foundation (Figure 16.13; radius = r_0). According to this analog, the equation of motion of the foundation can be given in the following form:

$$m\ddot{x} + c_x\dot{x} + k_x x = Q_0 e^{i\omega t} \tag{16.54}$$

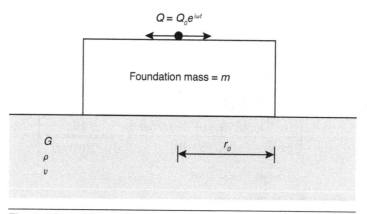

Figure 16.13 Sliding vibration of a rigid circular foundation

where

$$m = \text{mass of the foundation}$$

$$k_x = \text{static spring constant for sliding} = \frac{32(1-v)Gr_0}{7-8v} \qquad (16.55)$$

$$c_x = \text{dashpot coefficient for sliding} = \frac{18.4(1-v)}{7-8v}r_0^2\sqrt{\rho G} \qquad (16.56)$$

For sliding vibration:

$$D_x = \text{damping ratio in sliding} = \frac{0.288}{\sqrt{B_x}} \qquad (16.57)$$

where the dimensionless mass ratio (B_x) is given by:

$$B_x = \frac{7-8v}{32(1-v)}\frac{m}{\rho r_0^3} \qquad (16.58)$$

For rectangular foundations, the preceding relationships can be used by obtaining the equivalent radius r_0, or:

$$r_0 = \sqrt{\frac{BL}{\pi}}$$

where B and L are the length and width of the foundation respectively.

For the constance force excitation (i.e., Q_0 = constant), the resonant frequency f_m may be given as:

$$f_m = \left(\frac{1}{2\pi} \sqrt{\frac{32(1-v)Gr_0}{(7-8v)m}} \right) \sqrt{1-2D_x^2}$$

(16.59)

and for rotating mass-type of excitation:

$$f_m = \left(\frac{1}{2\pi} \sqrt{\frac{32(1-v)Gr_0}{(7-8v)m}} \right) \frac{1}{\sqrt{1-2D_x^2}}$$

(16.60)

Similarly, for constant force excitation, the amplitude of vibration at resonance is:

$$A_{x(resonance)} = \frac{Q_0}{k_x} \frac{1}{2D_x\sqrt{1-D_x^2}}$$

(16.61)

and for rotating mass excitation:

$$A_{x(resonance)} = \frac{m_1 e}{m} \frac{1}{2D_x\sqrt{1-D_x^2}}$$

(16.62)

where

m_1 = total rotating mass causing excitation, and

e = eccentricity of each rotating mass

For constant force excitation, the amplitude of vibration at a nonresonant frequency is:

$$A_x = \frac{\left(\dfrac{Q_0}{k_x}\right)}{\sqrt{\left[1-\left(\dfrac{\omega^2}{\omega_n^2}\right)\right]^2 + 4D_x^2\left(\dfrac{\omega^2}{\omega_n^2}\right)}}$$

(16.63)

For rotating mass excitation:

$$A_x = \frac{\left(\dfrac{m_1 e}{m}\right)\left(\dfrac{\omega}{\omega_n}\right)^2}{\sqrt{\left[1-\left(\dfrac{\omega^2}{\omega_n^2}\right)\right]^2 + 4D_x^2\left(\dfrac{\omega^2}{\omega_n^2}\right)}}$$

(16.64)

where

$$\omega_n = \sqrt{\frac{k_x}{m}} \qquad (16.65)$$

16.7 TORSIONAL VIBRATION OF FOUNDATIONS

Similar to the cases of vertical, rocking, and sliding modes of vibration, the equation for the torsional vibration of a *rigid circular foundation* (Figure 16.14) can be written as:

$$J_{zz}\ddot{\alpha} + c_\alpha \dot{\alpha} + k_\alpha \alpha = T_0 e^{i\omega t} \qquad (16.66)$$

where

J_{zz} = mass moment of inertia of the foundation about the axis $z - z$

c_α = dashpot coefficient for torsional vibration

k'_a = *static spring constant for torsional vibration* $= \dfrac{16}{3}Gr_0^3 \qquad (16.67)$

α = rotation of the foundation at any time due to the application of a torque $T = T_0 e^{i\omega t}$

The damping ratio D_α for this mode of vibration was determined as (Richart et al. 1970)

$$D_\alpha = \frac{0.5}{1 + 2B_\alpha} \qquad (16.68)$$

B_α = the dimensionless mass ratio for torsion at vibration $= \dfrac{J_{zz}}{\rho r_0^5} \qquad (16.69)$

For constant force excitation, the resonant frequency for torsional vibration is:

$$f_m = \left(\frac{1}{2\pi}\sqrt{\frac{k_\alpha}{J_{zz}}}\right)\sqrt{1 - 2D_\alpha^2} \qquad (16.70)$$

and for rotating mass excitation (see Figure 16.14):

$$f_m = \left(\frac{1}{2\pi}\sqrt{\frac{k_\alpha}{J_{zz}}}\right)\left(\frac{1}{\sqrt{1 - 2D_\alpha^2}}\right) \qquad (16.71)$$

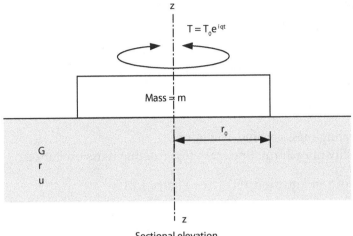

Sectional elevation

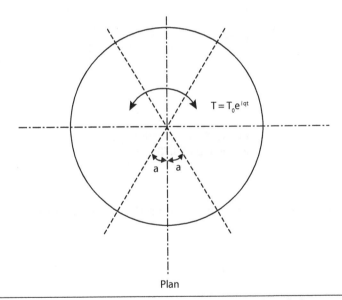

Plan

Figure 16.14 Torsional vibration of a rigid circular foundation

For constant force excitation, the amplitude of vibration at resonance is:

$$\alpha_{\text{resonance}} = \frac{T_0}{k_\alpha} \frac{1}{2D_\alpha \sqrt{1 - D_\alpha^2}}$$

(16.72)

For rotating mass-type excitation:

$$\alpha_{resonance} = \frac{m_1 e\left(\frac{x}{2}\right)}{J_{zz}} \frac{1}{2D_\alpha \sqrt{1-D_\alpha^2}} \tag{16.73}$$

where

m_1 = total rotating mass causing excitation

e = eccentricity of each rotating mass (for rotating mass excitation)

For the definition of x in Equation 16.73, see Figure 16.15.

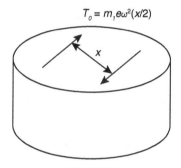

$T_0 = m_1 e\omega^2(x/2)$

Figure 16.15 Definition of x in Equation 16.73

For a rectangular foundation with dimensions $B \times L$, the equivalent radius may be given by:

$$r_0 = \sqrt[4]{\frac{BL(B^2 + L^2)}{6\pi}} \tag{16.74}$$

Example 16.7: A radar antenna foundation is shown in the figure on page 481. For torsional vibration of the foundation, it is given:

$$T_0 = 24.4 \times 10^4 \text{ N-m}$$

The mass moment of inertia of the tower about the axis $z - z = 13.56 \times 10^6$ kg-m^2

The unit weight of concrete used in the foundation = 23.68 kN/m^3

Continues

Example 16.7: *Continued*

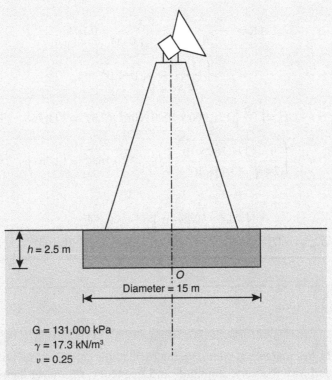

h = 2.5 m

O

Diameter = 15 m

G = 131,000 kPa
γ = 17.3 kN/m³
υ = 0.25

Calculate:

a. the resonant frequency for torsional mode of vibration
b. the angular deflection at resonance.

Solution:

a.

$$J_{zz} = J_{zz(\text{tower})} + J_{zz(\text{foundation})}$$

$$= 13.56 \times 10^6 + \frac{1}{2}\left[\pi r_0^2 h\left(\frac{23.58 \times 1000}{9.81}\right)\right]r_0^2$$

$$= 13.56 \times 10^6 + \frac{1}{2}\left[(\pi)(7.5)^2(2.5)\left(\frac{23.58 \times 1000 \text{ N}}{9.81}\right)\right](7.5)^2$$

$$= 13.56 \times 10^6 + 29.87 \times 10^6 = 43.43 \times 10^6 \text{ kg-m}^2$$

$$B_\alpha = \frac{J_{zz}}{\rho r_0^5} = \frac{43.43 \times 10^6}{\left(\dfrac{17.3 \times 1000}{9.81}\right)(7.5)^3} = 1.038$$

Continues

Example 16.7: *Continued*

$$D_\alpha = \frac{0.5}{1+2B_\alpha} = \frac{0.5}{1+(2)(1.038)} = 0.163$$

$$f_m = \left(\frac{1}{2\pi}\sqrt{\frac{k_\alpha}{J_{zz}}}\right)\sqrt{1-2D_\alpha^2}$$

$$k_\alpha = \frac{16}{3}Gr_0^3 = \left(\frac{16}{3}\right)(131,000\times1000\text{ N/m}^2)(7.5)^3 = 294,750\times10^6$$

$$f_m = \left(\frac{1}{2\pi}\sqrt{\frac{294,750\times10^6}{43.43\times10^6}}\right)\sqrt{1-(2)(0.163)^2} = 12.76\text{ cps}$$

b.

$$\alpha_{\text{resonance}} = \frac{T_0}{k_\alpha}\frac{1}{2D_\alpha\sqrt{1-D_\alpha^2}} = \left(\frac{24.4\times10^4\text{ N-m}}{294,750\times10^6}\right)\frac{1}{(2)(0.163)\sqrt{1-(0.163)^2}} = 0.257\times10^{-5}\text{ rad}$$

Reminder

- In *free undamped vibration*, the oscillations continue indefinitely and with the same amplitude and frequency; this never happens in reality. Some degree of damping is almost always present.

- $\omega = 2\pi f$

- $\omega_n = \sqrt{\dfrac{k}{m}}$ and $f_n = \dfrac{1}{2\pi}\sqrt{\dfrac{k}{m}}$ where f_n is the undamped natural frequency.

- Damping is modeled by a dashpot (where a piston is pushed into a viscous liquid). The larger the velocity, the larger the resisting force that is provided by the dashpot.

- In *free damped vibration*, the critical damping coefficient $c_c = 2\sqrt{km}$; the damping ratio $D = c/c_c$. When $D < 1$, oscillations take place at a frequency of $f_d = f_n\sqrt{1-D^2}$ with a gradual decay of the amplitude. The higher the damping, the lower the frequency. For $D \geq 1$, there is no oscillation.

- In *forced* vibration, the force can be in the form of a constant force (or moment or torque) or rotating masses. At resonant frequency f_m, the amplitude of the displacement (or rotation) reaches the maximum.

REVIEW EXERCISES

1. A foundation is supported by a spring as shown in Figure 16.2. Given: weight of the foundation $W = 24$ kN; spring constant $k = 12,000$ kN/m. Determine the natural frequency of vibration of the system.

   Answer: 11.14 cps

2. A machine foundation can be idealized to a mass-spring system as shown in Figure 16.2. Given: weight of machine and the foundation combined $= 400$ kN; spring constant $= 100,000$ kN/m. Determine the natural frequency of the undamped free vibration of this foundation.

   Answer: 7.88 cps

3. Refer to Review Exercise 2. What would be the static deflection z_s of this foundation?

   Answer: 4 mm

4. A foundation weighs 800 kN. The foundation and the soil can be approximated as a mass-spring-dashpot system as shown in Figure 16.5. Given: spring constant $= 200,000$ kN/m; dashpot coefficient $= 2340$ kN-s/m. Determine the following:
   a. Damping ratio
   b. Damped natural frequency

   Answer: 0.29, 7.54 cps

5. The foundation given in Review Exercise 4 is subjected to a vertical force $Q = Q_0 \sin \omega t$ in which $Q_0 = 25$ kN and $\omega = 100$ rad/s. Determine the amplitude of the vertical vibration of the foundation.

   Answer: 3.8×10^{-2} mm

6. A 20 m-thick sand layer in the field is underlain by rock. The ground water table is located at a depth of 5 m measured from the ground surface. Determine the shear modulus of this sand at a depth of 10 m below the ground surface. Given: void ratio $= 0.6$; specific gravity of soil solids $= 2.68$; angle of friction of sand $= 36°$. Assume that the sand is round-grained.

   Answer: 95,940 kPa

7. A layer of clay deposit extends to a depth of 15 m below the ground surface. The ground water table coincides with the ground surface. For the clay, given: void ratio = 1.0; specific gravity of soil solids = 2.78; plasticity index = 20%; overconsolidation ratio = 2; effective stress friction angle ϕ = 26°. Determine the shear modulus of this clay at a depth of 7.5 m.

Answer: 48,343 kPa

8. A concrete foundation (unit weight = 23.5 kN/m³) supporting a machine is 3.5 m × 2.5 m in plan and is subjected to a sinusoidal vibrating force (vertical) having an amplitude of 10 kN (not frequency dependent). The operating frequency is 2000 cpm. The weight of the machine and foundation is 400 kN. The soil properties are: unit weight = 18 kN/m³; shear modulus = 38,000 kN/m²; Poisson's ratio = 0.25.

Determine:
a. the resonant frequency of the foundation
b. the amplitude of vertical vibration at resonance

Answer: 672 cpm, 0.0368 mm

9. The concrete foundation (unit weight = 23.5 kN/m³) of a machine has the following dimensions (see Figure 16.11): L = 3 m; B = 4 m; height of the foundation = 1.5 m. The foundation is subjected to a sinusoidal horizontal force from the machine having an amplitude of 10 kN at a height of 2 m measured from the base of the foundation. The soil supporting the foundation is sandy clay. Given: G = 30,000 kN/m²; ν = 0.2; soil density ρ = 1700 kg/m³.

Determine:
a. the resonant frequency for the rocking mode of vibration of the foundation
b. the amplitude of rocking vibration at resonance
(Note: The amplitude of horizontal force is not frequency dependent. Neglect the moment of inertia of the machine.)

Answer: 827 cpm, 0.000186 radians

10. Solve Review Exercise 9 assuming that the horizontal force is frequency dependent. The amplitude of the force at an operating speed of 800 cpm is 20 kN.

Answer: 847 cpm, 0.00042 radians

11. Refer to Review Exercise 9.

    Determine:
    a. the resonant frequency for the sliding mode of vibration
    b. the amplitude for the sliding mode of vibration at resonance
    Assume the weight of the machinery on the foundation to be 100 kN.

    Answer: 605 cpm, 0.0752 mm

12. Repeat Review Exercise 11 assuming that the horizontal force is frequency dependent. The amplitude of the horizontal force at an operating frequency of 800 cpm is 40 kN. The weight of the machinery on the foundation is 100 kN.

    Answer: 747 cpm, 0.186 mm

13. A concrete foundation (unit weight = 23.5 kN/m³) supporting a machine has the following dimensions: length = 5 m; width = 4 m; height = 2 m. The machine imparts a torque T on the foundation such that $T_0 e^{i\omega t}$. Given: T_0 = 3000 N-m. The mass moment of inertia of the foundation is 75×10^3 kg-m². The soil has the following properties: v = 0.25; unit weight = 18 kN/m³; G = 28,000 kN/m².

    Determine:
    a. the resonant frequency for the torsional mode of vibration
    b. the angular deflection at resonance

    Answer: 756 cpm, 5.9 × 10⁻⁶ radians

Index

A

AASHTO soil classification system, 37, 39–40, 41, 46

ABAQUS, 6

Acceleration, 470

Acid sulphate soils, 1

Active state, lateral earth pressures, 225, 231, 232, 233–235, see also Lateral earth pressures

Activity, 37

Adhesion, earth retaining structures, 381, see also Earth retaining structures

Adhesive resistance, 239

Aeolian soil, 2

Air, 11, 12
  mass of, 11
  stresses and, 67
    normal, 65–72

Air content, 55, 62

Airflow, 6

AIR/W, 6

Allowable bearing capacity, 292, 305, 307, 326

Allowable pressures, 317, 318

Alluvial soil, 2

Aluminum, 32, 33

American Association of State Highway Transportation Officials (AASHTO) soil classification system, 37, 39–40, 41, 46

American Society of Civil Engineers, 5

Amplitude, 459, 461, 463, 482, 483, 285
  decrease in, 455
  low strain, 463
  maximum, 460

Amplitude of vibration, 466, 467, 468, 469, 470, 471
  at resonance, 460

Anchored sheet piles, 385, 395–399

Angle of internal friction, 187

Angle of shearing resistance, 187

Angular deflection at resonance

Angular distortion, 290, 291

Angular-grained soil, shear modulus for, 463, 464

Angular grains, 31

Angular resonant frequency, 462

Anisotropic soils, flow net in, 85–86

Anisotropy, 203

Applied normal stress, 323

Applied pressure, settlement in footings and, 310

Archimedes principle, 13

Artificial slopes, 421

AS, see Australian Standards

Atomic force microscope, 33

Atomic structure of clay minerals, 32, 33

At-rest state, 225, 226–230, 271, 273, 275

Attapulgite, 33

Atterberg, A., 34

Atterberg limits, 34–37, 38, 39, 41, 44, 256, 277

Auger drilling, 254

Australian Standards (AS), 27

Average degree of consolidation, 156

Average hydraulic gradient, 75

Average vertical strain, 141

Axisymmetric loading, 313, 315

B

Backfills
  compaction and, 49, see also Compaction
  inclined granular, 235–236
  shear failure, 181

Backhoe, 49, 252

Backpressure, 194, 195, 196, 197, 205, 208, 210, 216

Base circles, 425, 426

Base failure, 425

Basements, 299
  lateral earth pressures, 230, see also Lateral earth pressures

Batter pile, 341, 342

Bearing capacity
  allowable, see Allowable bearing capacity